ADVANCES IN ANIMAL WELFARE SCIENCE
1986/87

ADVANCES IN ANIMAL WELFARE SCIENCE 1986/87

edited by

Michael W. Fox, D.Sc., Ph.D., B.Vet.Med., MRCVS
Director, The Institute for the Study of Animal Problems
Washington, DC 20037, USA

and

Linda D. Mickley B.S.
Research Associate, The Institute for the Study of Animal Problems
Washington, DC 20037, USA

1987 **MARTINUS NIJHOFF PUBLISHERS**
a member of the KLUWER ACADEMIC PUBLISHERS GROUP
BOSTON / DORDRECHT / LANCASTER

Distributors

for the United States and Canada: Kluwer Academic Publishers, P.O. Box 358, Accord Station, Hingham, MA 02018-0358, USA
for the UK and Ireland: Kluwer Academic Publishers, MTP Press Limited, Falcon House, Queen Square, Lancaster LA1 1RN, UK
for all other countries: Kluwer Academic Publishers Group, Distribution Center, P.O. Box 322, 3300 AH Dordrecht, The Netherlands

ISBN-13: 978-94-010-7996-9   e-ISBN-13: 978-94-009-3331-6
DOI: 10.1007/978-94-009-3331-6

Copyright

Katherine Houpt
Dept. of Physiology
New York State Veterinary College
Cornell University
Ithaca, New York, USA

John A. Hoyt
The Humane Society of the United States
Washington, DC, USA

B.O. Hughes
Agricultural Poultry Research Centre
Roslin, Midlothian, Scotland

Frank Hurnik
Dept. of Animal Science
University of Guelph
Guelph, Ontario, Canada

Ron Kilgour
Ministry of Agriculture and Fisheries
Ruakura Animal Research Station
Hamilton, New Zealand

Devra Kleiman
National Zoological Park
Smithsonian Institution
Washington, DC, USA

Elizabeth A. Lawrence
Tufts University
School of Veterinary Medicine
Boston, Massachusetts, USA

Jenny Remfry
Universities Federation for Animal Welfare
Potters Bar, Herts, England

Bernard E. Rollin
Dept. of Philosophy
Colorado State University
Fort Collins, Colorado, USA

Andrew N. Rowan
Tufts University
School of Veterinary Medicine
Boston, Massachusetts, USA

Harry C. Rowsell
Canadian Council on Animal Care
Ottawa, Ontario, Canada

Joyce S.A. Tischler
The Animal Legal Defense Fund
San Francisco, California, USA

G. van Putten
Research Institute for Animal Production
"Schoonoord," Zeist, Holland

D.B. Wilkins
Royal Society for the Prevention
    of Cruelty to Animals
The Causeway, Horsham, Sussex, England

Thomas Wolfle
National Institutes of Health
Dept. of Health and Human Services
Bethesda, Maryland, USA

Foreword

This third volume of articles dealing with advances in animal welfare science and philosophy covers a wide variety of topics. Major areas of discussion include the ethics and use of animals in biomedical research, farm animal behavior and welfare, and wildlife conservation.

Three articles dealing with aspects of equine behavior and welfare cover new ground for this companion species.

An in-depth study of the destruction of Latin America's tropical rain forests links the need for conservation and wildlife protection with the devastating impact of the international beef (hamburger) industry, and also highlights serious welfare problems in the husbandry of cattle in the tropics.

Papers from a recent symposium at Moorhead State University, *Animals and Humans: Ethical Perspectives* have been included in this volume. Many of these are "benchmark" papers presenting the most up-to-date and documented evidence in support of animal welfare and rights. Articles opposing these position papers are included since they were part of the symposium, and because they provide the reader with a deeper understanding of the arguments given in support of various forms of animal exploitation. While there is no intent to endorse these views by publishing them, it should be acknowledged that without an open and scholarly exchange of opposing views, the possibility of constructive exchange and conflict resolution will remain remote.

Papers from this symposium opposing the use of animals in medicine were not available for inclusion in this volume. However, the papers by Robert Sharpe and Martin Stephens in Part I offer convincing views in opposition to the wholesale exploitation and suffering of animals in the name of medical progress.

We wish to express our gratitude to the Manuscript Review Committee and the twenty-nine contributors to this volume.

M.W. Fox
L.D. Mickley
Washington, DC

LIST OF CONTRIBUTORS

David J. Allan
Senior Lecturer, Pathology
Dept. of Medical Laboratory Science
Queensland Institute of Technology
Brisbane, Australia, 4000

Jack L. Albright
Professor
School of Agriculture
  and Large Animal Clinics
School of Veterinary Medicine
Purdue University
West Lafayette, IN 47907
USA

Donald J. Barnes
Director, Washington Office
The National Anti-Vivisection Soc.
112 N. Carolina Ave, SE
Washington, DC 20003
USA

R.W. Bell
Professor
Psychology Dept.
Texas Tech University
Lubbock, TX 79409
USA

Judith K. Blackshaw
Lecturer in Animal Behaviour
Dept. of Animal Sciences and
  Production
University of Queensland
St. Lucia, Brisbane,
Australia, 4067

S.G. Brazier
Graduate Student
Dept. of Animal Sciences
Texas Tech University
Lubbock, TX 79409
USA

Sharon E. Cregier
North American Editor
Equine Behavior Journal
University of Prince Edward Island
Charlottetown, PEI
Canada, C1A 4P3

Stanley E. Curtis
Professor
Dept. of Animal Sciences
College of Agriculture
1207 W. Gregory Dr
University of Illinois at
  Urbana-Champaign
Urbana, IL 61801
USA

Michael W. Fox
Director, Institute for the
  Study of Animal Problems
Scientific Director,
The Humane Society of the
  United States
2100 L St, NW
Washington, DC 20037
USA

Harry Frank
Associate Professor
Depts. of Resource and Community
  Science, Psychology
University of Michigan-Flint
Flint, MI 48503
USA

John W. Grandy
Vice President for Wildlife
  and Environment
The Humane Society of the
  United States
2100 L St, NW
Washington, DC 20037
USA

Linda M. Hasselbach
Dept. of Resource and
   Community Science
University of Michigan-Flint
Flint, MI 48503
USA

J.C. Heird
Assistant Professor
Dept. of Animal Science
Texas Tech University
Lubbock, TX 79409
USA

J.F. Hurnik
Professor
Dept. of Animal and Poultry Science
University of Guelph
Guelph, Ontario
Canada, N1G 2W1

Michael Hutchins
Curatorial Intern
Dept. of Mammalogy
New York Zoological Society
Bronx, NY 10460
USA

Dawn M. Littleton
Dept. of Psychology
Canisius College
Buffalo, NY 14209
USA

U.A. Luescher
Assistant Professor
Dept. of Clinical Studies
University of Guelph
Guelph, Ontario
Canada, N1G 2W1

Gary F. Merrill
Associate Professor
Graduate Program in Physiology
Bartlett Hall, Cook College
Rutgers University
New Brunswick, NJ 08903
USA

Linda D. Mickley
Research Associate
Institute for the Study
   of Animal Problems
2100 L St, NW
Washington, DC 20037
USA

Ashley Montagu
321 Cherry Hill Rd
Princeton, NJ 08540
USA

Jan Narveson
Professor
Dept. of Philosophy
University of Waterloo
Waterloo, Ontario
Canada, N2L 3G1

Harvey K. Nelson
Regional Director
U.S. Fish and Wildlife Service
Federal Building, Fort Snelling
Twin Cities, MN 55111
USA

Tom Regan
Professor
Dept. of Philosophy
North Carolina State University
Raleigh, NC 27695-8103
USA

William Robinson
Professor
Dept. of Biology
Northern Michigan University
Marquette, MI 49855
USA

Douglas Shane
Environmental Consultant
1805 Shallcross Ave
Wilmington, DE 19806
USA

Robert Sharpe
Director, Lord Dowding Fund
The National Anti-Vivisection Soc.
51 Harley St
London, W1N 1DD
England

Martin L. Stephens
Associate Director
Laboratory Animal Welfare Dept.
The Humane Society of the
    United States
2100 L St, NW
Washington, DC 20037
USA

Christen Wemmer
Assitant Director for
Conservation and Captive Breeding
Research and Conservation Center
National Zoological Park
Front Royal, VA 22630
USA

James A. Will
Professor
College of Agriculture
    and Life Sciences
119 Veterinary Science Bldg.
1655 Linden Dr
University of Wisconsin-Madison
Madison, WI 53706
USA

LIST OF CONTENTS

PART II
PROCEEDINGS FROM THE MOORHEAD STATE UNIVERSITY
CONFERENCE: "ANIMALS AND HUMANS: ETHICAL PERSPECTIVES"

Part I

ETHICS, WELFARE, AND LABORATORY ANIMAL MANAGEMENT

David J. Allan[1] and Judith K. Blackshaw[2]

Introduction

Animals have been used in medical research from as far back as 129-199 A.D. when Galen, a Greek medical scientist, used a pig for his experiments. In the sixteenth and seventeenth centuries, anatomical dissections were carried out on animals; Galvani used frogs in 1791 for his experiments and the Russian physiologist, Pavlov, carried out his famous dog experiments in the early 1900s. Since this time, large numbers of animals have been used in biomedical and other research. In 1963 the first edition of "The Guide for the Care and Use of Laboratory Animals" was published, and the United States Public Health Service began to require all recipients of grants in which animals were used to adhere to these guidelines. There is now worldwide interest in welfare issues, and the ethics of using animals for research has been raised in many countries.

The study of ethics is the treating of moral questions and is concerned with right and wrong. There will always be differing opinions on the ethics of animal use which may raise dilemmas for workers. In a practical sense this has often been dealt with by drafting codes for the care and use of experimental animals. Adherence to a code, however, does not exclude the experimenter from actively considering ethical and welfare issues.

Three important issues in laboratory animal management are the ethics of using the animals for experimentation, the welfare of the animals being used, and the scientific validity of the selected species and number to be used.

Ethical Considerations

Animal Rights. —A distinction should be drawn between the concepts of animal welfare and animal rights. Advocates of animal welfare acknowledge that we have a right to use animals according to our needs and emphasize the maintenance of high standards of animal management. Animal rights organizations promote the complete cessation of all aspects of human "exploitation" of animals (Singer 1976).

The philosophy of animal rights has developed through the thoughts of a number of modern philosophers who have become involved in public debate on the ethics of animal experimentation. These debates and discussions have

1

been used for criticizing many current practices involving animal use. Singer (1985) commented that although there were one or two nineteenth century thinkers who asserted that animals have rights, the serious political movement for animal liberation is a product of the 1970s. Several philosophers have written books on animal rights (e.g., Singer 1976; Regan 1983) which have become the basis for the animal liberation movement. Singer (1980, 1985) stated that from an ethical point of view, humans and nonhumans stand on an equal footing. That we use chimpanzees for experiments and yet would not do so with retarded humans of lower mental level is, in Singer's view (1976, 1980, 1985), an example of speciesism and is as indefensible as the most blatant racism. However, Singer (1980, 1985) is very careful to point out that he does not mean that animals have all the same rights as humans; he advocates equality of consideration of interests, not equality of rights. Where animals and humans have similar interests, those interests are to be counted equally, with no automatic discount just because one of the beings is not human. When Singer (1980) commented on animal experimentation he posed the question: What experiments on animals can be justified without speciesism? He answered: Only those experiments which would also be justified if performed on an orphaned, irreparably retarded human being at a comparable level of sentience and awareness. Exactly which experiments this criterion justifies depends on the extent to which one believes that it is permissible to sacrifice the interests of one human being to benefit another. Because Singer takes a broadly utilitarian view on these issues he can justify some experiments, but there are many that he cannot justify.

Regan (1983, 1985) regards himself as an advocate of animal rights and does not take Singer's utilitarian view. He believes there are fundamental differences between utilitarianism and the rights view. This is very apparent in the use of animals in science. For the utilitarian, whether the harm done to animals in pursuit of scientific ends is justified, depends on the balance of aggregated consequences for all those affected by the outcome (Regan 1983). The rights view is that an individual should never be harmed merely on the grounds that this will or might produce "the best" aggregate consequences. To do so is to violate the rights of the individual, and this is why the harm done to animals in pursuit of scientific purposes is wrong. So Regan (1985) is committed to total abolition of the use of animals in science, agriculture, sport, hunting, and trapping.

It is important to understand these views on the use of animals as they are an attempt to give a philosophical and ethical basis to the practice of using experimental animals. The most valuable outcomes of these discussions have been to heighten the awareness of the public and the animal researcher to the ethical issues involved when animals are used for research.

Animal Experimentation. —These divergent views raise the crucial issue of whether it is proper to use animals in scientific investigations and, if so, what ethical obligations we have to provide for their welfare.

Laboratory animals are presently used in a wide variety of experimental situations:

i. scientific research—which studies the basic life processes, the cause of diseases, and the investigation of new therapies and drugs, with an aim to prevent, cure, or alleviate human and animal suffering,

ii. educational processes,

iii. experimental work on animal breeding and nutrition,

iv. diagnostic testing for disease in man or animals,

v. development, production, and testing of commercial products, such as vaccines, antisera, shampoos, food additives, and cosmetics.

The British Veterinary Association has recently given a great deal of thought to the need for animal experiments (Seamer 1982) and supports the principle of the three R's.

i. *Replace* animals where possible, with valid alternatives.

ii. *Reduce* the number of animals needed for an experiment by the use of a planned design, so a statistically valid result can be obtained with the minimum number of animals.

iii. *Refine* the experiment and choose the most appropriate animal model.

Criticism of the use of animals for inessential experiments and the large numbers of animals used in experiments has led to questioning the obligations of the research worker to the animals. This has also stimulated the active consideration of alternatives and adjuncts which may be suitable to use in place of, or in conjunction with, some animal experiments.

Alternatives to Animal Experiments.—It is important to consider the alternatives which may be suitable to use in place of, or in conjunction with, animal experiments (Balls 1983).

i. A wide range of procedures could contribute to a reduction in the current reliance on animal experiments in toxicity testing (i.e., the LD50 test). These could include improved data storage, and more opportunity for the international exchange and use of this information. In the authors' opinion, needless repetition of toxicity tests occur and an international register of the toxicity of chemicals would help cut down the number of tests.

ii. Mathematical modelling using computer-assisted models of various biological processes are important adjuncts to the use of live animals. It must be recognized that development of these models is dependent on the amount and quality of the information used in their construction. This information may be derived from new animal experiments, and animals might also be needed to test the validity of the constructed model. Mathematical modelling is in the early stages of development but can be expected to increase in the near future.

iii. Use of vertebrate embryos or lower organisms has the advantage of acquiring rapid results on the general effects of chemicals on complete living organisms. Other alternatives include bacteria, algae, fungi, protozoans, coelenterates, plants, insects, echinoderms, and molluscs. Bacterial tests for mutagenicity are well established and are being used to make early decisions about new product development (quoted in Balls 1983).

iv. Man is the ultimate experimental animal in the final stages of testing any chemical intended for the use of man. Already human volunteers are widely used in the testing of cosmetics and toiletries. Occupational association has allowed the identification of many known human carcinogens. Cancers associated with occupations such as radiology and asbestos mining are examples of how man serves as the primary experimental subject.

v. In vitro techniques have been developed including the use of subcellular fractions, cell suspensions, tissue biopsies, tissue slices, whole organ perfusion, and tissue culture. Thus the classic Friedman-Lopham rabbit test for human pregnancy has been replaced by an in vitro technique (Stark 1983). The main potential advantage of in vitro methods is that information can be obtained in isolation, but this may be the main disadvantage as the animal body never works in isolation.

The trend towards alternatives will continue but they must be subjected to rigorous and critical independent assessment. Animal experiments and alternative methods will be used together for the foreseeable future.

Obligations of the Research Worker. —It is the humane use and care of all animal life that defines an ethical and conscientious society. When animals are used for research, there is a clear obligation on those using the animals, and on those deriving benefit from such use to:

i. provide humane care and treatment,
ii. minimize pain and discomfort,
iii. avoid unnecessary use of animals.

There is an obligation on those using animals to be accountable for their proper use. This can be achieved by institutional, funding agency, and editorial responsibilities, and periodic reviews by ethics committees.

The institution should have strict codes for the care and use of laboratory animals, and the funding agencies should require adherence to these ethical codes. Nevertheless, the ultimate responsibility for preventing animal pain and suffering lies with the individual scientist.

In the final stage, before they consider publishing the work, editors of journals should make sure that the welfare of the experimental animals has been taken into account.

It should be considered a privilege to use experimental animals, and so their welfare must be assured.

Welfare Considerations: The Animal and the Environment

The environment is central to laboratory animal management and the welfare of the animal must be considered throughout the breeding-holding phase and the experimental phase.

The Breeding-Holding Phase. —The laboratory animals used for experimental purposes have been selected and bred for many generations under laboratory conditions, and need a well-controlled environment to keep them healthy. The design of animal accommodations must take into account the variety of species held and the differing ages of the animals.

The environment influencing the animal during this phase has physical, nutritional, and general biological components.

i. physical environment

Factors which affect the health and welfare of the animals include noise, temperature, humidity, ventilation, and light/dark cycles.

It is known that noise can adversely affect animals which rely on hearing for courtship and mating behavior (Fletcher 1976). Under normal animal house conditions the ambient noise levels are 42-44 decibels and may rise to 54-68 decibels during feeding and cage cleaning. In most cases the duration of this higher level noise is not long enough to cause problems (Fletcher 1976). Sudden noises are often more of a problem than intensity. Auditory fire alarms with gradual buildup cause few of the problems seen with sudden onset alarms. Similarly, crashing feed can lids or other intrusive noises caused by workers in the room are likely to be stressful, affecting productivity and physiological stability.

The optimum temperature range for most laboratory animals is 19-23° Centigrade (Harkness and Wagner 1983), and ideally the temperature should not vary more than 2° Centigrade from an average level.

Control of humidity is necessary for the health and comfort of the animals and staff. A range of 30-70% saturation is acceptable for most animal houses (Harkness and Wagner 1983).

Recirculation of air in an animal house is not advisable as dust and microorganisms could be widely distributed. Room air changes with fresh or filtered air should be between ten and fifteen complete air changes per hour to achieve uniform ventilation, depending on the arrangement and size of the cages and the stocking density (Harkness and Wagner 1983; Lane-Petter 1976).

Direct sunlight is usually excluded from the animal house because it makes the control of lighting and temperature difficult. A lighting cycle of 12 hours daylight and 12 hours darkness ensures a constant breeding cycle throughout the year for most species.

ii. nutritional environment

While the nutritional requirements of some laboratory animals are fairly well documented (Coates 1976), knowledge about the nutritional require-ments of most nonhuman primates and many other species is largely or totally lacking and in need of research attention. Laboratory animals should have access to a clean, reliable water supply and wholesome, clean, nutritious palatable food on a regular basis to ensure the appropriate intake of protein, fat, carbohydrate, vitamins, salts, minerals, and fiber.

iii. biological environment

Major aspects of the biological environment are the control of communic-able disease, and the management of the social environment.

The control of disease organisms (parasites and pathogenic microor-ganisms) is a fundamental requirement for successful laboratory animal management and influences the welfare of each animal. Hygiene in the animal house is directed mainly at the prevention of infections and infestations in the whole animal population. This is achieved by the quarantine of incom-ing animals, restricted entry of people, the maintenance of physical barriers (e.g., flyscreens, air locks, and air curtains), adequate sterilization procedures, and the incorporation of design features that enable the easy and effective cleaning of fixed and moveable structures. Treatment of infection is an aspect of laboratory animal medicine involving specialized diagnostic and management procedures.

People, too, represent a very important aspect of the social environment of animals. The technician or caretaker should be recognized by colony individuals as a dominant member of the social hierarchy. Colonies that fail to recognize this role of the technician/caretaker are likely to inadvertently increase the stress of the colony as demonstrated by a multitude of conditions including increased susceptibility to disease, reduced fertility, and aggression.

The provision of an optimum social environment will ensure the maximum output from each unit of the breeding colony and will help maintain the good health and welfare of the growing animals. The housing and breeding systems selected for each species are the main factors determining the social environment for reproduction. For example, the permanently mated monogamous breeding system is suitable for rats and mice, but unsuitable for rabbits which are temporarily mated (Festing 1976). Design of the reproductive environment also needs an understanding of the behavioral characteristics of the different species. Rats may not litter on a nesting place even when supplied with nesting material, and the young may fall through a grid floor. The cage should have a solid floor and soft nesting materials to prevent possible injury or loss of the young. The social environment of growing animals is influenced by density, mixing of littermates, and sex. The failure to recognize this may jeopardize their health and even survival.

The Experimental Phase. — The welfare of animals used in the experimental environment is ultimately the responsibility of the research worker. Consideration must be given to handling and restraint, experimental procedures (e.g., anesthesia, blood collection), and euthanasia.

The techniques of handling and restraint of each species should be mastered so that the animals are not injured, unduly stressed, or able to inflict injury on the handler. Most experimental animals become accustomed to handling but special restraint methods are necessary for some species, e.g., support of the rear quarters of a rabbit will prevent kicking of the hind legs and so prevent fracture of its lumbar vertebrae or scratching of the handler.

Experimental procedures need to be planned in detail and implemented by personnel experienced in the techniques. Minimization of pain can be achieved by thoughtful selection of the experimental technique and the use of anesthesia, analgesic, or sedative agents where necessary. Drugs that should be favored for use in laboratory animals meet the following criteria: Simple equipment and easy manipulative technique for administration; a reliable action and a wide safety margin in the species being used; nonvolatile, nonexplosive, and nontoxic to the worker during storage and use; cost-effective. Injectable general anesthetic agents, for example, should be considered in preference to an explosive volatile agent such as ether for routine use in the laboratory.

Euthanasia of laboratory animals is carried out in three main circumstances: Culling of unwanted animals, relief of suffering in an individual animal, or as the end-point of an experiment. When undertaking euthanasia, the welfare of the animal and the personnel must be considered. To safeguard the welfare of the animal, the technique chosen should achieve quick, quiet, and painless death and should not induce fear, apprehension, or panic in the animal. Other animals should be protected from the sight, sound, or smell of the procedure. The welfare of the personnel is protected by adopting a procedure that is physically and chemically safe and aesthetically acceptable. Euthanasia should not be carried out in a public or communal area.

Conclusion

The ethics and welfare considerations in laboratory animal management are multiple and complex, involving animals in all phases of their life cycle. In an intensive laboratory animal colony, the welfare of individual animals within the group is of prime importance and is not adequately reflected by measures of overall productivity. Management decisions can be difficult in situations where individual and group welfare interests conflict, e.g., the culling of aged animals from a breeding colony. When planning animal management or experimental procedures, the well being of the animal house staff and research workers should not be overlooked.

Endnotes

[1] Senior Lecturer, Pathology, Dept. of Medical Laboratory Science, Queensland Institute of Technology, Brisbane, Australia, 4000.

[2] Lecturer in Animal Behaviour, Dept. of Animal Sciences and Production, University of Queensland, St. Lucia, Australia, 4067.

References

Balls, M. 1983. Alternatives to experimental animals. *Vet. Rec.* 113 (7): 398-401.

Coates, ME. 1976. The nutrition of laboratory animals. In: *The UFAW Handbook on the Care and Management of Laboratory Animals.* Edinburgh, London and NY: Churchill Livingston. pp. 27-56.

Festing, MF. 1976. Production methods. In: *The UFAW Handbook on the Care and Management of Laboratory Animals.* Edinburgh, London and NY: Churchill Livingston. pp. 57-73.

Fletcher, JL. 1976. Influence of noise on animals. In: *Control of the Animal House Environment.* Laboratory Animal Handbooks 7. London: Laboratory Animals Ltd. pp. 51-62.

Harkness, JE and Wagner, JE. 1983. *The Biology and Medicine of Rabbits and Rodents.* Philadelphia: Lea and Febiger.

Lane-Petter, W. 1976. The animal house and its equipment. In: *The UFAW Handbook on the Care and Management of Laboratory Animals.* Edinburgh, London and NY: Churchill Livingston. pp. 74-94.

Regan, T. 1983. *The Case for Animal Rights.* London, Melbourne and Henley: Routledge and Kegan Paul.

—. 1985. The case for animal rights. In: *In Defence of Animals.* Singer, P. ed. Oxford: Basil Blackwell Pub. Ltd.

Seamer, JH. 1982. B.V.A. Policy on animal experimentation. *Vet. Rec.* 110(11): 241-44.

Singer, P. 1976. *Animal Liberation: Towards an End to Man's Inhumanity to Animals*. London: Jonathan Cape Ltd.

—. 1980. Animals and humans as equals. *Anim. Reg. Stud.* 2: 165-74.

—. 1985. Prologue. Ethics and the new animal liberation movement. In: *In Defence of Animals*. Singer, P. ed. Oxford: Basil Blackwell Pub. Ltd.

Stark, DM. 1983. Developing replacements to whole animal experimental systems. *Calif. Vet.* 37(1): 89-90.

THE CRUEL DECEPTION[1]

Robert Sharpe[2]

With new legislation to replace the Cruelty to Animals Act 1876 near at hand, the powerful vested interest groups whose profits and livelihood depend on laboratory animals are stepping up their campaigns to ensure the survival of vivisection. Have the benefits really been so great, and can vivisection achieve major advances in our present state of health?

History shows (McKeown 1979) that the real reasons for the dramatic increase in life expectancy since the middle of the last century are improvements in nutrition, living and working conditions, hygiene and sanitation, with medical measures only having a relatively marginal effect. The reduction in Britain's death rate from the 1850s was almost exclusively due to the decline of the infections—mainly TB, bronchitis, pneumonia, influenza, whooping cough, measles, scarlet fever, diphtheria, smallpox, and the water- and food-borne diseases such as cholera, typhoid, diarrhea, and dysentery. Mortality for nearly all the infections was declining before, and in most cases long before, specific therapies became available.

The Evidence

Beginning with the airborne infectious diseases, it can be seen that deaths from *tuberculosis* were falling well before the introduction of specific measures to treat or prevent it (McKeown and Lowe 1976; see figure 1, page 10).

Specific drug treatment, introduced in 1947, is considered to have speeded up the decline in England and Wales, but not in America (McKinlay and McKinlay 1977). Whilst it is difficult to assess the contribution of BCG vaccination after introduction in Britain, some doubt surrounds its value under all circumstances (Weitz 1982), whilst the Netherlands had the lowest death rates from respiratory TB for any European country in 1957-59 and in 1967-69 despite its having no national BCG program (McKeown 1979).

Until recently, *pneumonia, bronchitis,* and *influenza* were grouped together in national statistics (McKeown 1979). The introduction of antibiotics does not seem to have made an impact on the declining death rate (McKeown 1979; see figure 2, page 10), but this is hardly surprising, because acute bronchitis and influenza are viral diseases for which antibiotics are ineffective. What the diagram does show is that mortality was declining before any specific medical treatment.

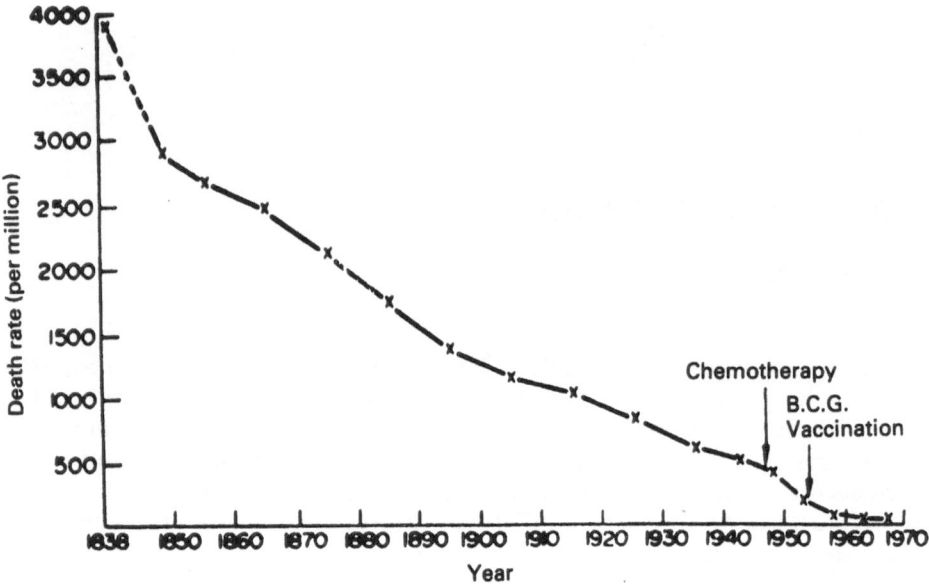

Figure 1. Respiratory tuberculosis: mean annual death rate: England and Wales.

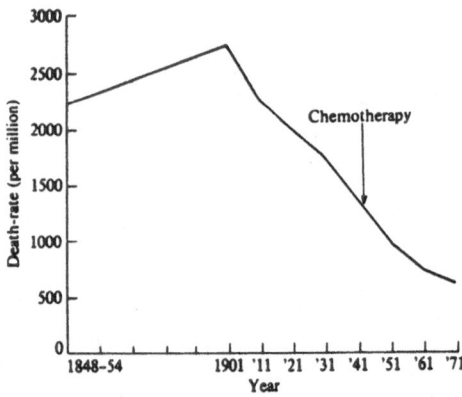

Figure 2. Bronchitis, pneumonia, and influenza: death rates (standardized to 1901 population): England and Wales.

In the United States, on the other hand (McKinlay and McKinlay 1977), statistics are available for pneumonia since 1900 and show that death rates were declining well before the introduction of sulphonamides or antibiotics (figure 3).

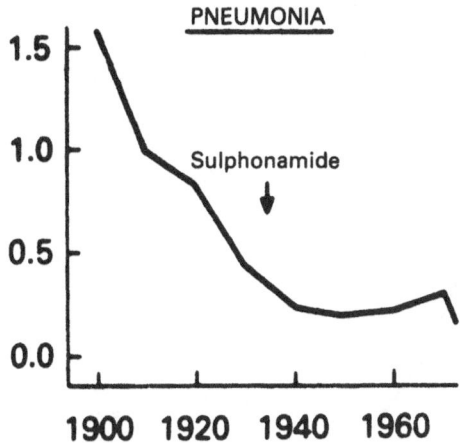

Figure 3. Fall in the standardized death rate per 1000 population in relation to specific medical measures. United States 1900-1970. (After John and Sonja McKinlay 1977.)

In England and Wales, pneumonia gained a place for itself in the national statistics from the early 1930s, and although it is known from clinical experience that antibiotics can treat bacterial diseases, it is difficult to judge their overall contribution from death rates which were already declining. Unhappily, pneumonia, bronchitis, and influenza are still major causes of sickness and death. Since there are many forms of influenza virus, vaccines against one form may be useless against another.

The decline in death rate from *scarlet fever* was well under way before the introduction of sulphonamides in 1935 and antibiotics in 1945 (McKeown and Lowe 1976; see figure 4, page 12).

The death rate from *whooping cough* in England and Wales has declined rapidly since the 1860s (McKeown 1979; see figure 5, page 12). The value and safety of immunization has been hotly debated and in West Germany, where the vaccine was not used nationally, the number of cases still declined (McKeown 1979).

Measles has declined in much the same way as whooping cough, in this case immunization being introduced in England and Wales only recently in 1968 (see figure 6, page 12).

The declining death rate from *diphtheria* in children under 15 years of age is illustrated in figure 7, page 13 (Porter 1971).

Although the sharp decline in death rates between 1860 and 1875 was not associated with any specific therapy, later declines coincided first with the introduction of antitoxin treatment and secondly with the introduction of a national immunization campaign. Had mortality from other common childhood infections remained the same or increased at the same periods then it would be tempting to assume that antiserum and immunization were mainly responsible for the decline in diphtheria deaths around 1900 and 1940.

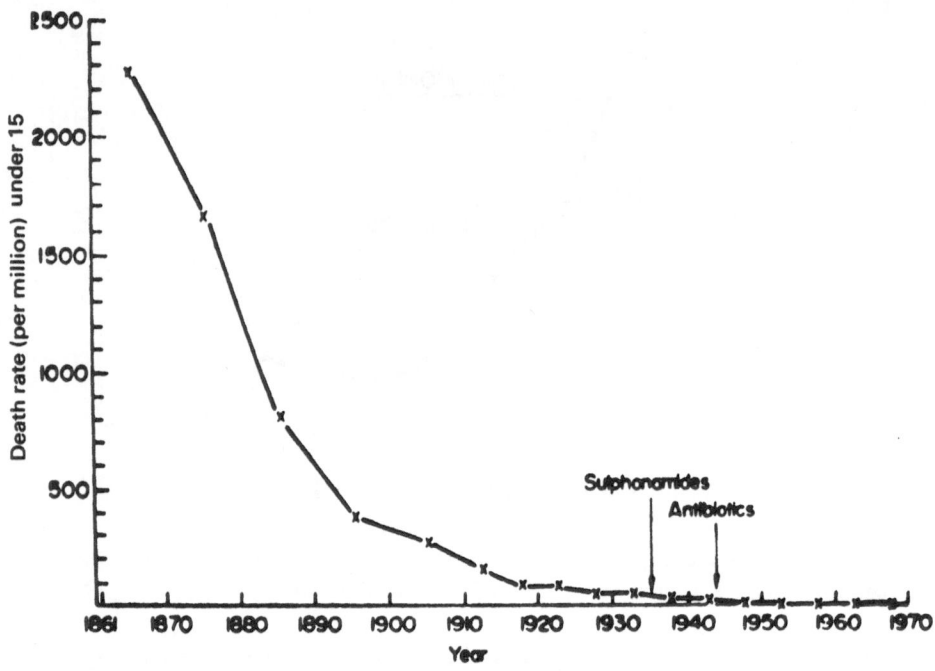

Figure 4. Scarlet fever: mean annual death rate in children under 15: England and Wales.

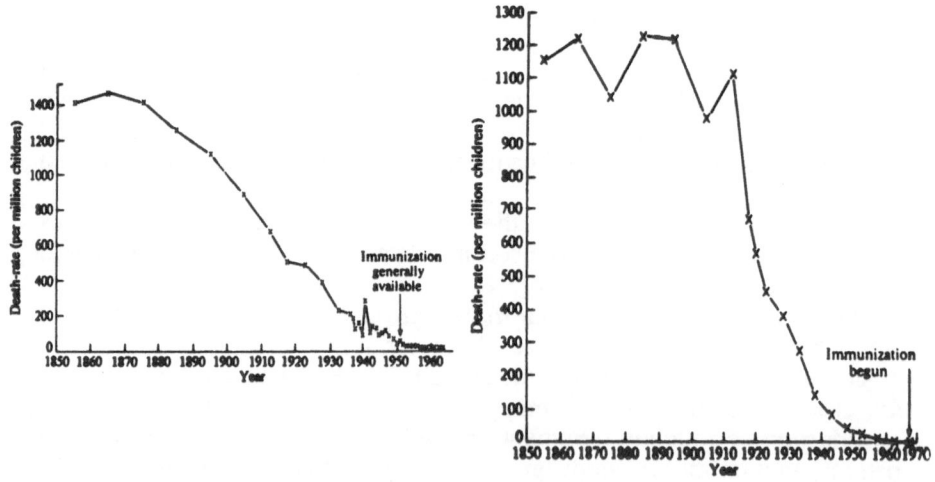

Figure 5. Whooping cough: death rates of children under 15: England and Wales.
Figure 6. Measles: death rate of children under 15: England and Wales.

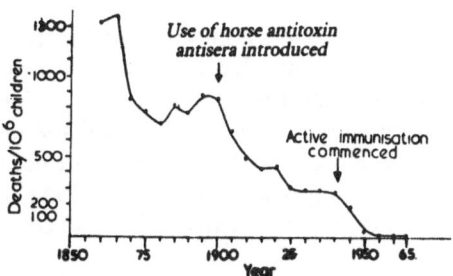

Figure 7. Deaths of children under 15 years attributed to diphtheria.

But deaths from measles and whooping cough, for instance, did decline over this period without any treatment or immunization, suggesting that other influences may also have been operating with diphtheria.

At the same time the value of antitoxin treatment had never been accepted generally (Porter 1971), whilst in the U.S. at least (McKinlay and McKinlay 1977), introduction of immunization did not produce major detectable changes in the already declining death rate. It is clearly not possible to assess accurately the contribution of these measures to the decline of deaths due to diphtheria in England and Wales.

The combined death rates from the commonest infections of childhood (scarlet fever, diphtheria, whooping cough, and measles) among children up to 15 years old, shows that around 90% of the total decline in mortality between 1860 and 1965 had occurred before the introduction of antibiotics, and widespread immunization against diphtheria (Porter 1971; see figure 8).

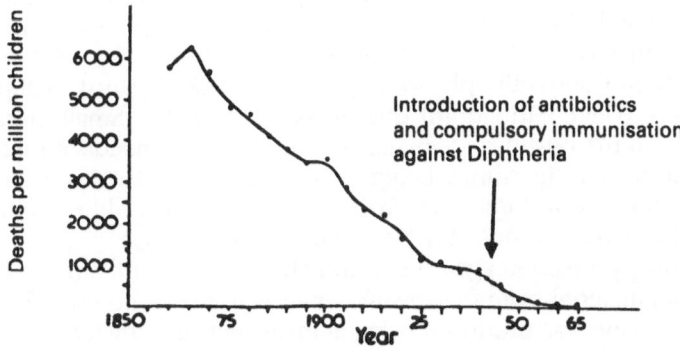

Figure 8. Deaths of children under 15 years attributed to scarlet fever, diphtheria, whooping cough, and measles.

Although the contribution of *smallpox* to the overall decline in death rate in England and Wales between the 1850s and 1970s was small (McKeown 1979), this is the one major disease for which vaccination was available before 1900.

Once again, it is not easy to assess accurately the contribution of vaccination, and historical analysis has shown that a combination of effects is likely to have been responsible (Hardey 1983; see figure 9). Lack of reliable information during earlier centuries coupled with simultaneous decline of many other infectious diseases, suggesting influences other than vaccination, make interpretation of the declining death rate for smallpox difficult. Vaccination was stopped in this country after 1940 when risks were considered to outweigh the benefits.

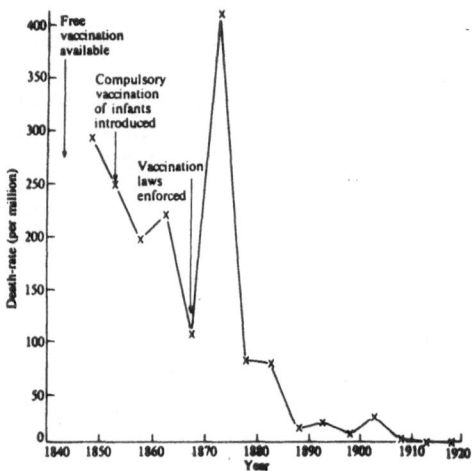

Figure 9. Smallpox: death rates: England and Wales.

Inoculation against smallpox had been practiced in the Far East for many centuries, long before Jenner developed a vaccine using cowpox in 1798. It is thought that Lady Wortley Montague (Porter 1971), wife of the British ambassador in Constantinople, introduced inoculation against smallpox into this country [United Kingdom] during the early 1700s. Small amounts of material from the pustules of smallpox sufferers were inoculated into those seeking protection, immunity being conferred against dangerous attacks. It had also been realized that risks from the inoculation could be reduced by making fluid from the pustules less virulent (Inglis 1963).

Developing protection against smallpox has travelled full circle. Centuries ago, human material from fellow sufferers was used. In more modern times smallpox vaccine was produced using animals, but recently techniques have been developed to produce the vaccine from human cells in culture.

Declining death rates from water- and food-borne diseases, such as cholera, dysentery, diarrhea, typhoid, and nonrespiratory tuberculosis, also made a major contribution (McKeown 1979; McKeown and Lowe 1976). *Cholera* was introduced into Britain from the continent of Europe during the nineteenth century. The last epidemic was in 1865 and from that time cholera was negligible.

The decline of cholera is a good example of how practical advances are not necessarily dependent on scientific understanding (Porter 1971). By careful observation John Snow deduced that contaminated water was the origin of the spread of cholera in a London outbreak. In 1854 he removed the handle of the Broad Street pump to prove his point and the outbreak stopped. Yet this was nearly 30 years before the discovery of the cholera bacterium.

The general decline of the *diarrheal diseases* began in the late nineteenth century. From the early 1900s, a rapid decline set in and by the time intravenous rehydration therapy was introduced in the 1930s to prevent dehydration, 95% of the decline had already occurred (McKeown 1979).

The decline of the enteric fevers—*typhoid* and *paratyphoid*, a major killer of the 19th century—was also rapid and began before the turn of the century. Specific treatment was not available until 1950, but by then mortality from enteric fever had almost been eliminated in England and Wales (McKeown 1979).

Tuberculosis can also arise by drinking milk from cows with bovine tuberculosis. Disease levels were already low by the time the first specific treatment, streptomycin, was introduced (McKeown 1979).

Consequently, deaths from nearly all the infectious diseases were declining before, and in most cases long before, effective therapeutic procedures became available. According to a detailed analysis by McKeown (see McKinlay and McKinlay 1977), the situation for the 19th century can be summarized:

> ...the decline in mortality in the second half of the nineteenth century was due wholly to a reduction of deaths from infectious diseases; there is no evidence of a decline in other causes of death. Examination of the diseases which contributed to the decline suggested that the main influences were: a) rising standards of living, of which the most significant feature was a better diet; b) improvements in hygiene; and c) a favorable trend in the relationship between some micro-organisms and the human host. Therapy made no contributions, and the effect of immunization was restricted to smallpox which accounted for only about one-twentieth of the reduction of the death rate.

Whilst for the 20th century:

> The main influences on the decline in mortality were improved nutrition on air-borne infections, reduced exposure (from better hygiene) on water and food-borne diseases and less certainly, immunization and therapy on the large number of conditions included in the miscellaneous group. Since these three classes (of infections) were responsible respectively for nearly half, one-sixth, and one-tenth of the fall in death rate, it is probable that the advancement in nutrition was the major influence.

And in the United States, American researchers J. and S. McKinlay of Boston University have concluded (McKinlay and McKinlay 1977):

> In general, medical measures (both chemotherapeutic and prophylactic) appear to have contributed little to overall decline in mortality in the United States since about 1900—having in many instances been introduced several decades after a marked decline had already set in and having no detectable influence in most instances. More specifically, with reference to those five conditions (influenza, pneumonia, diphtheria, whooping cough, and poliomyelitis) for which the decline in mortality appears substantial after the point of intervention —and on the unlikely assumption that all of this decline is attributable to the intervention—it is estimated that at most 3.5% of the total decline in mortality since 1900 could be ascribed to medical measures introduced for the diseases described here.

Declining Death Rates

Rene Dubos (Pasteur's biographer):

> ...by the time laboratory medicine came effectively into the picture the job had been carried far toward completion by the humanitarian and social reformers of the nineteenth century (*Mirage of Health*, 1959, New York).

Consequently, society's "control" of the infectious diseases rests primarily on efficient public health services and a good standard of living. Medical measures clearly played only a relatively minor role in increasing life expectancy. It must therefore be concluded that, *at most,* animal experiments had only a marginal effect on reducing the death rate, even on the unlikely assumption that animals were involved in developing all such measures.

Of course animals were not involved in all therapeutic advances, in truth many discoveries being initially made by human observation, only later to be retested on animals! Examples include the discovery of vitamins and the early inhalation anesthetics (which enabled surgery to advance rapidly), as well as many vital medicines such as morphine and digitalis (Koppanyi and Avery 1966; Talalay 1964; see also *Clinical Medical Discoveries* by M. Beddow Bayly, 1961, National AntiVivisection Society).

The Situation Today

According to national statistics the principal causes of death in today's Western society include heart disease, cancer, strokes, bronchitis and emphysema, alcohol-related problems, diabetes, and high blood pressure, most of which are very difficult or impossible to cure but all largely preventable, because they are often caused by our affluent and wasteful Western lifestyle. We suffer these illnesses because of faulty diet and because we smoke, eat and drink to excess, fail to exercise properly and allow our environment to be polluted by dangerous substances.

One example is smoking, which is considered to be the single most preventable cause of death in the United Kingdom, being responsible for 100,000 premature deaths a year (Russell 1982; *Brit. J. Addiction* 1984). Tobacco has been connected with various cancers, heart disease, bronchitis and emphysema, peptic ulcers, and is a hazard to the fetus (*Brit. Med. J.* 1983). Another example is faulty diet, which has been linked to heart disease, certain common cancers, strokes, diabetes, high blood pressure, and a host of other diseases (*New Scientist* 1983).

Consequently if *major* advances in our health are to be achieved the emphasis must be on prevention, since treatment is so difficult. Once again, it follows that *at best*, animal experimentation could only have a marginal impact, even assuming it to be a reliable, or indeed the only, method of research.

Are We Getting Healthier?

In recent years, with animal experiments carried out on a huge scale, has our overall health improved? In 1951, a 45-year-old man could expect to live an extra 26.4 years, whilst in 1981 he could a further 27.5 more years (*Social Trends* 1985). Despite this small increase, evidence from America suggests that this is more than counterbalanced by more years of serious disablement (Melville and Johnson 1982).

According to the General Household Survey (*Social Trends* 1984), there has been a progressive *increase* in the number of people reporting chronic sickness (defined as any long standing illness, disability, or infirmity) between 1972 and 1982. This is supported by Government figures which show an increase between 1967 and 1983 in the number of insured people incapacitated by sickness (*Social Security Statistics* 1984). In 1982 an astonishing 41% of British men aged between 45 and 64, and 42% of women in the same age group, reported being chronically sick (*Social Trends* 1985). In 1961, 233.2 million prescriptions were dispensed in the U.K. and this had risen dramatically to 389.2 million in 1983 (*Social Trends* 1985)! It doesn't look as if we are getting healthier.

Conclusion

This article has not addressed ethical issues or the frequently misleading nature of animal-based research. Nevertheless, analysis of disease trends has enabled the following conclusions to be reached: i) The dramatic increase in life expectancy since the middle of the last century is overwhelmingly based on improvements in public health; ii) The impact of animal experiments on such improvements in health can only have been marginal, *at best*, whilst major advances in the future must once again come from disease prevention; iii) Despite the enormous scale of animal experimentation in recent years, our overall health appears to be declining.

Of course there is a place for research, but conducted humanely without using animals. Nevertheless, whatever methods are used, much more could be achieved by effective disease prevention campaigns.

Endnotes

[1] Reprinted with permission from *Animals' Defender and Anti-Vivisection News*, May/June 1985.

[2] Director, Lord Dowding Fund, The National AntiVivisection Society, Ltd., 51 Harley St, London W1N 1DD, England.

References

Brit. J. Addict. 1984.79: 241-43.

Brit. Med. J. 1983. Nov. 26: 1570-71.

Hardey, A. 1983. *Med. Hist.* 27: 111-28.

Inglis, B. 1963. *A History of Medicine.*

Koppanyi, T. and Avery, MA. 1966. *Clin. Pharmacol. and Therap.* 7: 250-70.

McKeown, T. 1979. *The Role of Medicine.* Oxford: Blackwell.

McKeown, T. and Lowe, CR. 1976. *An Introduction to Social Medicine.* Oxford: Blackwell.

McKinlay, JB and McKinlay, S. 1977. *Health and Society.* (Millbank Memorial Fund). pp. 405-28.

Melville, A. and Johnson, C. 1982. *Cured to Death: The Effects of Prescription Drugs.* Secher and Warburg.

New Scientist. 1983. July 14 and Oct. 26.

Porter, RR. 1971. Presidential address at the Swansea Meeting of the British Association for the Advancement of Science. Sept. 3, 1971.

Russell, W. 1982. *Brit. Med. J.* March 13: 837.

Social Security Statistics. 1984. p. 26 (DHSS).

Social Trends. 1984. No. 14

—. 1985. No. 15.

Weitz, M. 1982. *Health Shock.* Hamlyn.

THE SIGNIFICANCE OF
ALTERNATIVE TECHNIQUES IN BIOMEDICAL RESEARCH:
AN ANALYSIS OF NOBEL PRIZE AWARDS[1]

Martin Stephens[2]

Introduction

The current controversy over the use of animals in research has put the biomedical and psychological communities on the defensive. Their main defense has been to emphasize the health benefits to humans from animal research (e.g., Miller 1985; Gay undated).

This defense has been challenged on scientific grounds in at least two ways. First, critical assessments of various fields of animal research have revealed that claims of health benefits are grossly exaggerated (Drewitt and Kani 1981; Kuker-Reines 1982; Giannelli 1985; Sharpe 1985; Reines 1986, undated; Stephens 1986a; Anonymous 1986). A second and complementary response is to argue that, whatever the benefits have been, the use of animals should be replaced to the fullest extent possible by alternative methods.

Alternatives include the three Rs of replacement, reduction, and refinement (Russell and Burch 1959), that is, methods that completely *replace* the use of animals in a procedure, that *reduce* animal use, and *refine* procedures so that pain, suffering, or deprivation are lessened (Stephens 1986b). Reduction and refinements are considered interim steps toward the ultimate goal of the complete replacement of laboratory animals with nonanimal methods.

The "alternatives approach" consists of developing and employing methods specifically designed as alternatives. The aim of the approach is to determine the extent to which alternatives can replace traditional uses of animals. This aim has an ethical and compassionate appeal that is being bolstered by recent scientific advances in developing alternatives (Stephens 1986b).

The alternatives approach was first discussed comprehensively in 1959 by Russell and Burch (1959). Before this, some alternative methods already had been developed but were employed almost exclusively for scientific reasons, with humane considerations being overlooked. The alternatives approach advocates both scientific and humane considerations.

Perhaps because of the recentness of this approach—indeed, of some of the alternatives themselves—animal advocates have largely overlooked the past achievements of alternative techniques in research. One of the few

historical examples that proponents of alternatives have cited is the role of
cell culture in polio research (Rowan 1979). The present paper documents
many others.

Table 1 lists methods that are considered here to be alternatives. Some
of these methods are readily classified according to the three Rs. For example,
human studies and physicochemical assays are replacements. However, other
methods can fall into more than one category. For example, the method of
using less sentient organisms is a replacement when microorganisms or plants
are employed instead of multicellular animals, while the method will usually
be a refinement when invertebrates are employed instead of vertebrates.

The importance of alternative methods in the history of biomedical research
can be inferred from Nobel Prize awards in medicine or physiology. These
awards are generally believed to recognize research "of the highest caliber,
the most enduring influence, and the most importance to biomedical science"
according to the National Academy of Sciences (1985). Before turning to an
analysis of these awards, we should note that any comparison of the historical
significance of alternative methods and traditional animal research is likely to
be biased against alternatives for several reasons, including the acceptance of
animal methods as the main paradigm of research, the historical paucity of
ethical and compassionate concern for laboratory animals, and the newness of
some alternative methods. Hence we should not expect alternative methods to
have outshined traditional animal methods in the history of biomedical research.

Table 1. Alternative Methods

---

1. Human Studies
   a. epidemiological
   b. clinical (observations as well as studies)
   c. post-mortem
2. In vitro studies
   a. studies of sub-cellular components
   b. short-term (less than 24 hr.) studies of cells or tissues
   c. studies of cells or tissue in culture
3. Mathematical and computer modeling studies
4. Studies of less sentient organisms
   a. vertebrate embryos
   b. invertebrates
   c. microorganisms
   d. plants
5. Physicochemical assays of biological substances
6. Miscellaneous methods
   a. naturalistic studies of animals
   b. clinical studies of animals
   c. studies of mechanical models

Methods

Seventy-six Nobel Prizes in medicine or physiology have been awarded between 1901, the inaugural year, and 1985. These awards were classified into two categories, alternative and nonalternative, depending on whether or not alternative methods made a major contribution to the research. The major contributions to projects in the non-alternative category were made by in vivo studies of vertebrates. Most projects clearly fell into one or the other category. It is recognized that the classification of some projects is arguable, but the conclusions presented below would hold regardless of how these few cases were judged.

The classification scheme examines only the prize-winning research itself, and does not consider the pre-existing foundation of biomedical knowledge necessary to conduce the research (see Comroe and Dripps 1976).

Sufficient information was available to classify all but two awards (those given in 1906 and 1924). When an award was divided among two or more research projects, it was classified in the alternative category as long as alternative techniques made a major contribution to at least one project. Eight awards were of this kind and are identified as such in the Results section.

Any award-winning project in the alternative category is a testament to the power of the techniques advocated by animal protectionists. However, studies that involved alternative techniques can be subdivided into two categories, namely, those that could have been conducted on intact vertebrates, but which were not, and those that necessarily could not. Although projects in the former subcategory hold more promise for animal advocates, projects in the latter category should also be welcomed by animal advocates as well as others for contributing to biomedical knowledge in ways that traditional animal methods could not. Examples of these projects include descriptive studies of molecular biology, in which sub-cellular components are researched.

It is recognized that this secondary classification, as with the primary classification, involves subjectivity in some cases.

Results

Surprisingly, fully two-thirds (50) of the Nobel Prizes fall into the alternatives category. Table 2 provides a brief description of these awards and lists the alternative techniques associated with each one.

I emphasize that the classification of a project in the alternative category does not mean that intact, vertebrate animals were not used. Rather, this designation indicates that alternative methods were the key, even though in vivo vertebrate methods also may have been involved and may have played an important role.

Nicolle's research on the spread of typhus, awarded the Nobel Prize in 1928, illustrates this distinction. Nicolle, a hospital physician, observed that people who touched new typhus patients or their clothes contracted typhus themselves. Yet once the patients were admitted to the hospital, contagion ceased. In his Nobel lecture, Nicolle noted:

I asked myself what happened between the hospital door and the sickroom. What happened was this: the typhus patient was relieved of his clothes and linen, and was shaved and washed. The agent of the contagion was therefore something attached to the skin, to his linen, and something of which soap and water rid him. This could only be the louse....

If it had not been possible to reproduce the malady in animals and consequently to verify the hypothesis, this simple determination would have sufficed to make clear the mode of propagation of typhus (quoted in Sourkes, 1966, p. 134).

Hence, Nicolle himself stated that his clinical observations were the key to identifying the louse as the agent of typhus transmission. This discovery, in turn, was the main reason for Nicolle's receipt of the prize (Sourkes 1963).[3]

Figure 1 shows the prevalence of alternative projects among the Nobel Prize winners throughout this century. The prevalence has increased steadily during the past 70 years. During the last 20 years, 19 of the 20 prizes were awarded to projects in the alternative category! This has resulted primarily from the increasing prominence of in vitro studies, including molecular and biochemical studies of sub-cellular components and studies of cells and tissues in culture.

The 50 studies in the alternatives category were further classified into those that could have been conducted on intact vertebrates and those that necessarily could not (see Methods). The projects fell about equally in both subcategories (24 vs. 26, respectively).

The award-winning projects in either subcategory form a diverse collection. Some projects were of practical value in the fight against diseases. Examples include the discovery of the insect vector of malaria (awarded the Nobel Prize in 1902), the discovery of the louse as the transmitter of typhus (1928), the discovery of penicillin (1945), the production of a vaccine against yellow fever (1917), the discovery of a hormonal treatment for prostate and breast cancer (1966), and the elucidation of the mechanism for familial hypercholesterolemia (1985). The successful cultivation of the polio virus in tissue culture (1954) paved the way for the development of effective polio vaccines.

Other award-winning projects using alternative methods made significant contributions to basic biology. Examples include the discovery that genes regulate chemical processes—the one gene, one enzyme principle (1958), the discovery of the double helix structure of DNA (1962), and the discovery of the interaction between tumor viruses and the genetic material of cells (1975).

Still other projects developed techniques that have rapidly become invaluable in biomedical research. These developments include radioimmunoassay (1977), restriction enzymes (1978), computer-assisted tomography (1979), and monoclonal antibodies (1984).

Let us examine the most recent, award-winning project as an example of the kinds of studies that constitute such highly regarded research. Brown and Goldstein were honored in 1985 for elucidating the fundamental mechanism of familial hypercholesterolemia (FH), a genetic disorder of cholesterol metabolism that predisposes carriers to coronary heart disease. Much of Brown and Goldstein's research was conducted in vitro. Fibroblasts from patients homozygous for FH were cultured and shown to have abnormal

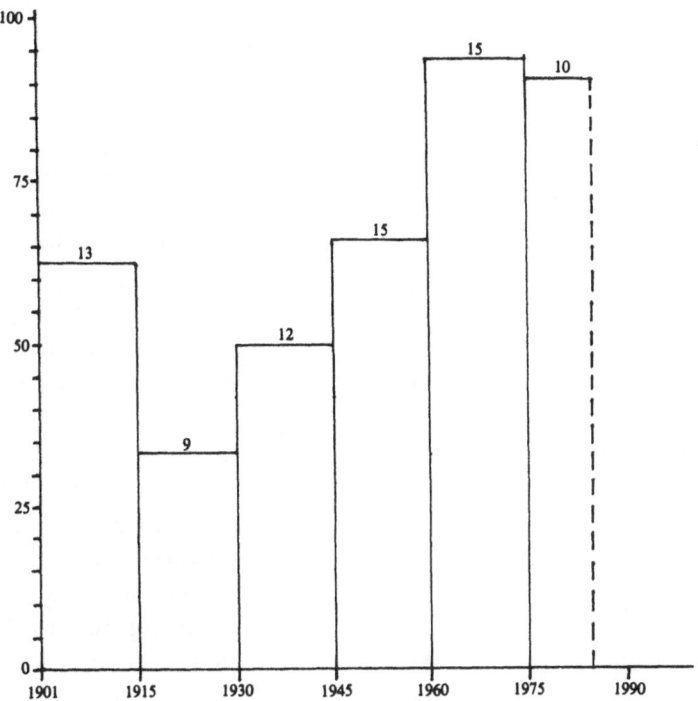

Figure 1. Percentages of twentieth century Nobel Prizes in medicine or physiology awarded to projects that involved major contributions from alternative techniques. (Numbers atop bars indicate sample sizes.)

cholesterol metabolism and cholesterol-associated receptors. The pathway of cholesterol through the cell, including the critical step of receptor-mediated endocytosis, was worked out.

Molecular studies on patient-derived material showed the correspondence between the structure and function of the receptors and also identified various mutations that led to defective receptors.

In addition to these in vitro studies, Brown and Goldstein's work also involved clinical studies. A young patient was homozygous for FH and therefore lacked functional cholesterol-associated receptors. The patient had very high cholesterol levels and severe coronary arteriosclerosis. A liver transplant seems to have corrected the receptor problem and a heart transplant replaced the damaged heart.

Brown and Goldstein's work, which has implications beyond cholesterol metabolism and heart disease, was recently reviewed in *Science* (Motulsky 1986). That comprehensive article mentions animals studies in only one line, and these studies were apparently not conducted by Brown and Goldstein.

Table 2. Research projects awarded Nobel Prizes that involved major contributions from alternative techniques.

Year	Winner	Technique[1]	Topic
1902	R. Ross	LSO	discovered insect vector of malaria (Anopheline mosquitoes) and other aspects of this disease
1903	N. Finsen	IV	treatment of diseases, especially lupus melgaries, with concentrated light radiation
1907	C. Laveran	H	role of protozoa in causing diseases
1908	E. Metchnikoff[2]	LSO, IV+	immunity (LSO = larval starfish and water fleas)
1909	T. Kocher	H	physiology, pathology, and surgery of the thyroid gland
1910	A. Kossel	IV	protein chemistry of cells, including nucleic substances
1911	A. Gullstrand	H, MM	dioptrics of the eye
1914	R. Bárány	H	physiology and pathology of the vestibular apparatus
1915-1918, 1921, 1925: No Prizes awarded			
1927	J. Wagner-Jauregg	M	malaria inoculation in treatment of dementia paralytica
1928	C. Nicolle	H+	work on typhus
1930	K. Landsteiner	IV	discovery of the human blood groups
1931	O. Warburg	IV	nature and mode of action of the respiratory enzyme in yeast
1933	T. Morgan	LSO or IV	role of the chromosome in heredity (fruit flies)
1935	H. Spemann	LSO	organizer effect in amphibian embryonic development
1937	A. von Szent-Györgyi	IV	biological combustion process, with special reference to vitamin C and fumaric acid

Table 2. Research projects awarded Nobel Prizes that involved major contributions from alternative techniques. *(continued)*

Year	Winner	Technique[1]	Topic
1940–1942: No Prizes awarded			
1944	J. Erlanger H. Gasser	IV	differentiated functions of single nerve fibers
1945	Fleming[2]	IV+	discovery of penicillin
1946	H. Muller	LSO	production of mutations by X ray (LSO = fruit fly)
1948	P. Müller	LSO	efficiency of DDT as a contact poison against several species of insects
1949	E. Moniz[2]	H	therapeutic value of a psychosurgical procedure in certain psychoses
1951	M. Theiler	IV+	vaccine against yellow fever
1953	H. Krebs F. Lipmann	IV	citric acid cycle and coenzyme A and its role in intermediary metabolism
1954	J. Enders F. Robbins T. Weller	IV+	cultivation of poliomyelitis viruses in tissue culture
1955	H. Theorell	IV	nature and mode of action of oxidizing enzymes
1956	A. Courand W. Forssmann D. Richards	H	heart catheterization and pathological changes in the circulatory system
1958	G. Beadle E. Tatum	LSO	genes regulate chemical processes (bread mold)
	J. Lederberg	LSO	genetic recombination and the organization of the genetic apparatus of bacteria
1959	S. Ochoa A. Kornberg	IV	mechanisms of the biological synthesis of RNA and DNA
1962	F. Crick J. Watson M. Wilkins	IV, PC	molecular structure of nucleic acids and its significance for the transfer of information in living material

Table 2. Research projects awarded Nobel Prizes that involved major contributions from alternative techniques. *(continued)*

Year	Winner	Technique[1]	Topic
1963	J. Eccles A. Hodgkin A. Huxley	IV +	ionic involvement in the excitation and inhibition of nerve cell membranes
1964	K. Bloch F. Lynen	LSO, IV	mechanism and regulation of cholesterol and fatty acid metabolism (LSO = yeast)
1965	F. Jacob A. Lwoff J. Monod	LSO	genes that control activity of other genes (LSO = bacteria and viruses)
1966	C. Huggins[2]	H +	hormonal treatment for cancer of prostate and breast
1967	G. Wald[2] K. Hartline	IV (+ ?) LSO, H (+ ?)	chemical and physiological visual process in the eye (LSO = Limulus)
1968	M. Nirenberg R. Holley H. Khorana	IV IV IV	interpretation of the genetic code and its function in protein synthesis
1969	M. Delbruck A. Hershey S. Luria	LSO/IV LSO/IV LSO/IV	replication mechanism and genetic structure of bacterial viruses
1970	B. Katz[2]	IV (+ ?)	transmitters in nerve terminals and the mechanism of their storage, release, and activation
1971	E. Sutherland, Jr.	IV	mechanisms of the action of hormones
1972	R. Porter G. Edelman	IV IV	chemical structure of antibodies
1973	K. von Frisch K. Lorenz N. Tinbergen	Mi[3] Mi Mi	organization and elicitation of individual and social behavior patterns
1974	A. Claude G. Palade C. de Duve	IV IV IV	structural and functional organization of the cell
1975	R. Dulbecco D. Baltimore H. Temin	IV + IV IV	interaction between tumor viruses and the genetic material of cells

Table 2. Research projects awarded Nobel Prizes that involved major contributions from alternative techniques. *(continued)*

Year	Winner	Technique[1]	Topic
1976	B. Blumberg[2]	H	new mechanisms for the origin and dissemination of infectious disease
1977	R. Yalow	IV	development of radio-immunoassay and the principles underlying it
	R. Guillemin A. Schally	Mi[4]	hypothalamic hormones
1978	W. Arber H. Smith D. Nathans	LSO/IV LSO/IV LSO/IV	discovery and application of restriction enzymes (LSO = bacteria and viruses)
1979	A. Cormack G. Hounsfield	MM, H MM, H +	development of the X-ray diagnostic technique, computer-assisted tomography
1981	R. Sperry[2]	H +	functions of the cerebral hemispheres
1982	S. Bergstrom B. Samuelsson J. Vane	IV, PC IV, H + IV +	biochemistry and physiology of prostaglandins
1983	B. McClintock	LSO	discovery of mobile genetic elements (LSO = corn)
1984	C. Milstein G. Kohler[5]	IV	development of a technique for monoclonal antibody formation
1985	M. Brown J. Goldstein	IV, H	cholesterol biochemistry and hypercholesterolemia

[1] H = human studies, IV = in vitro studies, MM = mathematical modeling, PC = physicochemical techniques, LSO = studies of less sentient organisms (vertebrate embryos, invertebrates, microorganisms, and plants), and Mi = miscellaneous. + = in vivo studies of nonhuman vertebrates (non-embryos) also involved.
[2] Award shared with researcher(s) whose work involved major contributions from non-alternative methods.
[3] All three researchers conducted naturalistic, ethological studies.
[4] Biological material (hypothalami) derived from slaughterhouse animals.
[5] Award shared with N. Jerne for his theoretical contributions.

Sources:
Sourkes, T.L. 1966. *Nobel Prize Winners in Medicine and Physiology, 1901-1965*. New York: Abelard-Schuman.
Science. 1966-1984. Various articles on Nobel Prize winners.
Garfield, E. 1985. The 1984 Nobel Prize in Medicine is awarded to Niels K. Jerne, Cesar Milstein, and Georges J.F. Kohler for their contributions to immunology. *Current Contents*, 11 November.

Discussion

The number of major contributions that alternative techniques have made to Nobel Prize-winning research is astonishing. Most of the Nobel Prizes in medicine or physiology were awarded before the alternatives approach was first articulated in 1959, and this approach has yet to be embraced as a guiding principle of biomedical research. Undoubtedly, few if any award-winning projects used alternative techniques out of concern for animals. Yet the adoption of alternative methods is welcome regardless of its underlying motivation. The adoption of alternatives solely for the sake of science increases expectations of what can be achieved if alternatives are adopted for the sake of animals as well.

As striking as the results are, even more awards would have gone to projects that used alternatives techniques if not for the traditional emphasis on in vivo vertebrate studies in biomedical research. For example, many animal researchers were skeptical of tissue-culture systems in the early days of this technique's existence. If not for this skepticism, tissue culture "might have been used to discover many of the vitamins, amino acids, and hormones" according to the National Academy of Sciences (1985). Tissue culture could have been used to discover the hormone insulin, for instance. Even human studies could have yielded this discovery. Yet the researchers who discovered insulin used traditional in vivo methods on dogs. They were awarded the Nobel Prize in 1923. This byproduct of tradition is often regarded as a triumph of animal research, yet other techniques could have done the job.

The results presented above contrast with statements made by defenders of traditional animal research. In the foreword to the latest compendium of the health benefits of animal research (Gay undated), William Raub of the National Institutes of Health (NIH) praises animal research and downplays alternative methods:

> Research with laboratory animals has been so integral to the progress of biomedicine that it is difficult to exaggerate the contribution. Virtually every medical innovation of the last century—and especially the last four decades— has been based to a significant extent upon the results of animal experimentation. Had laboratory scientists studied only relatively simple living systems such as invertebrates, microorganisms, and cell cultures or had clinical scientists lacked avenues of inquiry apart from human experimentation with all its necessary ethical constraints, mystery would reign in many areas where invaluable knowledge now exists (Raub, undated).

Clearly, the public is receiving a biased view of biomedical research from the animal research industry.

This industry has also provided a biased view of the importance of traditional animal experiments in Nobel Prizewinning biomedical research. The National Society for Medical Research (NSMR), a now-defunct industry group that defended animal research, identified all award-winning projects from 1901 to 1977 that used animals (NSMR, undated). At first glance, NSMR's list is impressive, totalling 36 prizes (out of a potential 58). However, NSMR's analysis is misleading for several reasons. First, although public outcry against the use of animals in research focuses primarily on vertebrates (indeed, mostly mammals and birds), NSMR's list includes several projects on invertebrates, including mosquitoes (1902), starfish and water fleas (1908), fruit flies (1933, 1958), and horseshoe crabs (1967/Hartline).

Second, NSMR's analysis did not distinguish between in vivo studies and alternative studies of animals. The latter include in vitro studies (1953, 1963, 1967/Wald, 1970, 1977), ethological studies (1973), and biochemical studies using normally discarded material from slaughterhouses (1977).

Third, no assessment was made of the importance of intact vertebrates in research projects that also involved alternative techniques. Hence, several projects are listed for which alternative techniques were the key feature (1928, 1945, 1951, 1966/Huggins, and 1976/Blumberg).

When the foregoing points are taken into consideration, less than 25 prizes remain on NSMR's list—a modest total indeed.

Nobel Prizes in medicine or physiology were also analyzed by the National Academy of Sciences (1985), in an assessment of the value of research employing various "model systems." The Academy concluded that studies of microorganisms and invertebrates (that is, less sentient organisms), as well as "lower vertebrates," have made great strides in our understanding of biology and medicine. Unfortunately, the report noted, "the proportion of NIH resources that supports research in this area may be small in comparison to the resources dedicated to research with mammals" (p. 75).

One of the Academy's conclusions was remarkably supportive of alternative techniques:

> Proposals for the study of invertebrates, lower vertebrates, microorganisms, cell- and tissue-culture systems, or mathematical approaches should be regarded as having the same potential relevance to biomedical research as proposals for work on systems that are phylogenetically more closely related to humans (p. 75).

Researchers as well as funding agencies should heed this advice. Unfortunately many researchers currently consider alternative methods to be "adjuncts" to traditional animal research. The present paper suggests that this somewhat pejorative label should be dropped, and alternative methods should gain higher prominence in the researcher's armamentarium against biomedical ignorance.

This gain in prominence was predicted by Sir Peter Medawar, a Nobel Laureate in medicine or physiology. Although Medawar defended animal research and argued that its abolition would seriously impair research, he added:

> ...this does not imply that we are for evermore, and in increasing numbers, to enlist animals in the scientific service of man. I think that the use of experimental animals on the present scale is a temporary episode in biological and medical history, and that its peak will be reached in ten years' time, or perhaps even sooner (Medawar 1972).

Medawar's words are especially relevant to his own field—immunology. Medawar used intact vertebrate animals in his prize-winning research, yet immunology is currently benefitting greatly from in vitro work (Garfield 1985; National Academy of Sciences 1985).

An expanded role for alternative techniques can transform biomedical research from an animal-centered enterprise to a human-centered one. Human-centered research would emphasize sophisticated clinical, epidemiological, and post-mortem studies, as well as tissue culture, mathematical modeling, and physicochemical studies of human material or human-derived data. Technical advances, such as positron emission tomography, are expanding the scope of sophisticated, non-invasive clinical studies.

Conclusion

Alternative techniques have made a major contribution to some of the twentieth century's most significant biomedical research. In some cases these techniques have substituted for the use of vertebrates; in other cases they have added to our biomedical knowledge in ways that were not feasible or practical using vertebrates.

The results provide a firm basis for expecting researchers to increase their reliance on alternative techniques and decrease their reliance on traditional animal research methods.

The ultimate goal of the alternatives approach should be human-centered research (as discussed above), with animal studies limited to naturalistic and clinical situations.

Acknowledgements

I extend my thanks to J. McArdle for suggesting that this analysis be undertaken and for commenting, along with A.N. Rowan, on an earlier draft of this manuscript. I also thank S. O'Berri, G. Lugbill, and C. Blair for secretarial assistance.

Endnotes

[1] This paper is an expanded version of an analysis presented in *Alternatives to Current Uses of Animals in Research, Safety Testing, and Education.* (Washington, DC: The Humane Society of the United States, 1986.)

[2] Dr. Stephens is Associate Director, Laboratory Animal Welfare, The Humane Society of the United States, 2100 L St, NW, Washington, DC 20037.

[3] Despite the insight that Nicolle gained from his clinical observations, it is arguable that he would have been awarded a Nobel Prize without having demonstrated the louse's role using animals. However, this is largely a sociological issue, not a biomedical one. It testifies to the influence that Koch's postulates held among researchers (see Kuker-Reines 1982, p. 9). Today, the greater availability of alternative techniques often provides non-traditional means of demonstrating biological relationships, if such demonstrations are deemed necessary. We need to be more imaginative in our studies and less reliant on animals.

References

Anonymous. 1986. Cocaine research on animals. *Reverence for Life*. March-April: 8-9.

Comroe, JH. and Dripps, RD. 1976. Scientific basis for the support of biomedical science. *Science*. 192: 105-11.

Drewett, R. and Kani, W. 1981. Animal experimentation in the behavioral sciences. In: Sperlinger, D. ed. *Animals in Research: New Perspectives in Animal Experimentation*. New York: John Wiley and Sons. pp. 175-201.

Gay, WI. ed. Undated. *Health Benefits of Animal Research*. Washington, DC: Foundation for Biomedical Research.

Giannelli, MA. 1985. Three blind mice, see how they run: A critique of behavioral research with animals. In: Fox, MW and Mickley, LD. eds. *Advances in Animal Welfare Science, 1985-86*. Washington, DC and Dordrecht, The Netherlands: The Humane Society of the United States and Martinus Nijhoff, respectively. pp. 109-64.

Kuker-Reines, B. 1982. *Psychology Experiments on Animals: A Critique of Animal Models of Human Psychopathology*. Boston: New England Anti-Vivisection Soc. (NEAVS).

Medawar, P. 1972. Quoted in Rowan (1979).

Miller, NE. 1985. The value of behavioral research on animals. *Amer. Psych.* 40(4): 423-40.

National Academy of Sciences. 1985. *Models for Biomedical Research: A New Perspective*. Washington, DC: National Academy Press.

National Society for Medical Research. Undated. *What do these Nobel Prize winners have in common?* (Pamphlet). Washington, DC: NSMR.

Raub, W. Undated. Foreword. In: Gay, WI. ed. *Health Benefits of Animal Research*. Washington, DC: Foundation for Biomedical Research.

Reines, B. 1986. *Cancer Research on Animals: Impact and Alternatives*. Chicago: National Anti-Vivisection Soc. (NAVS).

—. Undated. *Heart Research on Animals: A Critique of Animal Models of Cardiovascular Disease*. Jenkintown, PA: American Anti-Vivisection Soc. (AAVS).

Rowan, AN. 1979. *Alternatives to Laboratory Animals: Definitions and Discussion*. Washington, DC: Institute for the Study of Animal Problems.

Russell, WM. and Burch, RL. 1959. *The Principles of Humane Experimental Technique*. London: Methuen and Co.

Sharpe, R. 1985. The cruel deception. *Animals' Defender and Anti-Vivisection News*. May/June: 43-46.

Sourkes, TL. 1966. *Nobel Prize Winners in Medicine and Physiology 1901-1965*. New York: Abelard-Schuman.

Stephens, ML. 1986a. *Maternal Deprivation Experiments in Psychology: A Critique of Animal Models*. Jenkintown, PA, Chicago, Boston : AAVS, NAVS, NEAVS, respectively.

—. 1986b. *Alternatives to Current Uses of Animals in Research, Safety Testing, and Education: A Layman's Guide*. Washington, DC: The Humane Society of the United States.

SOCIALIZED VS. UNSOCIALIZED WOLVES (*CANIS LUPUS*) IN EXPERIMENTAL RESEARCH[1]

Harry Frank [2,5], **Linda M. Hasselbach**[3], and **Dawn M. Littleton**[4]

The antiquity of man's interest in wild animal behavior is documented in upper Paleolithic cave drawings, which faithfully depict not only the color and form of the animals he hunted, but also their gait, posture, and social expression. Indeed, Geist (1978) suggests that man's capacity to survive the decline of Pleistocene megafauna was largely dependent on his ability to record and predict patterns of game migration and adapt his weaponry to changing predator-defense behaviors. With the rise of systematic agriculture, animal behavior occupied a less central role in human affairs, but remained a topic of continued fascination. Much of Aristotle's *Biological Treatises,* for example, is devoted to topics familiar to any contemporary student of behavioral biology: locomotion, reproduction, behavioral organization, and so forth.

In this century, the maturation of the behavioral sciences and the rise of environmental awareness have elevated the study of animal behavior to a professional discipline and have catapulted animal behavior researchers into the ranks of Nobel Laureates. Until recently, however, animal behavior research, whether conducted in the wild or in large, natural-setting enclosures, has been essentially an *observational* enterprise. Only in the last few decades have the precision and control of *experimental* methodology been brought to bear on the behavior of wild species (e.g., Leyhausen 1979).

Even in the arena of observational research, the question of whether or not one can draw legitimate inferences about wild animals from the behavior of tame specimens is not new, and among those who work with the gray wolf (*Canis lupus*) debate remains lively. Those who advocate use of unsocialized animals argue that cross-species social attachments may create behavioral abnormalities, and those who take the opposing view argue that socialization reduces stress and permits greater expression of natural behavior patterns.

In the experimental setting human contact is both more frequent and more intimate than in observational research, and the issue therefore assumes even greater importance. The present paper discusses two experimental studies of wolf information processing, one of which was conducted with unsocialized animals and one of which was conducted with socialized animals, and examines the both the management and methodological consequences of these approaches.

In 1980 the first author advanced a model of wolf and dog information processing suggesting that natural selection has favored evolution in the gray wolf (*Canis lupus*) of two concurrent systems of information processing. The more recently acquired system was characterized as "cognitive" and was suggested to have evolved in response to pressures that accompanied the rise of group hunting and that are relaxed under conditions of domestication. Accordingly, it was hypothesized that wolves should perform better than domestic dogs (*Canis familiaris*) on experimental *problem-solving* tasks, which call into play such complex cognitive capacities as insight into means-ends relationships, serial organization of behavior, cognitive mapping, imagery, and foresight. The more primitive "instinctual" system was characterized as a repertoire of closed behavioral programs governing behaviors that (1) exhibit little plasticity and (2) are elicited by very specific stimulus configurations. In contrast, the model pointed out that domestic dogs have been selected (incidentally or intentionally) for tractability, that is, (1) behavioral plasticity, which is reflected in high behavioral variability, and (2) responsiveness to a broad range of stimuli. It was therefore hypothesized that domestic dogs should perform better than wolves on experimental *training* tasks, in which to-be-learned behaviors are typically acquired by some variant of operant conditioning.

In a series of studies conducted in 1980 and 1981, Frank and Frank (1982, 1983, 1984,1985) administered an age-graded series of learning tasks to four Eastern timber wolf (*C. l. lycaon*) pups and four Alaskan Malamute (*C. familiaris*) pups. Performance of the wolf pups was also compared with data reported by J.P. Scott, J.H. Fuller, and their associates (e.g., Scott and Fuller 1965; Scott et al. 1967) for large samples of domestic dogs. Training tasks were distinguished from problem-solving tasks on the basis of three criteria: (1) Cues were arbitrarily selected by the experimenter; (2) reinforcement was administered by the experimenter; (3) the to-be-learned behavior had no perceptible, functional connection with the reinforcement. As predicted, wolves performed better than dogs on seven of seven problem-solving measures, and dogs performed better than wolves on five of seven measures of training-task performance.

In presenting a summary of these results at a European mammalogy conference, Frank and Frank (1984) reported that by six weeks of age it was evident that the wolf pups had been only partially socialized to humans and that by 20 weeks of age were nearly as wary of the experimenters as pups reared without human contact. One of the conference participants (M. Kiley-Worthington, personal communication, August 1982) suggested that the wolf pups' relatively poor training-task performance might reflect differences in socialization, rather than differences in trainability. Basically, she argued that the problem-solving and training tasks differed not only in the information-processing capacities they tapped, but also in social context. By definition, the training tasks required that environmental feedback be mediated by a human training agent. In contrast, feedback in the problem-solving tasks resulted directly from the animals' transactions with the test apparatus. Insofar as socialization to humans might involve sensitization to human behavioral cues, therefore, the incompletely socialized wolf pups may have been operating at a comparative disadvantage in the training situation, much like a nearsighted child trying to learn to read.

To eliminate this possible source of confounding, the present authors repeated the training-task experiments in 1983 using an intensively socialized group of wolf pups. Except for the two measures in which the test hypotheses failed in the initial study, the 1983 study replicated the original results. Nevertheless, behavioral differences between our maternally reared (1980) and hand-reared (1983) wolf pups were profound—and in many ways much more instructive than their formal, quantitative similarities in test performance.

Rearing Regimens

1980.—The four pups used in the initial study, one male and one female from each of two litters, were acquired at 11 days (± 24 h) of age from the Carlos Avery Game Park, Forest Lake, Minnesota. They were fostered on a highly socialized, mature maiden wolf approximately 67 days after ovulation. To promote lactation and maternal behavior, the foster mother was administered a series of progesterone injections beginning 10 days before the pups were introduced. Pups and foster mother were housed in 1.8 × 1.2 m den-box inside a 3.7 × 5.8 m barn. The barn and a 3.7 × 1.5 m outdoor pen were erected on a single 3.7 × 7.2 m concrete slab.

E. Klinghammer (personal communication, May 1980) had advised the experimenters that it is difficult to socialize wolf pups who interact with adult wolves, so a rather complicated routine was established by which two pups were with the experimenters at all times until approximately six weeks of age. Pairs were exchanged with the foster mother at mealtimes, and sleeping schedules were rotated and staggered (1) so that each pup spent one night with each experimenter followed by two nights with the foster mother and (2) so that no two pups spent consecutive nights together. When sleeping with an experimenter, pups were confined to an 1.2 × 2 m plywood pen furnished only with a mattress. By six weeks of age, individual differences in socialization were apparent, so from six to nine weeks of age only the less socialized pups slept with the experimenters. At nine weeks of age the overnight socialization periods were discontinued, and contact hours were reduced to approximately six per day plus the time spent in the testing situation.

Although the foster mother lactated for seven weeks, production was minimal, and the pups were hand fed four times a day until four weeks of age. Solid food was introduced at three weeks of age, and the transition from a milk-replacer formula to a meat and kibble diet was complete by Day 38, though nonnutritive suckling on the foster mother persisted for a short time thereafter.

After six weeks of age, when experimental testing began, the pups were allowed occasional forays into the half-acre wooded enclosure which surrounded both the home barn and the 9.8×9.8 m plywood testing arena. During these sessions they interacted freely with two adult male canids, an Alaskan Malamute and a Malamute-wolf hybrid, who were maintained in the same enclosure.

At 10 months of age the four pups were returned by air to the Carlos Avery Game Park.

1983. — The wolf pups used in the 1983 replication study were five males and three females acquired at 8 days (± 12 h) of age from the Ross Park Zoo, Binghamton, NY. Two adult females tended the pups from birth until six days of age, when the pups were removed from the den by the zoo staff, so it is uncertain whether they represent two litters or just one. It is known that at least three died shortly after birth, and 11 is an unusually large number for one litter. However, only one of the two females lactated, and she is known to have given birth to 13 pups in a single litter.

These pups were housed in a 3×4 m air-conditioned laboratory at the University of Connecticut's Biobehavioral Sciences complex from nine days of age to 47 days of age, when construction was completed on our testing facility at the Spring Manor Poultry Farm (Mansfield Depot, CT). Their new accommodations included a 20×22 m outdoor pen and an indoor kennel area (two 3.3×3.3 m kennels and a 3.3×6.6 m switching run) that occupied one end of a renovated 22×6.3 m chicken coop. Most of the remaining floor space was given over to testing apparatus.

For the first three weeks the pups were bottle fed six times daily (midnight, 4:00 a.m., 8:00 a.m., noon, 4:00 p.m., and 8:00 p.m.). To facilitate the 4:00 a.m. feeding and to ensure virtual round-the-clock human contact, one of the authors slept in the lab next to the 1×1 m monkey cage in which the pups were confined from approximately 1:00 a.m. to 7:30 a.m. At three weeks of age the 4:00 a.m. feeding was discontinued and the pups were left alone during the night. By this age they were too large to confine in the monkey cage for extended periods and were therefore allowed overnight run of the laboratory. Ten days later (Day 31) solid food was introduced, and within a day or two the pups had established themselves on a six-hour feeding cycle. All bottle feeding was discontinued at 39 days of age, though formula was provided in pans until the pups were 47 days old. At 50 days of age the number of daily feedings was reduced to three, at 75 days of age to two, and at 123 days to one.

Since the independent variable in our study was degree of socialization, the pups were not allowed contact with any canids until 6 weeks of age, and every effort was made to maximize their exposure to people. From their first day at the University of Connecticut, therefore, the pups were handled, cuddled, fed, romped with and petted by a host of volunteers from the university community and their families. Faculty members, graduate students, secretaries, friends of the project personnel, and children all served as surrogate packmates. Peak contact hours were during daytime feedings, especially noon and 4:00 p.m., but it was not unusual for project personnel to find themselves supervising visitors at 8:00 a.m. or 10:00 p.m. After the pups were transferred to the off-campus testing site, daytime visitation was confined to weekends, but security guards from the nearby Mansfield Training Center were frequent nocturnal visitors.

Because the University of Connecticut maintains a large population of both wild and domestic canids, stringent health precautions were observed until the pups received their first series of routine immunizations. Project personnel maintained a complete change of clothing at the laboratory, which left the building only to be laundered. Exposed skin areas were washed with a laboratory scrub compound before handling pups. Volunteers were required to bring clothing that had been machine washed and immediately sealed in plastic bags. In addition, all volunteers wore laboratory coats that were laundered with chlorine bleach and transported to and from the laboratory sealed in plastic bags. No outside footwear was permitted in the animal room. Volunteers and project personnel wore either socks laundered in chlorine bleach or rubber sandals that were immersed in a dilute solution of bleach (30:1) several times a day. Even visitors who did not handle the pups were required to wear laboratory coats and observe the same footwear procedures as the project staff and volunteers. Despite these precautions, three of the eight contracted Corona's Virus (diagnosed by electron microscopy) at 17 days of age and required aggressive support (subcutaneous injections of Ringer's Lactate, dextrose intubation, anti-emetics, and antibiotics for secondary infection). The three survived the infection, but one of the others subsequently died of an intestinal blockage 10 days after swallowing the tip of a pacifier nipple.

At 7 months of age, the pups were donated to Wolf Haven in Tenino, WA.

Differences

From the animals' frame of reference, the most pervasive difference between the 1980 and 1983 studies was the 1983 pups' reduced susceptibility to stress resulting from human proximity. From the experimenter's frame of reference, the important differences lie in the behavioral *consequences* of these differential stress levels, which often involve complex interactions and which may therefore be impossible to anticipate and difficult even to recognize without explicit, empirical comparison.

I. Management and Maintenance

In conducting research with sentient wild species, management and maintenance present problems that seldom arise in work with domesticated species. Many of these problems, however, can be eliminated or reduced if humans are accepted as a natural feature of their social environment. In addition to our colony of seven wolves, for example, the University of Connecticut also maintained a resident pack of unsocialized wolves. Because the resident pack's enclosure was an unpartitioned open field, their fear and aggressiveness made it impossible to perform even the most basic maintenance on a routine basis. In contrast, the present authors were able to disinfect the kennel area, remove feces, and inspect the fenceline daily and perform repairs or modifications on the physical plant as needed, complicated only by the wolves' obstreperous curiosity and the necessity of keeping unattended tools and equipment out of their reach. Other management differences had a more significant impact on our operation and merit discussion in greater detail.

Feeding.—The earliest notable differences between the 1980 and 1983 pups was in feeding behavior. The maternally reared (1980) litter were, from their first day out of the den, difficult to feed. While still on the bottle, any slight deviation in formula, stiffness or orifice opening of the nipple, temperature of the formula, position in which the pup was held, etc., was likely to result in rejection of the nipple. The only difficulty with the 1983 litter was obtaining nipples that were stiff enough to prevent the pups from aspirating formula in their enthusiasm.

We feel that the most plausible explanation for this difference surrounds the complex motivational components of feeding behavior. In the course of nursing, a pup satisfies hunger needs, receives tactile and thermal stimulation, and satisfies non-nutritive sucking needs. To the extent that all of these were satisfied by human caretakers in the 1983 litter, the bottle-feeding situation may have acquired secondary reinforcement value, in addition to the primary reinforcement associated with reduction of hunger. In the maternally reared 1980 litter, non-nutritive needs were largely satisfied by the foster mother wolf. The overall satisfaction derived from hand feeding may have been further reduced by the small but continuously available volume of milk produced by the foster mother. Alternatively, we also recognize the possibility that the stimulus pattern that elicits sucking behavior in wolf pups may have become more canalized in the 1980 pups, who nursed on their biological mothers until 11 days of age, than in the 1983 litter, who were removed from the den at 6 days of age.

After weaning, the feeding differences persisted. In order to ensure that the maternally reared pups ate anything, they were fed two at a time in individual dishes in the experimenter's home. They were fed three times a day from four weeks of age to nine weeks and twice a day until 15 weeks of age. Considerable time and effort were expended to overcome their indifference to food, which was seldom consumed in less than 45 minutes. After 20 weeks of age the pups were too unmanageable to be brought indoors and four rations were placed in different locations in the home barn. The hand-reared pups were, in comparison, almost insatiable. Indeed, they were so eager to eat that at five months of age it became necessary to confine them to their kennel at night so that the authors would not be mobbed when setting out their morning rations. Feedings seldom lasted longer than three to five minutes. At 20 days of age, when body weights were first recorded for both groups, the maternally reared pups averaged 3.9 pounds and the hand-reared pups 3.2 pounds. By 30 days of age both groups averaged 5.25 pounds, and the group averages remained essentially identical until 12 weeks of age. Although average weights for the two groups were thus very similar during the period of most rapid growth, the hand-reared pups came from smaller stock and may therefore have realized more of their growth potential. Furthermore, after 12 weeks of age the growth curve for the maternally reared pups flattened out, while the curve for the hand-reared pups continued to accelerate until they reached the capacity of the scale.

Health care.—The health care advantages of working with socialized wolves are discussed in detail by Albert, Goodmann and Klinghammer (in press) and by Klinghammer and Goodmann (in press). The explicit contrasts discussed below between health care delivery problems confronted with the maternally and hand-reared pups are not intended to pre-empt their recommendations, but, rather, to underscore them.

Injections were not really a problem with either group prior to two months of age. After their routine immunizations the 1980 litter had few health problems and therefore required no injections until health certificates were required for shipment to Minnesota when they were approximately 10 months old. Although the four pups were, by this time, confined to the home barn, the operation required more than an hour. One pup at a time was chivvied into the paddock and the guillotine door lowered. Then the first author cornered the pup and squeezed him against the fence fabric with his leg so the veterinarian could administer the injection from outside the pen. The author still bears prominent scars. The hand-reared (1983) pups received injections almost weekly from 17 days of age until seven months of age. Once the syringes were prepared, a typical round of booster shots for the entire group of seven required about 15 or 20 minutes. Pups were allowed into the testing area one at a time, and the delays were occasioned primarily by pups who had already received their shots crowding to get back in for additional attention.

Oral medications can be a much more difficult enterprise than injections. The usual procedure is to wrap pills in ground meat, cheese, or some other palatable carrier or to incorporate it into the animals' food (which should be attempted only after it is established that the medication does not produce a taste aversion). Liquid medication is injected into the mouth with a hypodermic syringe. For the 1980 pups, pills were always introduced in balls of canned cat food and, if accepted, subsequently mixed with their daily food. This, of course, was possible only so long as feeding was conducted individually and carefully monitored to ensure that every pup consumed his entire ration. Less palatable preparations were administered in liquid form, which often assumed the character of rodeo bulldogging. Group feeding made it impossible to incorporate medication into the 1983 pups' food, but they accepted unconcealed pills with good grace, even after we began daily heartworm prophylaxis, and liquid medication was received from a syringe with the same uncritical acceptance as bottle-fed formula. Indeed, especially palatable preparations (e.g., Strongid-T) were so eagerly consumed that we occasionally substituted such medications for food rewards administered during experimental testing.

Body weights for both groups were recorded several times a week from 20 days of age until they exceeded the scale's capacity (30 pounds) at approximately 100 days of age. The 1980 pups, however, began to show fear of the scale at about four or five weeks, and keeping them on the weighing pan long enough to obtain an accurate reading simply became so stressful for both the pups and the experimenters that it was not always possible to obtain reliable recordings at every weighing. Since illness in a young pup is often reflected in body weight before other symptoms are manifest, the present authors contrived to make weighing a pleasurable activity for the 1983 pups. From three to seven weeks of age they were placed on the scale every two or three days and petted, stroked, and scratched until they either fell asleep or showed signs of wanting to leave. Within a week or two the pups would actively compete to occupy the weighing pan, and even after they became so heavy that the scale would tip over with a grating, metallic clatter when a pup tried to mount it, they showed no aversion to the scale. Weighing was discontinued only when the pups became too large to be completely supported by the pan. The pups in figures 1a-f range from 22 to 99 days of age. It should be noted that the 1980 pups were weighed far more often than the 1983 pups and therefore had more frequent exposure to the scale. The difference between the two groups' orientation to it must consequently be attributed to the positive reinforcement that human contact and attention constituted for the socialized pups.

The range of health care procedures endured by the 1983 litter was considerably greater than that experienced by the 1980 litter. In part, this was because the former were more tolerant, in part because they permitted (demanded, actually) more frequent intimate contact, thereby facilitating detection of health problems, and in part because the 1983 pups were reared in an environment that exposed them to more canine pathogens. Besides the medications mentioned above, several of the pups were treated for urinary tract conditions (*Pseudomonas aeruginosa*) and (*Proteus sp.*) for which vaginal and penile sheath lavages were prescribed. It is inconceivable that such a procedure would have been tolerated by the maternally reared pups, but the hand-reared pups accepted the treatment with absolute equanimity. The greatest test of their tolerance, however, came at 13 weeks of age. By this time their performance in perceptual tasks and difficulties negotiating new objects in their enclosure had led us to suspect that some of the pups might have juvenile cataracts, which is not uncommon in wolves weaned on commercial milk replacer before two weeks of age. The entire litter was therefore administered an ophthalmic examination by Dr. Herbert J. Van Kruiningen of the University's Pathobiology Department. Apart from the fact that Dr. Van Kruiningen was a total stranger to the pups, the examination required that they be manually restrained and an unfamiliar instrument be placed within a few inches from their eyes. The photograph in figure 2 (page 42) demonstrates the placidity with which this activity was accepted.

Logistics.—In addition to health care, one of the most difficult problems in working with wolves is logistics—getting the animals from one place to

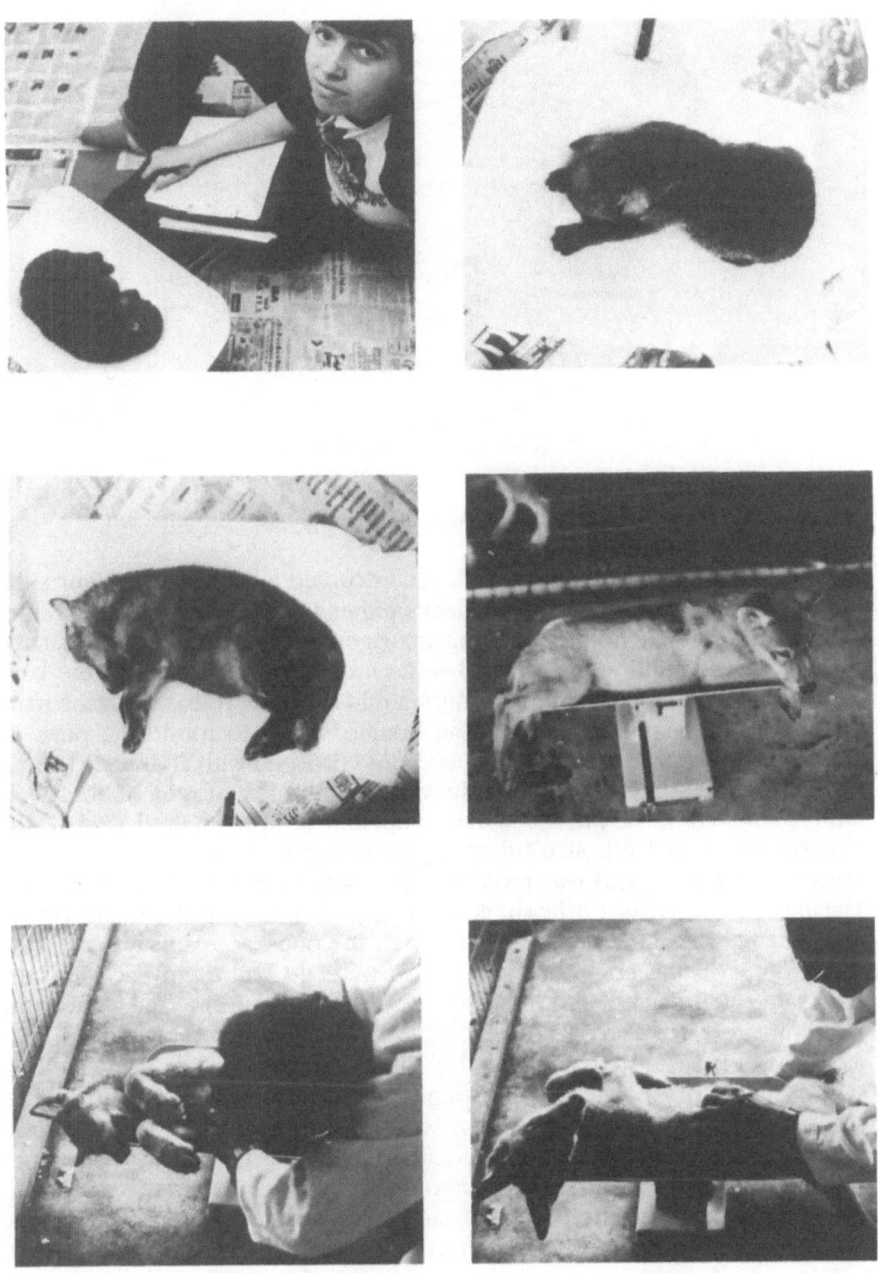

Figure 1a-f. Socialized wolf pups on baby scale from 22 to 99 days of age.

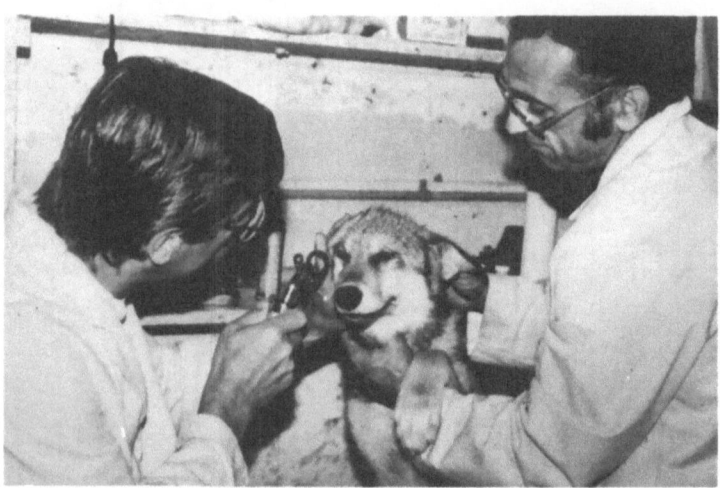

Figure 2. Socialized wolf pup undergoing ophthalmological examination.

another. Although the 1980 research site occupied a half-acre enclosure, the pups spent most of their first 20 weeks either in the experimenters' house or confined to their home barn. This management practice was necessitated by their speed, agility, and general reluctance either to return to the barn or to approach the experimenters. After a mid-morning release into the main compound, it frequently required an hour or more to round the pups up and return them to the barn so that we could proceed with afternoon testing. "Recapture" was often possible only with the aid of the foster mother wolf, who would chase down a wayward pup and jaw-pin it until one of the experimenters arrived. At 67 days of age one pup remained loose in the enclosure all night and was recovered the next morning when we put the Malamute and hybrid on leash, and she approached to greet them. By 119 days of age, it became necessary to restrict the pups' access to water during their romps in the enclosure and to use water to bait them back into the barn. After 130 days of age, the pups would avoid entering the barn if the experimenters were inside the enclosure, and it was necessary to rig lines to the paddock gate so that it could be shut by remote control during feeding. Within a few days, however, the pups adopted the strategy of feeding in shifts so that only one or two pups were in the paddock at any one time. Thereafter they were permanently confined to the barn until early November, just after the testing program was completed, when the foster mother worked the exterior latch on the barn door and released the pups. They remained at large for several weeks, and our efforts to recapture them are instructive.

First, new fencing was erected to connect the barn with the testing arena and the perimeter fence of the compound. This created an irregular quadri-lateral enclosure (approximately 20 × 20 m) with a single, narrow entrance between the barn and the perimeter fence. A roll of 6-foot welded wire fence fabric was secured to the barn, so that someone in the barn could—in principle—slip out, unroll the 6-m length of wire and fasten it to the fence, thus confining the pups in the small enclosure. For several days the pups were fed inside this enclosure, and on the first absolutely windless day one experimenter ensconced himself in the barn, from which he could watch the small enclosure through a knothole and rush out to unroll the wire as soon as all of the pups began to eat. The pups approached the barn warily, entering the small enclosure no more than two at a time and only for a few seconds. Eventually, the boldest male entered the enclosure alone and carried the four food dishes out one at a time.

The trap was finally sprung only by a most devious ploy. A ladder was placed against the outside of the perimeter fence at the entrance to the small enclosure. After several days one experimenter concealed himself in the woods some 200 m beyond the fence. The pups were fed in the small enclosure, and as soon as they began to eat, the other experimenter sounded a whistle, and the first experimenter approached the fence on cross-country skis, keeping the barn between himself and the pups. Skis are almost silent in deep snow, and he was able to reach the fenceline undetected, kick off the skis, climb the ladder, drop into the compound, unroll the wire and secure it.

All of this elaborate and time-consuming effort was in dramatic contrast to the procedures required to return the socialized, 1983 pups to their kennel—we called them.

In addition to the routine logistical demands involved in management of any animal colony, experimental research imposes other requirements. Animals must be moved to and from the testing site, placed in and removed from test apparatus, and in some cases kept isolated from others who have not yet been tested. In 1980 this was accomplished by first rounding up the entire group, confining them to the paddock, cornering the pup who was to be tested, carrying him to the test arena, and then carrying him back to the home barn after testing. These experimental logistics were particularly onerous with two of our tests.

At 15 weeks of age the pups were administered a visual discrimination (oddity) task using the Wisconsin General Test Apparatus (WGTA). The WGTA test chamber was a 1×1 m box mounted on a 1-m-high table. Pups entered and exited the chamber through a rear door. After testing, the 1980 pups generally refused to leave the chamber voluntarily, and it was necessary to reach in and extract them as gently as possible. This often produced panic, and one experimenter still carries a deciduous canine tooth embedded in his forearm.

In 1983 our only problem was holding back the mob of yet-untested pups and limiting admission to the testing area to a single participant. Once out of the kennel the testee was usually content to follow or precede the experimenter to the apparatus. Although these pups were, like their predecessors, sometimes reluctant to return to the compound when testing was complete, this was generally expressed in the form of submission and prolonged play solicitation, rather than defensive aggression. Figures 3a and 3b illustrate the procedures required to get the 1983 pups into and out of the test apparatus itself.

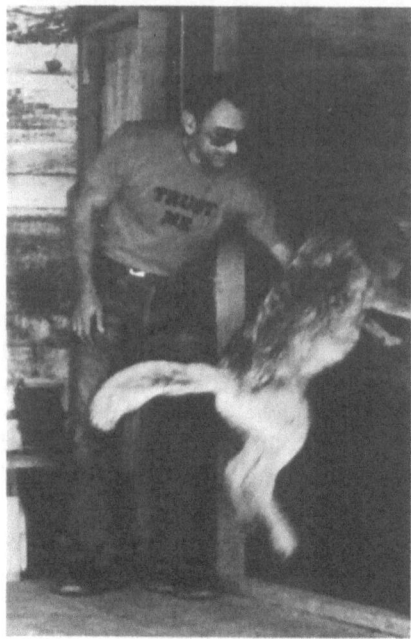 

Figure 3. a. Socialized wolf enters testing chamber of Wisconsin General Test Apparatus on author's signal.
Figure 3. b. Socialized wolf leaves Wisconsin General Test Apparatus when called by author (not shown).

A test of intermodal (visual vs. auditory) cue discrimination was administered at 22 weeks of age and employed the WGTA test chamber as a start box, equipped for this experiment with a plexiglass guillotine door in front, through which pups were released into the runway of a T-maze. If an auditory signal (whistle) was presented just prior to release, one exit of the T would be found open; if a visual signal (a flashing light) was presented, the other exit would be found open. After each trial it was expected that the pup would return to the start box, receive food reward for a correct choice, and begin the next trial. This procedure, which worked so well with Scott and Fuller's domestic dogs (1965, pp. 238-42), proved completely unworkable with the 1980 pups. As soon as a pup discovered the open exit he either bolted for the wall of the test arena and attempted to scramble over, or, more typically, simply did his best to avoid recapture, leading the experimenters on a frantic chase that might last several minutes. Small pens therefore had to be constructed outside each arm of the T, and pups had to be carried from the pen to the start box after each trial.

Figures 4a and 4b illustrate typical behavior of the 1983 pups in performing the cue-discrimination task. As with the visual discrimination (WGTA) task, our only problem was containment of those not scheduled for immediate testing. Because pups now weighed nearly 50 pounds, however, debarring all but one from the test area was more difficult than it had been at 15 weeks. On one occasion the stampede of volunteers bowled over the third author, and three pups managed to crowd into the start box simultaneously.

 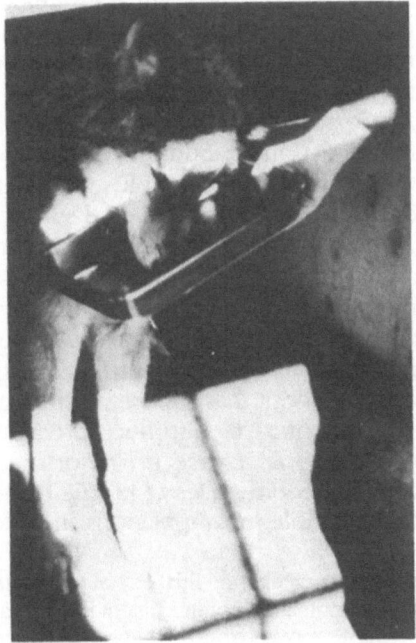

Figure 4a. Socialized wolf exits T-maze after correct choice and gallops back to start box. Figure 4b. Socialized wolf returns to start box and accepts food reward for correct response.

Transportation and shipping are another logistical problem confronting those who conduct research with wolves. When the 1980 wolves were shipped to Minnesota, they had to be immobilized with injections of ketamine hydrochloride and promazine HCl just to get them into their shipping kennels. In transit to the airport and while awaiting departure, respiration and other vital signs required continuous monitoring, and arrangements were made with airport personnel to keep the kennels in a quiet and relatively secluded location until they could be stowed aboard the aircraft. The 1983 pups were administered a half-strength dosage of oral acepromazine maleate several hours before flight time and were perfectly content to sit in the middle of a busy cargo bay at Bradley International Airport, with the doors to their kennels either closed or open, while the authors, cargo handlers, and casual passersby entertained them.

II. Methodological Considerations: Error Variance and Confounding

The design of any scientific experiment always requires the experimenter to balance costs of labor, equipment, and materiel against the quality and reliability of data. At some point, the cost of controlling extraneous variability almost inevitably exceeds the additional precision that such control confers. In our experience with socialized and unsocialized wolves, this trade-off does not arise. The comparative ease of managing socialized wolves not only reduces the overall expenditure of labor, but also the ratio of time spent in management activity to time spent in actual research activity. Furthermore, the use of socialized participants, far from "contaminating" data with "socialization artifacts," can reduce extraneous variation, both systematic (i.e., confounding) and random (i.e., error). Extraneous variability may arise from three major sources (plus interactions among the three): task variables, environmental variables, and subject variables.

Task variables. — Generally speaking, the easiest source of extraneous variability to control is that associated with the experimental task itself. This means not only that all participants must be presented the same puzzle, apparatus, or stimulus materials, but that all the experimental procedures must likewise be constant from participant to participant. In the 1980 study, circumstances that can be attributed to lack of socialization compelled the experimenters to introduce procedural deviations that might ordinarily be expected to increase error variance. At 10 weeks of age, the pups were administered a series of puzzle boxes that required performance of increasingly complex manipulations in order to extract a food dish:

> Unfortunately, [the final day of testing] fell on July 4, and testing was punctuated by a combination of intermittent thunder and neighborhood fireworks that distracted the first two [wolf pups] and convinced us *to postpone testing the others until after the holiday weekend* (Frank and Frank 1985, p. 269, emphasis added).

Thus, the last two wolf pups addressed the final and most complex problems after a two-day hiatus which the others did not experience. Such interruptions do not invariably influence performance but can in principle either enhance it by reducing satiation or diminish it by reducing the effects of continuous practice. Although testing in 1983 was also subject to such external auditory distractions as heavy truck traffic, unexpected visits by maintenance personnel and animal care inspectors, and the operation of nearby construction equipment, voice or touch of a familiar human was always sufficient to reassure the socialized pups and draw their attention back to the task at hand. Thus, Thus performance was never so disrupted that we considered postponement of testing or other departures from our standardized testing procedures.

A similar deviation in our 1980 testing procedures occurred when one pup successfully escaped the test arena during the cue-discrimination test and remained at large for several days before she was recaptured and testing could resume. In 1983, our only logistical management problem involved pups breaking *in* to the test area, not breaking *out*.

Environmental variables. —Control of environmental variation that is unrelated to the experimental task itself is a more difficult and subtle problem. It is more difficult because the sources of variation can be almost infinite and may often be inherent in the research setting. For example, control of such factors as illumination, ambient temperature and windborne olfactory stimuli may simply be beyond the ability or economic resources of the experimenter. They are more subtle because they may interact with individual differences. In the example discussed above, the decision was made to postpone manipulation testing because it was felt that experimental deviation would introduce *less* error variance than the *differential* effects of auditory distraction on labile and phlegmatic individuals.

The most dramatic illustration of environmental lability in the 1980 pups was during visual discrimination on the WGTA. The experimenters had asked a colleague to photograph the session. To avoid distraction, the photographer used a telephoto lens from the upper story of the experimenters' house. Although she was 30 m from the test arena, concealed by a screened window and heavy curtains, the pups ignored the experimental task altogether and stared fixedly at the window until the photographer left. The socialization regimen imposed on the 1983 pups inured them to the presence of human strangers, and on occasion we were actually able to use unfamiliar personnel to assist with test administration.

Environmental stress interfered with test performance only once in 1983. One pup could not be habituated to isolation in the WGTA test chamber and tried relentlessly to dig, squeeze and chew his way out the moment he was confined. This problem would have been intractable with unsocialized pups, and we would have faced the choice of either eliminating his data or increasing error variance by recording refused attempts as failures. In the present instance, however, the pup relaxed and applied himself to the experimental task as soon as the third author climbed into the box and sat with him. (The author sat facing *away* from the stimulus display to ensure that she neither cued nor reinforced correct responses.)

Subject variables. —Sources of performance variation that derive from individual differences are the least accessible and often the least tractable. Furthermore, as indicated above, temperament and personality factors can interact in complex and unpredictable ways with both task and environmental factors. In the present study, however, the most salient differences surround motivation and appear directly related to socialization.

In the 1980-81 study the only hypotheses that were not confirmed were the predictions that the Malamutes would perform better than the wolves on the visual discrimination (WGTA) and cue-discrimination (T-maze) tasks. In

visual discrimination learning, wolf pups actually performed better than the
Malamutes in the initial discrimination (though the differences were not
statistically significant), but the dogs performed significantly better in reversal
learning ($P \leqslant .025$; Frank and Frank 1984). Differences were in the expected
direction on the T-maze but were only marginally significant ($P = .067$; Frank
and Frank ibid.) and therefore equivocal. Food reinforcement was employed
in both of these experiments, and the chronic indifference of the maternally
reared (1980) pups toward food proffered at feeding time (never a problem
with the perpetually greedy Malamutes) was also apparent in the testing
situation. Food rewards were frequently ignored or consumed half-heartedly
and often had to be offered in smorgasbord variety to maintain any overt
interest whatsoever. In this connection, the latent learning experiments of
the 1930s (e.g., Tolman and Honzik 1930) demonstrated that animals may
acquire to-be-learned behaviors but not *perform* them until performance is
reinforced, i.e., until motivated by adequate incentive. This consideration
raises the possibility that the 1980 pups' relatively poor performance on the
two discrimination tasks might have reflected insufficient motivation, rather
than inability to learn. This interpretation gains plausibility when it is noted
that the significant difference between wolves and Malamutes in reversal
learning on the WGTA was largely attributable to the wolves' frequent nonper-
formance or perseverative choice of the same stimulus object trial after trial.

The gustatory enthusiasm of the hand-reared (1983) pups offered a clear
opportunity to eliminate the possible confounding of motivational factors.
Despite the 1983 pups' eager participation in the cue-discrimination task and
avid consumption of food reward (see figures 4a and 4b), *none* of them
reached criterion, thereby providing unambiguous support for the original
hypothesis. Much to our surprise, however, they performed significantly
better than *either* the Alaskan Malamutes ($P \leqslant .05$) or the 1980 wolves ($P \leqslant
.005$) on the WGTA reversal learning task and significantly better than the
Malamutes on the initial discrimination task ($P \leqslant .005$). They likewise performed better than the 1980 wolves in the initial discrimination phase of
the experiment ($\overline{X}_{1980} = 82.50$; $\overline{X}_{1983} = 44.1$), but the difference was not significant due to the immense performance variability of the 1980 pups. This
variability was inflated by two of the pups, who exhibited only mild interest
in the task and hovered just below criterion until the last few days of testing.

Conceptual interpretation of these results is beyond the scope of this
paper, but the methodological implications are clear: Performance differences
between wolves and Malamutes in the 1980-81 study were confounded by
motivational factors, viz. differential incentive value of food reward, which
was eliminated in the 1983 study by the use of hand-reared wolves who
were thoroughly socialized to humans and as readily motivated by food
reinforcement as their domestic congeners.

Acknowledgements

The authors wish to thank the University of Connecticut School of Agriculture, which provided the testing site, the Pathobiology Department for unstint-

ing consultation and laboratory services, and the Farm Department for services too numerous to list. We also wish to thank Mrs. Elaine Werboff of Storrs, CT, for her generous contribution of essential building materials.

Endnotes

[1] This study was funded by grants from the Faculty Research and Development Fund of the University of Michigan-Flint and from the Office of the Vice-President for Research, University of Michigan.

[2] All of the authors participated equally in the socialization program described in the present paper and are listed alphabetically.

[3] Department of Resource and Community Science, University of Michigan-Flint, Flint, MI 48503.

[4] Department of Psychology, Canisius College, Buffalo, NY 14209.

[5] Associate Professor, Departments of Resource and Community Science, Psychology, University of Michigan-Flint, Flint, MI 48503. *To whom reprint requests should be sent.*

References

Albert, C, Goodmann, PA, and Klinghammer, E. n.d. Health care of wolves in captivity. In: Frank, H. ed. *Man and Wolf: Advances, Issues and Problems in Captive Wolf Research*. The Hague: Junk Publishers. In press.

Frank, H. 1980. Evolution of canine information processing under conditions of natural and artificial selection. *Zeitschr. fur Tierpsychol.* 53: 389-99.

Frank, H. and Frank, MG. 1982. Comparison of problem-solving performance in six-week-old wolves and dogs. *Anim. Behav.* 30: 95-98.

—. 1983. Inhibition training in wolves and dogs. *Behav. Proc.* 8: 363-77.

—. 1984. Information processing in wolves and dogs. Acta Zoo. Fenn. 171: 225-28.

—. 1985. Comparative manipulation-test performance in ten-week-old wolves (*Canis lupus*) and Alaskan Malamutes (*Canis familiaris*): A Piagetian interpretation. *J. Comp. Psych.* 99(3): 266-74.

Geist, V. 1978. *Life Strategies, Human Evolution, Environmental Design: Toward a Biological Theory of Health*. New York: Springer-Verlag.

Klinghammer, E and Goodmann, PA. n.d. Socialization and management of wolves in captivity. In: Frank, H. ed. *Man and Wolf: Advances, Issues and Problems in Captive Wolf Research*. The Hague: Junk Publishers. In press.

Leyhausen, P. 1979. *Cat Behavior: The Predatory and Social Behavior of Domestic and Wild Cats*. Tonkin, BA, trans., New York: Garland STMP. (Original work published 1956).

Scott, JP and Fuller, JL. 1965. *Genetics and the Social Behavior of the Dog*. Chicago: Univ. Chicago Press.

Scott, JP, Shepard, JJ and Werboff, J. 1967. Inhibitory training of dogs: Effects of age at training in basenjis and Shetland Sheepdogs. *J. Psych.* 6: 237-52.

Tolman, EC and Honzik, CH. 1930. Introduction and removal of reward, and maze performance in rats. *University of California Publications in Psychology.* 4: 257-75.

HUMAN/FARM ANIMAL RELATIONSHIPS[1]

Jack L. Albright[2]

Introduction

A recent educational pamphlet entitled "We Need Farm Animals"[3] was issued from an enlightened self-interest group and stated:

> Humans and domestic animals have mutually benefited each other for centuries. Each contributes to the well-being of the other; humans provide the feed and housing for animals, and animals produce milk, eggs, fiber, meat, draft power, and recreation for humans. Animals give us improved nutrition, better health, and companionship. Animals help us enjoy greater prosperity and an improved standard of living.

Fourteen important contributions arising from farm animals were discussed.[4] At the top of the list was the category of "Needs of Humans":

> Animals fulfill a basic need of humans—a desire to be wanted and accepted and to have something rely on us for attention and care. This may explain why most farmers feel so close to their farm animals. Animals also serve as companions and pets and contribute to our recreational needs. Contributions of animals as companions are significant even though it is difficult to quantify the pleasure and emotional value. Animals used in sports such as horse racing provide jobs and income for many people and entertainment and recreation for millions.

A final statement summarized and concluded:

> ...Many people of the world are supported almost completely by animals because they live in areas unsuitable for crop production.[5] Animals have proven to be vital and necessary for the nutrition, by-products, economic benefits, and companionship that they provide. Animal products are excellent complements to fruits, vegetables, and cereals, and they are needed to ensure a complete and balanced diet. Many additional products that come from animals often are not recognized as being of animal origin, but they contribute very much to our lives. The contributions of livestock are so great that animals are essential to the well-being of humans in all parts of the world.

Human/Farm Animal Interactions

At a conference on the Human/Animal Bond, Kilgour (1983) reported as follows:

> Humans keep animals, birds or insects for food, apparel, traction, cartage, sport, entertainment, scientific research, and industrial uses. Considerable control is exercised over mate selection, reproduction, maternal care and rearing, group size and composition, feed offered, living quarter space and design, etc. The "domestic contract" or "trade-off" is not entirely one-sided. Animals are protected from the worst exigencies of climate, seldom are without adequate food, and are comparatively healthy and protected from predators. In a sense the current concerns raised by animal welfarists are part of a re-examination of the terms of the domestic contract or the trade off implicit in "domestication." It is appropriate that this contract should be re-examined each few decades. Recent intensification in the animal industries and the growth of automation make it important for society to re-evaluate its stance to domestic animals.

Farm animals are also being kept as pets, as evidenced that there are more than 100 pigs now living in residential areas in the United States (Schwadel 1986).

Homo sapiens is one of the few species to enter into extended and complex social relationships with other species. In some cases, such as the milker with the cow, the shepherd and the dog, and the rancher with a horse, this may involve staking one's food supply and income, well-being, and even life on the trust and success of a close relationship. In the social contract between caretaker and animal, it seems obvious that the handler should acknowledge and accept appropriate behavior.

Human/Animal Combinations.—There are various combinations of human beings and farm animals. This paper attempts to evaluate those few studies of humans handling farm animals within a prescribed environment. Personality traits of dairy farmers and livestock people as determined by the Eysenck Personality Inventory (Eysenck 1977) need further study (Seabrook 1974; Arave and Brown 1979). Seabrook's sample size was small (20 herds) and these herds were criticized for having low yields while Arave and Brown's questionnaire did not go far enough.

The classical model is the human/animal relationship. Occasionally there are others. Different pathways are proposed as follows:

$$\text{Human} \longrightarrow \text{Human/Animal}$$

These models will be illustrated through two accounts. At the state mental institution herd, Medical Lake, Washington, they had an increasing number of dairy calves becoming ill with unusual behavioral symptoms. The herd manager investigated and questioned the various inmates including the calf feeder. The problem continued. As a last resort the herd manager decided to observe the various inmates and their visits to the calf barn. Finally, one particular inmate after dinner went past the garbage cans back of the pharmacy where he picked up a handful of assorted pills and headed for the calf barn.

Here the larger scene unfolded. In front of each calf, the inmate noted that the calves did not look well. Each night he would treat each calf to some unknown drug prescribed best to humans. The mystery was solved and this episode illustrates the importance of human-to-human observation and interaction with the drastic effects of indiscriminate dosage upon an animal's behavior and well-being.

Human ————► Animal/Animal

The other incident took place on sabbatical leave at the New Zealand Ruakura Animal Research Station. One evening after hours near the large animal learning maze, there was considerable commotion with the large group of tethered dogs barking incessantly. Then there would be quiet, followed by whistling and then barking. The resident mynah birds hidden in a grove of trees nearby had learned to imitate the human shepherd's whistles (calls) for herding sheep. Each evening the mynah birds would rehearse and deliver a chorus of whistles, driving the dogs to distraction.

Domestication and Tameness. —According to Murphey et al. (1981), domestication efforts have not been totally successful. Even after millennia of husbandry, behavioral control through forced compliance and confinement is still practiced. Physical restraints imposed upon livestock can result in lifelong dependence on the captors, with consequent cultural changes bringing about human dependence on the captives. Price (1984), in a recent review on domestication, searches for clues and behavioral aspects of animal domestication. In his study, he reminds us that:

> ...(perhaps) the first requirement for the successful domestication of any species is that man, the domesticator, have a recognized need or desire that can only be satisfied by controlling, protecting, and breeding a certain population of animals. So far, few species have been domesticated for other reasons (e.g., food production, companionship). During domestication some traits have become more or less frequent and conspicuous. During domestication body size has generally increased for some species (e.g, horse, rabbit) and has decreased for others (e.g., cattle, sheep). Some phenotypic characteristics (e.g., body color) of certain domesticated species have become more variable during domestication, whereas other traits (e.g., tameness) have become less variable.

Also according to Price (1984):

> Darwin suggested that domestication is more than taming, that it includes breeding animals in captivity, is goal-oriented, may occur without conscious effort on the part of man, increases fecundity, may bring about the atrophy of certain body organs, enables animals to achieve greater plasticity, and is facilitated by subjugation to man, the domesticator. Some contemporary definitions postulate that domestication is a condition in which the breeding, care, and feeding of animals are more or less controlled by man. This definition implies that a population of animals is rendered domestic by exposure to the captive environment and by the institution of certain management practices.

Farm animals selected and domesticated by man are very social by nature (Kilgour 1983). They are characterized by a dominance hierarchy, where man is able one way or another to establish dominance over every member of the animal group. Quite the opposite of the way successful cowmen and milkmaids handle milk cows through TLC (Tender Loving Care), the Fulani culture in northern Nigeria has made a general social virtue of the aggressive dominance that is vital in the husbandry of their cattle. A high level of aggressive, assertive, combative behavior, both verbal and physical abuse, is not only accepted but demanded in social relations with one human male to another and to their cattle (Lott and Hart 1977).

An important point often overlooked is that the temperament of farm animals is influenced by the temperament of their caretakers (Stricklin and Kautz-Scanavy 1984). With high-producing dairy cattle, a cow's fear of an aggressive handler may upset her if she has a nervous temperament. Temperament in cows is quite an important aspect of cowmanship as one undisciplined cow can unsettle the entire herd. (Burnside et al.[1971], reported that problem dairy cows—slow milkers or bad temperaments—culled and sold represent 3.6, 2.1, 2.1, and 1.9% of all Ayrshire, Guernsey, Holstein, and Jersey disposals, respectively. Of the problem cows, 42% of the Ayrshire, 10% of the Guernsey, 29% of the Holstein, and 37% of the Jersey cows were categorized as having bad temperaments.) Schmidt and Van Vleck (1974) categorized the different dairy breeds for temperament as follows: Ayrshire (nervous), Brown Swiss (docile, stubborn), Guernsey (docile), Holstein (docile), and Jersey (somewhat nervous).

Some farmers buy other farmer's culls knowing that they will become top quality cows in their herd. Therefore the handling of the cow has a lot to do with her temperament. Even in large herds, dairymen will have favorites among herd members. Well-adjusted cows show ready movement into milking parlors with or without grain as an inducement. Cows will often readily approach good herdsmen. It has been suggested that the true test of cowmanship is whether the cows exhibit approach behavior and come to the herdsman in the pasture (best) or turn away as he approaches (worst) or just stand still (neutral) when he comes closer (Albright 1978, 1981a).

Spatial Reactivity of Animals. — The spatial reactivity of animals to intruders with appropriate behavior terminology has been compiled by Waring (1985):

Animals exhibit specific reactive distances. For example, an approaching intruder first causes an animal to become attentive when it reaches the "investigative distance." The animal detected the approaching intruder when it reached the "perceptive distance." The distance at which an animal first begins to flee from an intruder is the "flight distance." The distance the fleeing animal then places between itself and the intruder before ceasing flight is the "withdrawal distance." When inhibited from fleeing, an animal first becomes defensive toward an approaching intruder at the "aggressive distance." The space around a resource defended by an individual or group is a "territory." The minimum distance tolerated between individuals under normal conditions is the "individual distance"; this creates a "personal space" around the individual ("group space" in the case of groups). The "strike distance" is the extent of a stationary animal's reach for inflicting physical harm on an intruder; in birds, it is often called "peck distance." A "submissive distance" is where an individual first shows cringing or other submissiveness when nearing a dominant. And the maximum distance an individual wanders from members of its social group under normal conditions before starting to return is called the "social distance." Reactive distances and their magnitude are influenced by the environmental context, physiological and psychological state of the animal, intensity of stimulation, experience, etc. Reactive distances are evident under free-ranging as well as captive conditions, and can be important to anyone trying to approach, manipulate, or manage animals.

Approach-Avoidance Relationships.—Undomesticated ungulates and domesticated cattle have paradoxical tendencies both to approach and to avoid humans (Hediger 1955). Murphey et al. (1981), have reviewed approach avoidance responses. Approach (investigating a person lying on the ground) and avoidance (flight distance) were studied among 525 cows—25 animals observed in 21 Brazilian herds of *Bos taurus, Bos indicus,* and *B. taurus* × *B. indicus* breeds. (Seven herds were compared in their investigatory responses to a human and a ball. Breed differences were evident for approach and avoidance behavior which had little relationship with one another. When reactions to the person and ball were compared, the responses were also breed specific and negatively correlated. Age took precedence over breed in investigating the human in a predator-prey context.) One observer (Murphey et al. 1980, 1981) stationed himself in or near each herd for 30 minutes recording the animal's general activities while allowing them to become accustomed to his presence, after which he tested the flight distance ("approach ability" or "unconceded distance") of individual cows. He used a split-image range finder and careful pacing (approximately one stride per second) to learn how close he could walk toward her before she fled or Murphey could touch her head. Murphey et al. (1981) found that there are breed differences in flight distances and that dairy cattle have less flight distance than do beef breeds. Earlier, dairy cattle were estimated to have a flight distance to man of 12 feet and beef cows of 16 feet (McFarlane 1976). In high-producing cows or in the country with the highest milk production per cow in the world—Israel—claims are made for their Israeli Friesian cow's degree of tameness and their zero flight distance (McFarlane 1976; Albright 1978). In the world's record milk producer Indiana Holstein Beecher Arlinda Ellen's case with the Beecher family waiting on her almost continuously throughout her life, it is easier to explain her zero flight distance, overall tameness, temperament, and willingness to approach family and stranger alike.

The fact that all of the dairy stock were more approachable by the human than were beef cattle in the Murphey study should not be too surprising. Dairy breeds have undergone considerable behavior-genetic selection to facilitate milking and they tend to be handled differently and to have had more contact with humans during their development than have beef cattle. This also makes dairy cattle more dependent upon humans than their beef counterparts. Beef animals tend to be better mothers than dairy cattle (Selman et al. 1970a, b) and come fairly close to fitting Kilgour and Dalton's (1983) behavioral definition of cattle:

> Cows are large, hairy ruminants living in herds which roam over a large area of grassland. The female withdraws from the herd to produce one (rarely two) precocial young which soon stand and suckle four to six times a day. They "lie-out" away from the dam for at least the first week of life during the daytime. These lying-out patterns together with head threats act to set greater distances between cows in the herd than sheep in a flock. In small herds, straightline social dominance ranks are found and these are stable over many months...

Stockmanship. — Stockmanship can be defined as "knowing the individual behavior of every animal in one's charge, and having the ability to recognize small changes in the behavior of any animal or all of the animals collectively" (Seabrook 1977).

According to Hollier (1979), there is an interaction of the pig with its physical surroundings, other pigs in its group, and with the person responsible for looking after its needs. In order to achieve better performance of growth efficiency, the pig is placed in an environment where it is dependent upon humans for most of its requirements. More of the physical aspects of the environment such as temperature and the basic social requirements such as stocking density and feed space just now are being understood. There is insufficient attention to the importance of having a trained person looking after the pigs. Developing a code of behavior and personal discipline in an approach to looking after intensively kept animals is going to become increasingly important in maximizing response in pig performance.

Good stock people exhibit three characteristics compared to the untrained person (Hollier 1979):

> Firstly, they are perceptive to conditions from the animals' point of view and have developed the discipline of trained observation of the health, comfort, and welfare of the pigs as a method of assessing ongoing performance levels.

> Secondly, they take the trouble to ensure that as soon as they see something is wrong it is **put right immediately.**

> Thirdly, they organize pig flow through the buildings by planning their production line. This means that at all times input, output, and inventory of pigs on hand are as closely balanced to the physical limitations of the building as is practically possible.

Observation is a key ingredient in stockmanship with a willingness on the part of the stockman to correct the conditions causing the deviation from normal behavior pattern.

According to Anderson (1974):

> Husbandry or stockmanship, the relation between man and his animals, although commonly recognized as being an important item in terms of man himself, particularly in the responsiveness of his animals, has received meager, if any, attention from the investigator. If indeed animal response is influenced by human behavior, what are the factors involved and can they be measured and disciplined so that chance alone is not the mediator?

In terms of job satisfaction there are five basic categories of human needs, and, as a general principle, the satisfaction of each group serves as a prerequisite to the next group. These groups are (1) biological needs, (2) safety and security needs, (3) need for affection, belonging, love, (4) the need for esteem and (5) self-actualization needs or full potential (Maslow 1970; Curtis 1983).

The Stockman's Personality and Milk Yields. — English dairymen and their cows were observed in a series of studies (Seabrook 1971, 1972a,b, 1973, 1974, 1977, 1980, and 1984). They defined good stockmanship as the "knowledge of the behavior of individual cows in the herd and the ability to notice deviations from normal behavior." From a study of about 50 herds of similar size (50-80 cows), composition, facilities, and management, Seabrook found that milk yield differences were accounted for by two factors, the level of concentrates fed and the herdsman. Of the 20% differences in milk yield between farms, concentrate levels between herds accounted for only about 25% of yield differences. This led to a long-term study in 20 herds of the herdsman and his personality. The frequency of recorded comments of various subjects of verbal and non-verbal signals, the frequency of human displacement activities like head scratching, yelling, cursing, and the interactions between herdsman and cows were recorded across all seasons and times of day. ("The herdsman considered it to be a study of cow behavior!") Stockmanship is best exercised when a high proportion of time is spent in contact with the cows, yet on one-man units, up to 60% of the working hours may be spent on non-cow contact. Such jobs causing high frustration should, whenever possible, be mechanized and simplified. Human annoyance can reach high levels just before meals, and if non-contact jobs are done then, the work is generally poor and careless. Human fatigue reaches its peak during milking when one-third of the cows remain to be milked, irrespective of herd size. Displacement activity levels peak at this point and the quality of decision making drops. The best human personality type for single unit dairy farms was characterized. A self-reliant, confident, introvert, quiet, reserved, non-sociable ("grumpy") person with cows can easily out-produce (eight herds with 5,191 liters) a similar person lacking confidence (six herds of 4,535 liters), while a confident extrovert ("cheerful Charlie") tends to have only average production achievement (4,629 liter average in six herds). A sound relationship is based on communication as well as confidence.

Some competent cowman talk to their cows when they are under stress. They use a pleasant voice but at times display the necessary dominance. The good communicator is somewhat placid, rather than excitable, and he reinforces good behavior by pleasant words and touch (contact comfort) with his cows. Seabrook (1977) also listed three other rules: patience, consideration for the needs of the cow, and consistency.

In the case of a good stockman, the animals do well and production is high, whereas the poor stockman can reduce productivity, although he apparently does all the jobs that are expected of him. The classic example of this effect is illustrated by the experience of Hampshire farmer, Rex Patterson, whose herds Martin Seabrook studied. Rex Patterson classified his tenants as "stockmen or milk extractors" in terms of the yield they obtained from the many dairy herds he owns. As a result he and his manager found that a good stockman and his attitude towards his cows will obtain up to a 20% increase in milk yield over a poor stockman on the same farm (Kiley-Worthington 1977).

Herdsmen in High-Producing Herds.—The personality of herdsmen was assessed in large, high-producing herds in the United Kingdom and North America (Reid 1977). The assessment of the 25 herdsmen in the study showed 17 of them to be of the confident introvert category. Some of the traits which the herdsmen in this study had in common were: instant recognition of each animal in the herd; a high percentage of their cows approached them; the average number of hours worked was 63 (because of their nature and interest, many herdsmen were only content when spending over 60 hours a week with their cows); 23 were married, all possessed a motor car, only one herdsman had further education; few had interests in community, church, or sports but several had gardening as a hobby. (Each of them in the United Kingdom grew flowers. In particular, roses, gladioli, chrysanthemums, dahlias, and wallflowers were grown by the herdsmen and these particular species require special treatment at specific times of the year, and like cows, respond to feeding). They (21 subjects) had a pet or pets, and almost everyone had kept rabbits, guinea pigs, hamsters, or mice as pets when they were children. Many had hand milked cows before the age of 10, most had few close friends while in school. Reid (1977) summarized the high-production herdsman as obtaining a higher percentage of the milk yield which her genetic capability permits than others would obtain from the same cow in similar large herd conditions (85 to 130 cows). The high-production herdsman achieved this by constant attention to the behavioral pattern and performance of each individual cow within the herd. His ambitions are complementary to the best interests of the dairy industry and his employer, who also tended to work long hours averaging 76 with a minimum of 40 and a maximum of 90. The herdsman recognizes that hoping without working creates illusions.

Opportunities for encouraging cows to associate the herdsman with pleasant feelings will occur during handling before calving, at calving time, in the collecting yard, and during milking. This relationship can be reinforced by feed rewards, patting the cow, tone of voice, and approaching the cow.

Account needs to be taken of the role of the human as a calf substitute (Seabrook 1977).

Effects of Human Handling of Chickens (Hens and Chicks). — Hughes and Black (1976) of Scotland found that handling caused stress and reduced egg production but only in hens not accustomed to it. Regular handling had no effect, indicating the hens had habituated to it. Depressing or stimulating effects of handling may be finely balanced with inherent fearfulness as a controlling factor. Irregular handling depresses performance, while habituation occurs during regular handling. Once the initial fear responses have waned, the extra stimulation provided by regular handling may enhance ability to adapt to other novel and stress-inducing stimuli.

There is little agreement about the effects of handling upon chicks' growth rates. Some authors (McPherson et al. 1961; Reichman et al. 1978) concluded that handling immature broilers and pullets had no effect, whereas Freeman and Manning (1979) found that regular handling decreased growth in chicks of a layer strain. Thompson (1976) and Gross and Siegel (1979) found increased growth following handling in broilers and layers, results consistent with findings in rats (Ruegamer et al. 1954; Weininger 1956; Levine 1962). Different handling regimes, methods, strain, sex and age differences may help to explain these inconsistencies (Jones and Hughes 1981).

The effect of regular handling (twice-daily) on growth and gain-to-feed ratios in male and female chicks of layer (two strains) and broiler strains were examined from hatching to three weeks of age in six batches of 160 birds each by Jones and Hughes (1981). Growth was significantly enhanced by regular handling in broilers and the females of the layer strains, and gain-to-feed ratios were generally greater in the handled birds. There were no significant treatment effects on growth or gain-to-feed ratios in males of the layer strains. Males had higher relative weight gains and gain-to-feed ratios than females.

The improved performance of the handled broilers agrees with the findings of Thompson (1976), but conflicts with those of McPherson et al. (1961) and Reichman et al. (1978), who found no effects of handling on growth. The birds were handled once weekly in the latter two studies, whereas Jones and Hughes' (1981) birds were handled twice-daily and those of Thompson were handled once a day for 15 days from hatching. The birds used by McPherson et al. (1961) and Reichman et al. (1978) were one and nine weeks old, respectively, before stimulation began, whereas Jones and Hughes' birds were handled from the first day of life. It is likely that the inconsistencies can be explained in terms of either differences in intensity of stimulation or the existence of a sensitive period.

The chicks of Freeman and Manning were handled twice-daily for five days per week for three weeks; a little less stimulation than that perceived by chicks handled by Jones and Hughes (1981). The interval on weekends producing an irregular regime may be an important factor. Strain differences may also account for this disagreement but Freeman and Manning weighed only a small number of chicks and did not distinguish sexes.

In a United States study by Gross and Siegel (1982), chicks were habituated to human beings (socialized) by being talked to, offered food, and handled gently within an environment with a minimum of noise. After seven weeks of socialization, the birds were challenge exposed with *Escherichia coli*. When compared with ignored groups, the socialized birds showed more than a 60% reduction in the prevalence of death and pericarditis. Furthermore, small flocks of socialized birds were more uniform in their response to *E. coli* than were similar nonsocialized flocks. Socialized chickens also had improved feed efficiency and increased antibody response. Socialization was also applied easily to larger flocks of chickens. Socialization of chickens to their handlers by being talked to, offered food, and handled gently results in increased feed efficiency, growth rate, uniformity of responses to all tests, resistance to stressors, antibody response to antigen, blood protein, and increased resistance to a wide variety of infectious agents (Gross 1983).

Responses of Pigs to the Presence of Humans.—Twelve commercial one-man pig farms of medium to high productivity were selected for the study of Hemsworth et al. (1981). A large integrated company controlled the farms providing pigs, feed, weekly management and twice-weekly veterinary advice. The 12 Dutch farms had very similar inputs apart from the ability of the stockman.

Two different behavioral tests were conducted on 1,225 and 480 pregnant sows, respectively, to examine their behavioral response towards humans: Sows displayed a significantly ($P > 0.05$) greater withdrawal response to the approaching experimenter's hand (Test 1) and a significantly ($P < 0.05$) lower approach behavior towards the stationary experimenter (Test 2) at farms in which the average total number of piglets born per sow per year was low.

The achievement and maintenance of a good human/animal relationship requires an animal's understanding of the type and nature of signals released by humans. In a follow-up experiment in the United States by Hemsworth et al. (1983), two experiments were conducted to evaluate the nature of several common signals. Both experiments were a 2 × 2 factorial of signal involving 48 and 44, 10-12 week-old pigs. The nature of the signals was evaluated by quantifying the approach behavior of the pigs to the experimenter in four three-minute tests over an eight-day period. Non-approach, squat posture (closer to the pigs) and bare hands by the experimenter were associated with an increase in the approach behavior of the pig. Therefore, naive pigs appear to interpret these signals to be nonthreatening in nature. There were also significant differences in litters on approach behavior.

These experiments demonstrate that the nature of signals released by humans can be identified. The effects of handling and stimulating on the behavior of young pigs were studied by Grandin et al. (1983). Pigs (24 four and a half-week-old Hampshire-sired crossbred) from five litters were placed in either a "stimulating" or a "nonstimulating" environment. The "nonstimulating" environment consisted of placing two pigs in each of six 1.22 m × 1.22 m nursery pens with plastic-coated expanded-metal floors. Lighting and temperature in the room were constant and the pigs were not handled except for adding feed to the self-feeders once-daily and cleaning the pens every third day. The "stimulating" environment consisted of 12 pigs placed together in one outdoor pen with a concrete floor and adjoining house bedded with straw. These pigs were handled and played with for at least 15, and often, 30 minutes daily and also provided with objects ("toys") with which to play (e.g., plastic milk crate, garbage can, chains, cloth strips, dirt, stones, newspapers, cardboard boxes, ropes and twine). The objects were changed daily. At the end of the nine-week trial, approach times to either a strange man or a novel object (red feeder standing on end) were measured in a 2.74 m wide octagonal pen with 1.22 m high white plywood walls and brown plastic carpeting on the floor. There was a three minute time limit. Resulting data were: Approach man: "stimulated," 59.5 seconds; "unstimulated," 100.3 seconds. Approach novel object: "stimulated," 49.8 seconds; "unstimulated," 83.5 seconds. Differences among litters were also apparent. One stimulated pig vocalized during the tests, and she and her littermates were slower to approach the man regardless of rearing environment (125.9 seconds vs. 60.9 seconds). Observations indicated that the pigs would stop playing with an object unless it was changed often, and the pigs played with some objects longer than others. If an object (e.g., bowling ball) became contaminated with manure, the pigs tended to avoid it.

Previous studies in the farm animal area have been limited from behavioral tests (swine), to once-daily (chicken) as well as close twice-daily interaction between the cowman and individual cows in the herd. These innovative, one-of-a-kind, creative studies attract a great deal of interest and comment; however, they are difficult to duplicate and repeat. Also, with more reliance placed upon machinery and controlled environments, currently there is less emphasis upon animal handling and husbandry-like skills. Thus, there is less time spent per food producing animal except in the case of the companion farm animal species such as the horse, goat, and dog.

Humans Are a Part of the Problem and Solution. —Other human/farm animal relationship studies include work on early experience (Albright 1981b; Donaldson 1970; Donaldson et al. 1971, 1972, and 1974). They showed social isolation during an early developmental period increased later milk production in Holstein dairy cattle.

Behavior modification is a powerful tool and it has been used to train dairy cattle by operant conditioning methods (Wisniewski 1977; Wisniewski and Albright 1978a, b). Through proper training of the cows and milking parlor operator, cows were induced to cooperate within the system instead of being forced to conform. Data collected by Wisniewski (1977) from 12,222 individual cow observations at milking times during a one year period using a double-five herringbone parlor system showed that routinely only 2.8% of the cows entered the milking parlor voluntarily or unassisted. Stated another way, the strategy of 97.2% of the cows was to wait for the milker to come and to coax them through the doorways into the milking stalls. Training of Holstein heifers and cows to enter the parlor using operant conditioning (with a training stimulus of either the door opening, flashing lights, or a buzzer) with negative reinforcement (shock prod) resulted in peak performance observed on day 7 when 99.2% of the cows entered the parlor by themselves. Performance of the cows decreased after the training period (64.9% entering unassisted vs. 80.7%) but this represented a 200-fold improvement over the pre-training period (2.2%). Untrained and partially trained cows followed the fully trained cows into the parlor.

In the above studies by Wisniewski, the leadership-followership behavioral trait is used to advantage. The urge to follow in a species being as great as the urge to lead causes group activity and movement to and from pastures, into and out of chutes, milking parlors and electronic feeders.

The domestication and taming process illustrate the close human/animal bond whereby "imprinting" (Albright 1982) as well as restraint, handling, training, and exhibiting animals takes place.

Those interested in the care and welfare of farm animals are concerned by what animals (and not humans) would choose when presented the opportunity. One such early test to determine how animals think, anticipate, and react was developed by Krushinsky (1965). With pigeons, hens, crows, rabbits, cats, and dogs, the reinforcement principle was utilized by placing food in one of two feeding bowls which stood side by side at the gap behind a screen. When an animal started to eat the food, both bowls were moved in straight lines in opposite directions. After covering a distance of 20 cm. the bowls disappeared behind non-transparent flaps. To solve the problem, the animal had to move around the screen on the side behind where the feeding bowl had disappeared. Experimental animals had success rates of: pigeons (7%), hens (52%), crows (86%), rabbits (27%), cats (86%), and dogs (89%).

Krushinsky's technique was adapted to farm animals so as to provide a slow-moving trolley carrying food within a large animal learning maze. It disappeared into an A-frame tunnel and, after certain periods of time, reappeared at the other end (Albright et al. 1982). Three dairy cows were tested. When the food trolley moved, they demonstrated a "startle" reaction (staring at the object with front feet firmly planted) which may have inhibited the learning process. It took them two weeks before they learned to continue eating from the moving food box and anticipate the reappearance of the trolley. After our human presence was removed, success came about by observing the cows with a TV monitor. The cows had become conditioned and dependent upon humans expecting us to remove them from the testing area after the food disappeared. They explored and solved the problem only after we were out of sight. Since cows learn from each other, each cow and later each bull were put in the testing arena separately. Earlier, another 10 of 11 cows "startled" when the food object started moving and they did not learn how to solve the problem.

None of the eight dairy bulls (four Jersey and four Friesian) learned to anticipate and had a larger startle response than cows. They stopped eating once the food box was set in motion. They showed a startle reaction when the trolley was 9 to 10 m away from them.

In addition, 12 out of 15 pigs were able to "extrapolate" from a known situation within three days, and six out of seven rams tested in four days.

Summary

Humans and farm animals contribute to the mutual well-being of one another. Humans provide the care, feed, and housing for animals that produce food, by-products, fiber, work, recreation, and also improve human nutrition, health and companionship. Although the bonding and relationship between handler(s) and livestock may be important, it has been difficult to study and to quantify. After proper bonding, each time an animal is handled it will be more tame and pliable. Significant relationships between personality characteristics of dairymen and milk production per cow were found in England and verified in North America. In surveying 50 one-man dairy herds of 50 to 80 cows, 11% more milk and greater willingness of cows to enter the milking area were obtained by those dairymen classed as confident introverts than by confident extroverts. Training of Holstein heifers and cows to enter the milking parlor using operant conditioning resulted in peak performance observed on day 7 when 99.2% of the cows entered the parlor by themselves as compared to the pre-treatment period of 2.2%. A Dutch study strongly suggests that the reproductive performance of pigs is associated with the relationship between the stockman and breeding stock. Later U.S. pig research demonstrated that handling and the nature of signals released by humans can be quantified. In Scotland, growth in broilers and layers was significantly enhanced by regular handling. Socialization of chickens to United States handlers by being talked to, offered food, and handled gently results in increased feed efficiency, growth rate, and resistance to stressors.

Endnotes

[1] Invited lecture for Symposium I: Applied Ethology, Midwest Animal Behavior Society Meeting, Southern Illinois University at Carbondale, IL, April 28, 1984.

[2] Professor of Animal Sciences,Dept. of Animal Sciences, School of Agriculture and Large Animal Clinics, School of Veterinary Medicine, Purdue University, West Lafayette, IN 47907. Also, Purdue Center for Applied Ethology and Human/Animal Interactions. This investigation is part of the Indiana contribution to NCR-131 Animal Care and Behavior and NC-119 Improving Dairy Herd Management Practices.

[3] Leaflet "We Need Farm Animals" is available from the American Society of Animal Science, 309 W. Clark St, Champaign, IL, 61820.

[4] Topics listed and discussed were on the Needs of Humans; Human Nutrition; Enjoyment of Eating; Recycling of Nutrients; Use of Crop Residues; Use of Industrial By-products; Use of Noncrop Land; Economic Contribution; Use of Feed Grains; Provide Fibers; Animal By-products; Food Storage; Power (Draft); and Human Research Applications. A summary statement and five references for additional information are included.

[5] Elsewhere in the pamphlet a vital, appropriate statement is made: "Due to inadequate rainfall, rocks and rough terrain, only about one-fifth of U.S.A. land is suitable for cropland. Worldwide the figure drops to one-tenth."

References

Albright, JL. 1978. The Behavior and Management of HighYielding Dairy Cows. BOCM SILCOCK Dairy Conference at Heathrow, London, England, Jan. 30. Booklet, 43 pp.

—. 1981a. The human-cow interaction. In: *Dairy Sci. Handbk.* 14: 357-62. Clovis, CA: Agriservices, Inc.

—. 1981b. The effects of early experience upon social behavior and milk production in dairy cattle. Presented at the Applied and Companion Animal Ethology Symposium, Animal Behavior Society, University of Tennessee, Knoxville, June 25. Mimeo 4 pp.

—. 1982. Behavioral responses to management systems—dairy. In: Woods, WR. ed. *Proceedings of the Symposium of the Management of Food Producing Animals.* Purdue University, West Lafayette, IN. Vol. 1:139-65.

Albright, JL, Kilgour, R and Wittlestone, WG. 1982. The Krushinsky Apparatus: A test for self awareness in farm animals. *Indiana Acad. Sci.* 91: 595-96.

Anderson, GW. 1974. Old wine in a new skin or animal behavior in the modern animal science curriculum. *J. Anim. Sci.* 39: 441-45.

Arave, CW and Brown, JE. 1979. Do livestock people have different personalities? *Hoard's Dairyman.* 124: 344-45.

Burnside, EB, Kowalchuk, WB, Lambroughton, DB and MacLeod, NM. 1971. Canadian dairy cow dispersals. 1. Differences between breeds, lactation numbers and seasons. *Can. J. Anim. Sci.* 51: 75-78.

Curtis, SE. 1983. What constitutes animal well-being? Int. Cong. on Animal Stress. Sacramento, CA. July 6-8, 27 pp.

Donaldson, SL. 1970. The effects of early feeding and rearing experiences on social, maternal and milking parlor behavior in dairy cattle. Ph. D. Dissertation, Purdue Univ., W. Lafayette, IN.

Donaldson, SL, Albright, JL, Black, WC and Harrington, RB. 1971. Effects of early feeding-rearing regimes on adult cattle behavior. *J. Anim. Sci.* 33: 194. (Abstr.).

Donaldson, SL, Albright, JL and Black, WC. 1972. Primary social relationships and cattle behavior. *Proc. Indiana Acad. Sci.* 81: 345-51.

—. 1974. Early feeding and rearing experiences and their effects upon behavior and milk production. *Int. Dairy Cong.* 19: 26-27.

Eysenck, HJ. 1977. *You and Neurosis.* Glasgow: William Collins Sons & Co. Ltd. 223 pp.

Freeman, BM and Manning, AC. 1979.Stressor effects of handling on the immature fowl. *Res. in Vet. Sci.* 26: 22326.

Grandin, T, Curtis, SE and Greenough, WT. 1983. Effects of rearing environment on the behavior of young pigs. *J. Anim. Sci.* 57 (Suppl. 1): 137. (Abstr.)

Gross, WB. 1983. Chicken-environment interactions. Research, management, behavior and well-being of farm animals. Presented at the Conference on the Human-Animal Bond. Minneapolis, MN. June 13-14. (Abstr.)

Gross, WB and Siegel, PB. 1979. Adaptation of chickens to their handler and experimental results. *Avian Diseases.* 23: 708-14.

—. 1982. Socialization as a factor in resistance to infection, feed efficiency, and response to antigen in chickens. *Am. J. Vet. Res.* 43: 2010-12.

Hediger, H. 1955. *Studies of the Psychology and Behavior of Captive Animals in Zoos and Circuses.* New York: Criterion Books.

Hemsworth, PH, Brand, A and Willems, P. 1981. The behavioral response of sows to the presence of human beings and its relation to productivity. *Livest. Prod. Sci.* 8: 67-74.

Hemsworth, PH, Gonyou, HW and Dzuick, PJ. 1983. Human communication with animals. *J. Anim. Sci.* 57(Suppl. 1): 137-38. (Abstr.)

Hollier, D. 1979. *Aspects of Swine Ecology.* Spring Green, WI: Gateway Press. 180 pp.

Jones, RR and Hughes, BO. 1981. Effects of regular handling on growth in male and female chicks of broiler and layer strain. *Brit. Poul. Sci.* 22: 461-65.

Kiley-Worthington, M. 1977. *Behavioural Problems of Farm Animals.* Oriel Press Ltd: Stockfield, Northumberland. 134 pp.

Kilgour, R. 1983. The role of human-animal-bonds in farm animals and welfare issues. Proc. Conf. on the Human-Animal Bond. Minneapolis, MN. pp 1-25.

Kilgour, R and Dalton, C. 1983. *Livestock Behavior: A Practical Guide*. St. Albans, U.K.: Granada Publ. Ltd. 320 pp.

Krushinsky, LV. 1965. Solution of elementary logical problems by animals on the basis of extrapolation. *Progress in Brain Research. Vol. 17. Cybernetics of the Nervous System*. Weiner, N and Schade, JP. eds. New York: Elsevier Pub. Co. pp. 280-308.

Levine, S. 1962. Psychophysiological effects of infantile stimulation. In: *Roots of Behaviour*. Bliss, EL. ed. New York: Harper and Brothers. pp. 246-59.

Lott, DF and Hart, BL. 1977. Aggressive domination of cattle by Fulani herdsmen and its relation to aggression in the Fulani culture and personality. *Ethos*. 5(2): 174-86.

Maslow, AH. 1970. *Motivation and Personality*. New York: Harper and Row.

McFarlane, IS. 1976. Social behavior in the herd. *Dairy Sci. Hndbk*. 9: 63-66.

McPherson, BN, Gyles, NR and Kan, J. 1961. The effects of handling frequency on 8-week body weight, feed conversion and mortality. *Poul. Sci*. 40: 1526-27.

Murphey, RM, Duarte, FA and Torres Penedo, MC. 1980. Approachability of bovine cattle in pastures: Breed comparisons and a breed × treatment analysis. *Behav. Genet*. 10: 17383.

—. 1981. Responses of cattle to humans in open spaces: Breed comparisons and approach-avoidance relationships. *Behav. Genet*. 11: 37-48.

Price, EO. 1984. Behavioral aspects of animal domestication. *Quart. Rev. Biol*. 59(1): 1-32.

Reid, NS. 1977. To endeavor to correlate common factors which may exist in herdsmen in high yielding dairy herds. *Nuffield Farming Scholarship Report*. Feb. 21 pp.

Reichman, KG, Barram, KM, Brock, IJ and Standfast, NF. 1978. Effects of regular handling and blood sampling by wing vein puncture on the performance of broilers and pullets. *Brit. Poul. Sci*. 19: 97-99.

Ruegamer, WR, Bernstein, L and Benjamin, JD. 1954. Growth, food utilization, and thyroid activity in the albino rat as a function of extra handling. *Science*. 120: 184-85.

Schmidt, GH and Van Vleck, LD. 1974. *Principles of Dairy Science*. San Francisco: W.H. Freeman and Co. p. 39.

Schwadel, F. 1986. Some pets not only live high on the hog but really are hogs. *The Wall Street J*. 66(95): 1, 12.

Seabrook, MF. 1971. Stress and the cowman. *Dairy Farmer*. 18(10): 47, 49, 73.

—. 1972a. A study to determine the influence of the herdman's personality on milk yields. *J. Agric. Labour Sci*. 1: 45-49.

—. 1972b. A study on the influence of the cowman's personality and job satisfaction on milk yields of dairy cows. *J. Agric. Labour Sci*. 1: 49-93.

—. 1973. Fit the cowman to the job. *Dairy Farmer*. 20(b): 22, 25, 53.

—. 1974. A study of some elements of the cowman's skill as influencing the milk yield of dairy cows. Ph. D. Thesis, Univ. Reading, Reading, England.

—. 1977. Cowmanship. *Farmers Weekly Extra*. 25(87): 3-26.

—. 1980. The psychological relationship between dairy cows and dairy cowmen and its implications for animal welfare. *Int. J. Stud. Anim. Prob*. 1: 295-98.

—. 1984. The psychological interaction between the stockman and his animals and its influence on performance of pigs and dairy cows. *Vet. Rec*. 115: 84-87.

Selman, IE, McEwan, AD and Fisher, EW. 1970a. Studies on suckling in cattle during the first eight hours post partum. I. Behavioural studies (dams). *Anim. Behav*. 18: 276-83.

—. 1970b. Studies on suckling in cattle during the first eight hours post partum. II. Behavioural studies (calves). *Anim. Behav*. 18: 284-92.

Stricklin, WR and Kautz-Scanavy, CC. 1984. The role of behavior in cattle production: A review. *Appl. Anim. Ethol*. 11: 359-90.

Thompson, CI. 1976. Growth in the Hubbard broiler: Increased size following early handling. *Dev. Psychobiol.* 9: 459-64.

Waring, GH. 1985. Spatial reactivity of animals and appropriate ethological terminology. Proc. Midwest Animal Behavior Society Meeting. Miami Univ., Oxford, OH. March 23.

Weininger, O. 1956. The effects of early experience on behaviour and growth characteristics. *J. Comp. and Physiol. Psy.* 49: 1-9.

Wisniewski, EW. 1977. Behavioral modification of milking parlor entrance order in dairy cattle trained by operant conditioning methods. Ph. D. Dissertation, Purdue Univ., W. Lafayette, IN.

Wisniewski, EW and Albright, JL. 1978a. Parlor entrance behavior of dairy cattle trained to enter a herringbone parlor with conditioning methods. Proc. Int. Symp. on Milking Machines. Louisville, KY: National Mastitis Council, Inc., Washington, DC. p. 460-66.

—. 1978b. Training heifers and cows to enter a milking parlor by operant conditioning methods. *World Cong. on Ethol. Appl. to Zootechnics.* 1(E): 67.

CONTRIBUTION TO A CONCEPT OF BEHAVIORAL ABNORMALITY IN FARM ANIMALS UNDER CONFINEMENT

U.A. Luescher[1] and J.F. Hurnik[2]

Introduction

Farm animals housed in close confinement often engage in activities that do not occur with animals maintained in traditional and more complex environments. Many of these activities consist of species-typical motor patterns directed towards unsuited or inappropriate objects, or performed as vacuum activities. For example, piglets fed from a trough from day 2 to day 21 after parturition display much nosing of penmates and ear sucking (DeBoer and Hurnik 1984). Similarly, confined veal calves in crates may lick their pelage excessively, or, when housed in groups, may suck the naval area of penmates; laying hens and broilers often engage in feather pecking and cannibalism. A list of reports in the literature concerning such behavior is given in Fox (1984). These behaviors are not adaptive, that is, they do not contribute to species-typical development, maintenance, or reproduction and they may even result in physical damage to the performer or its pen- or cagemates (Tschanz 1982). It is widely agreed that the occurrence of these abnormal behaviors is indicative of environmental inadequacy and animal suffering (Sambraus 1981; Fox 1984). Because abnormal behavior is believed to be an important criterion in evaluating animal housing systems and management practices, there is a need for a general concept of behavioral abnormality in farm animals that would facilitate making judgments about the acceptability of given production systems, and to predict effects of environmental changes on animal welfare (Duncan 1983).

This paper contributes to the development of a concept of behavioral abnormality by comparing the situation of animals in an artificial environment to an experimental learning situation. The concept is based on behavioristic theory developed by Seligman (1970) and Staddon and Simmelhag (1971). The concept provides a novel perspective of behavioral normality and abnormality.

To account for the fact that some stimulus-response associations are more easily established than others by animals in learning experiments, Seligman (1970) offered the so-called concept of "preparedness." He hypothetically postulated an ease-of-learning continuum. On one end were associations phylogenetically predisposed with high probability, which the author called "prepared" associations. On the other end were associations predisposed with low probability, which he called "contraprepared" associations. Between the two extremes is a middle range of so-called unprepared associations (those traditionally studied in learning laboratories).

Staddon and Simmelhag (1971) developed an entirely new concept that relates the nature of learning processes to evolutionary theory. They regard learning:

> ...as the outcome of two independent processes: a process of variation that generates either phenotypes in the case of evolution, or behavior in the case of learning; and a process of selection that acts within the limits set by the first process.

Thus, learning is considered to be in part a selection process analogous to, but separate from, phylogenetic selection. The ontogenetic selection process is controlled by what they call principles of reinforcement. They claim that "reinforcement acts only to eliminate behaviors that are less directly correlated with reinforcement than others."

The Staddon/Simmelhag theory takes into consideration the effect of genetics on behavior and on learning, and thus bridges a serious conflict between behaviorists and ethologists.

The present concept, developed on the basis of these theories, provides insight into the nature of environmentally-induced abnormal behavior in farm animals, and defines conditions required for its prevention. According to Hurnik et al. (1985), animal well-being can be defined as "a state or condition of physical and psychological harmony between the organism and its surroundings." Provision of appropriate conditions (those allowing normal behavior) is therefore considered an essential prerequisite for animal well-being.

Phylogenetic Adaptation

The process of phylogenetic adaptation results from an interplay between mutation increasing the range of genetically determined characteristics and the environment selecting those characteristics that give their carriers some advantage over noncarriers. Phylogenetic adaptation provides a defined range of behaviors in which an animal is able to engage.

The behavioral repertoire of a species often is subdivided into functional systems, such as ingestive, sexual, or eliminative behavior. Each functional system contains subcategories of behaviors, which will be called functional units of behavior. Behaviors within a functional unit are characterized by basically similar motor patterns and similar functions, and presumably are promoted by similar motivational factors.

The range of behaviors in each functional unit is determined by phylogeny. For example, the functional system of ingestive behavior includes the functional units of eating and drinking. The functional unit of eating may embrace such behaviors as eating grain, grazing grass, or chewing roots. Strong selection pressure on a functional unit of behavior during phylogeny results in a high degree of fixation and leaves little room for ontogenetic shaping. Weak selection pressure, in contrast, results in a broad range of possible behaviors within a functional unit, and leaves much room for learning. Behaviors with a high degree of fixation often are called instinctive; those with very low degree of fixation, learned.

Performance of a given behavior in general is preceded by the establishment of a stimulus-response association. When an animal is motivated, e.g., to sleep or to groom, it will search for appropriate releasing stimuli, such as a suitable shelter or an object for scratching on, according to genetically predisposed associations between sleeping and shelter, or grooming and a rough vertical surface. Such an association between action pattern and releasing stimulus thus is an integral part of a behavior. Therefore, each functional unit of behavior contains a phylogenetically determined range of stimulus-response associations.

Like Seligman's (1970) notion of "preparedness," the present concept suggests that the stimulus-response associations (and thus the behaviors) in a given functional unit are not predisposed in the genotype of an animal to be established with equal likelihood. Rather, they are characterized by their differential probability of being established. Phylogeny not only determines the range of possible stimulus-response associations contained within each functional unit of behavior, but also the probability of the associations relative to each other. As a result of phylogenetic adaptation, the stimulus-response associations that are most successful and presumably most rewarding in a

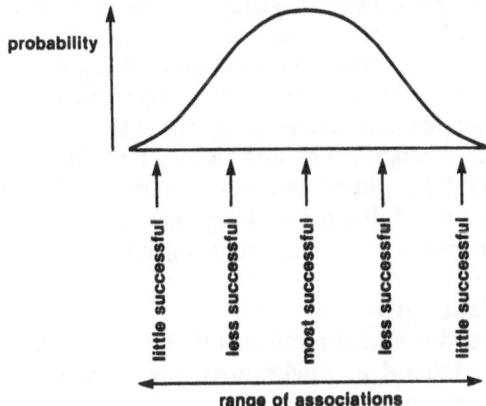

Figure 1. Genetically predisposed probability distribution for stimulus-response associations contained in one functional unit of behavior.

species' natural environment are generally predisposed with highest relative probability. This situation can be represented, for example, by a normal probability distribution as in figure 1.

The normal distribution has been chosen to represent real distributions for the sake of simplicity. On the abscissa, stimulus-response associations contained within one functional unit of behavior are arranged. For example, within the functional unit of roosting in hens, associations may range from between roosting and a thin twig to roosting and a very thick branch.

Different stimuli for associations of a functional unit, such as elevated perch, wooden slats, and wire floor for roosting in hens; or roots, straw, and tails of penmates for eating in pigs, usually vary with regard to several or many of their characteristics. To arrange associations to such stimuli along one axis is a simplification. As a matter of fact, they should be arranged along several independent axes, resulting in a multidimensional diagram.

The ordinate is a scale for the relative probabilities for the various associations within a functional unit. For instance, the association between roosting and a branch of a certain diameter (e.g., 2 inches), is predisposed in a hen's genotype with higher probability (and thus lies closer to the median of the distribution), than is the association between roosting and either a very thin or a very thick branch. Or the association between eating and roots is predisposed in the genotype of a pig with higher probability than the one between eating and straw, or eating and other pigs' tails.

Accordingly, in an artificial environment, provision of stimuli similar to the ones preferably responded to in the natural environment should facilitate the establishment of stimulus-response associations. Further, responses to these stimuli should be more probable than responses to less natural stimuli. Mees and Metz (1983) have obtained results that support this hypothesis (figure 2). In their experiment, young calves were given milk from pails with or without rubber teats. Frequency of sucking was higher in the group that had pails with teats. In both groups, if the pails were removed when empty, calves sucked on objects or ears of penmates. If empty pails were not removed, calves continued to suck on them. Frequency of overall sucking (including sucking on pails as well as other objects) was higher in the group that had pails with teats than in the group that had pails without teats. If the pails were removed, frequency of sucking dropped in both groups to below the levels exhibited by either group when the pails were present. This indicates a higher probability for sucking on a rubber teat than for sucking on a plain pail or even non-food related objects.

Ontogenetic Adaptation

In analogy to Staddon and Simmelhag (1971), behavioral adaptation during ontogeny can be defined as environmentally-controlled selection from a range of stimulus-response associations. The range as well as the probabilities of these associations are determined by phylogenetic processes, as explained earlier. In a given environment, some associations are more "directly corre-

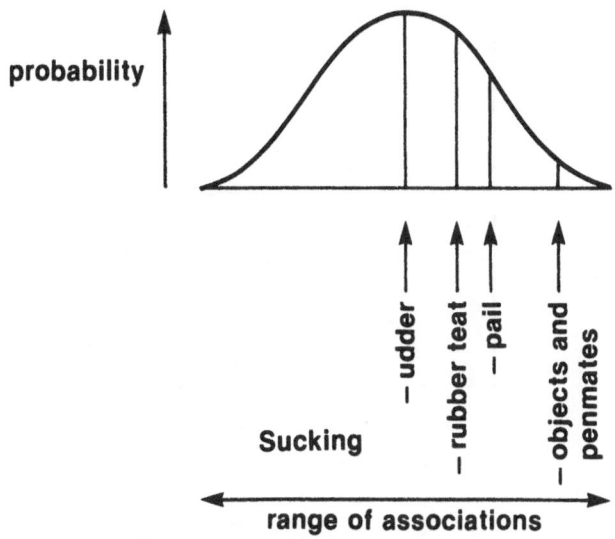

Figure 2. Functional unit of sucking in calves: Genetically predisposed probability distribution for stimulus-response associations.

lated with reinforcement than others" (Staddon and Simmelhag 1971), because the behaviors resulting from these association are most successful. The probability of these associations thus increases, while the probability of other associations is reduced.

This also applies to the natural environment of a species. In the natural environment, associations that are genetically predisposed with highest relative probability generally are reinforced the most. The genetically predisposed distribution of association probability and the phenotypic one resulting from ontogenetic adaptation of a given species in its natural environment are illustrated in figure 3 for eating in pigs. The medians of the two curves coincide, but their variances differ.

When an environment is provided that does not contain the natural stimuli appropriate to associations near the median of the probability distribution, these associations cannot then be behaviorally manifested, and thus are no longer reinforced. Other associations, for which stimuli are provided, are reinforced since the behaviors resulting from them become relatively successful (figure 4). If these associations lay sufficiently close to the median of the probability distribution, reinforcement of these associations will effectively suppress others, and the animal is considered to have adapted to the new situation.

As the deviation of the environment from natural becomes more pronounced, the genetically predisposed probability of the associations reinforced in the given environment is low, and the resulting behaviors are less rewarding. In accordance with Staddon and Simmelhag (1971), reinforcement of some stimulus-response association reduces the probability of those that

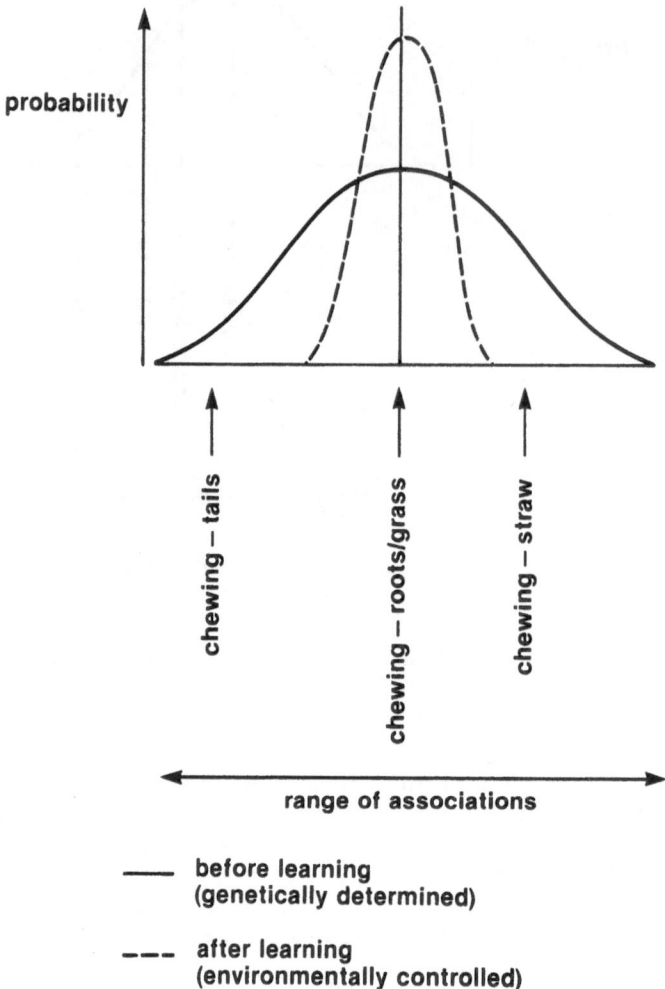

Figure 3. Functional unit of eating in pigs: Phenotypic probability distribution for stimulus-response associations in the natural environment, as compared to the genetically predisposed distribution.

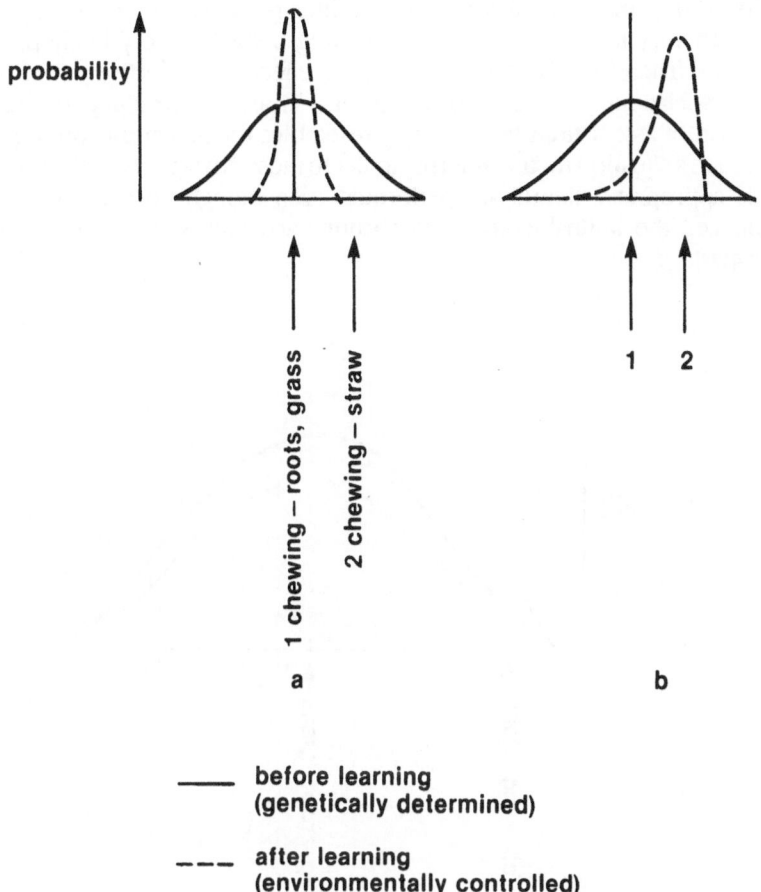

Figure 4. Functional unit of eating in pigs: Phenotypic probability distribution for stimulus-response associations in a semi-natural environment, as compared to the genetically predisposed distribution.

are not reinforced. Weak reinforcement of an association thus only slightly reduces the probability of non-reinforced associations. This situation is schematically illustrated in figure 5.

If the environment reinforces only stimulus-response associations predisposed with very low probability, the resulting behaviors very likely provide little reward. Therefore, the environment is not effective in selectively reinforcing certain associations and suppressing others. Thus, the resulting probability distribution of associations very strongly resembles the genetically predisposed one. In such a situation motor patterns similar to those characteristically displayed towards appropriate stimuli are performed even though no such stimuli are present, i.e., the animal engages in vacuum activities which often become stereotyped.

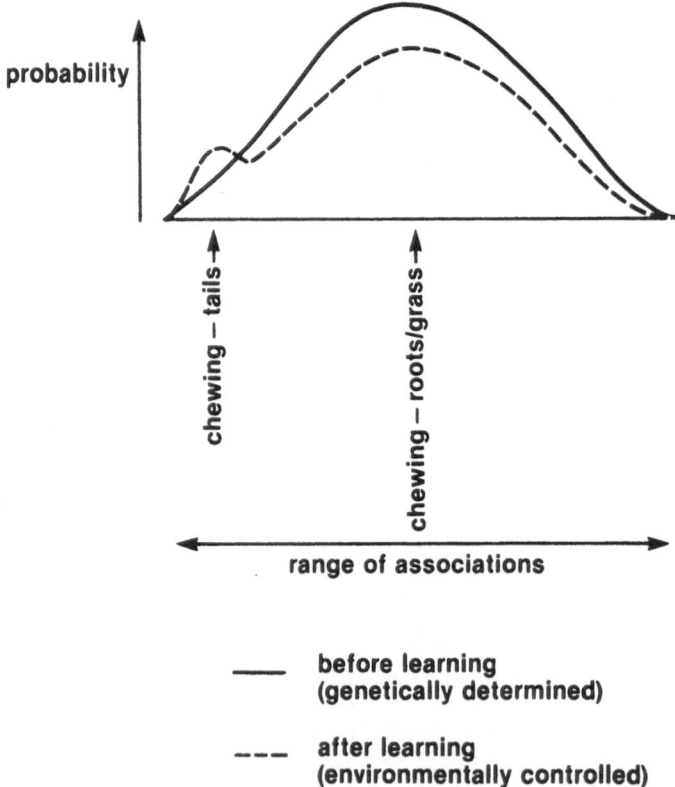

Figure 5. Functional unit of eating in pigs: Phenotypic probability distribution for stimulus-response associations in an environment which strongly deviates from natural, as compared to the genetically predisposed distribution.

The question, under what circumstances some functional units (for which only highly inappropriate stimuli are provided) can be replaced by others (for which there are more appropriate stimuli) is a complex issue which is not further addressed in this paper. It is maintained, however, that in most cases such replacement seems not possible, and that at any point in time an animal engages in some behavior of a functional unit with given probability.

Normal and Abnormal Behavior

The proposed concept provides an opportunity to define normal and abnormal behavior in statistical terms.

> Behavior can be considered normal if its underlying stimulus-response association is genetically predisposed with sufficiently high probability in relation to others within the same functional unit of behavior, i.e., if it lies within certain limits of the genetically determined probability distribution.

In contrast,

> Behavior can be considered abnormal if the underlying association is genetically predisposed with low probability, i.e., if it lies outside these limits (figure 6).

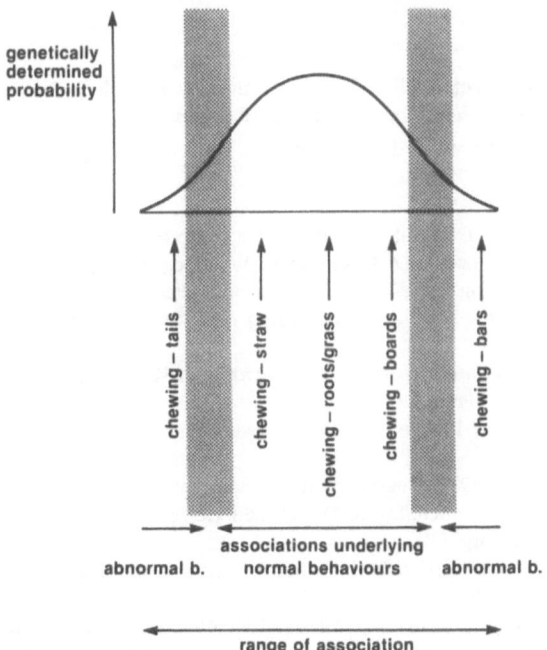

Figure 6. Functional unit of eating in pigs: Normal and abnormal behavior as defined statistically with reference to the genetically predisposed probability distribution for stimulus-response associations.

Conclusions

(1) Behavior generally is characterized by a motor pattern and its relation to the environment. Thus, a stimulus-response association is a determinative characteristic of any behavior related to stimuli in the environment.

(2) Stimulus-response associations are predisposed with a genetically determined probability, defined in relation to the probabilities of other associations within the same functional unit of behavior.

(3) The genetically predisposed distribution of association probabilities is determined by natural selection and domestication and can to some extent be influenced by selective breeding.

(4) Behavior resulting from stimulus-response associations predisposed with low relative probability (i.e., behavior directed toward inappropriate objects) is always abnormal, even if its motor pattern very strongly resembles that resulting from an association predisposed with high probability (i.e., behavior directed toward appropriate objects).

(5) Living conditions for farm animals are appropriate only if the environment reinforces the establishment of stimulus-response associations that are predisposed with sufficient probability.

(6) Ontogenetic adaptation (learning) results in a phenotypical probability distribution of stimulus response associations that deviates from the genetically predisposed distribution.

Endnotes

[1] Assistant Professor, Dept. of Clinical Studies, University of Guelph, Guelph, Ontario, Canada, N1G 2W1. *Address to which reprint requests should be sent.*

[2] Professor, Dept. of Animal and Poultry Science, University of Guelph.

References

DeBoer, S and Hurnik, JF. 1984. Automatic device for group rearing of piglets: feeding variations. 34th Annual Conference, C.S.A.S., Aug. 19-22, Winnipeg, Manitoba.

Duncan, IJ. 1983. Assessing the effect of housing on welfare. In: Baxter, SH, Baxter, MR, and MacCormack, JA. eds. *Farm Animal Housing and Welfare.* Dordrecht, The Netherlands: Martinus Nijhoff.

Fox, MW. 1984. *Farm Animals: Husbandry, Behavior and Veterinary Practice.* Baltimore, MD, USA: University Park Press.

Hurnik, JF, Webster, AB and Siegel, BP. 1985. *Dictionary of Farm Animal Behavior.* University of Guelph.

Mees, AM and Metz, JH. 1983 Saugverhalten von KalbernBedurfnis und Befriedigung bei verschiedenen Trankesystemen. Tagung Dt. vet. med. Ges. e.v., Fachgruppe Verhaltensforschung, 16.-19.11.1983, Freiburg, i.Br.

Sambraus, HH. 1981. Abnormal behavior as an indicator of immaterial suffering. *Int. J.Stud. Anim. Prob.* 2: 245-48.

Seligman, ME. 1970. On the generality of laws of learning. *Psych. Rev.* 77: 406-18.

Staddon, JE and Simmelhag, VL. 1971. The "superstition" experiment: A reexamination of its implications for the principles of adaptive behavior. *Psych. Rev.* 78:3-43.

Tschanz, B. 1982. Verhalten, Bedarfsdeckung und Schadenvermeidung bei Tieren. Tagung Nutztierkommission Schwiezer Tierschutz, Int. Ges. fur Nutztierhaltung, 23.4. 1982, Bern.

THE PSYCHOLOGY AND ETHICS
OF HUMANE EQUINE TREATMENT

Sharon E. Cregier[1]

Introduction

The effect on animals of man-induced stressors, such as the disruption of herd bonds, stabling, medication procedures and the like, has been the subject of increasing investigation. Obvious and shocking abuses against animals, bullfighting, certain training practices in the racehorse industry, and rodeo events such as wild horse races, steerbusting or calf-roping, are readily recognized and have, in some instances been stopped. (Steerbusting refers to roping, from horseback, of running cattle in such a manner as to flip the animal backward or jerk it down, knocking the wind out of the animal and occasionally breaking ribs, vertebrae, and neck.)

However, the bulk of incidents of abuse is in commercial associations between man and animal (Fraser 1982). These are the animals whose troubles are not obvious but who suffer pain and discomfort nonetheless. The abuses may be unwitting and remediable by education, or deemed to be without remedy. One such major abuse occurs during the transport of animals.

Transport: A Model of Abuse

Transport procedures involve bond and territorial disruption, forced driving, crowding, subjection to poisonous fumes from petrol, diesel or battery acids; sudden restraint; and subjection to erratic vehicular movement. For the utilized animals in agribusiness and elsewhere, transport causes more episodes of stress than any other common husbandry practice (Fraser 1982).

Horses, by nature neophobic, fragile, timid, and the most defenseless of hooved domestic animals, are especially distressed by transport (Sambraus 1981). Their distress is manifested by active and powerful resistance at the loading site, halter and hobble abrasions incurred during transport, vocalizations, kicking, choking from fear (sometimes to death), dehydration, and orosthenia (energetic mouthing activities; Fraser 1984), such as aggressive biting of companions in transit or rapid and erratic ingestion of feed during transit.

To control these symptoms of resistance by the horse, an extensive array of coercive paraphernalia has been developed to assist in the transport process: electric prods, whips, ropes, halters with steel or chain nosebands, crushes, hobbles, and punishment administered via electric shock from the driver's cab

77

should the animal "misbehave" during transport. Although the use of these coercive devices actually exacerbates the situation which they are supposed to contain, and often poses increased danger to the handler and animal, their popularity as selling items in livestock periodicals remains unabated.

Current research has demonstrated conclusively that horses suffer pain and discomfort, and are in danger of injury and death, when they must travel in the trailers and trucks that almost all haulers of livestock now use (Cregier 1982). These conveyances force the equine passenger to travel facing the engine, a position which—on loading, braking, during traffic maneuvers, and unloading—violates the basic attributes of equine physiology and psychology.

When the conventional conveyance moves off, the horse is thrust backward off its forequarters. When the conveyance brakes, the horse is pitched forward at the original rate of speed of the conveyance. The largest number of horse trailer injuries seen by one veterinarian are those to the horse's head, throat, and chest (F. Horney, personal communication, 1980).

In an attempt to mediate between the two opposing actions and protect its vulnerable head from injury on braking, the horse travels with most of its weight thrust toward its hindquarters, an area of its anatomy designed by nature to carry only 30-35% of its weight for sustained periods. In order to maintain the increased weight thrust on its hindquarters, the horse will abduct a hindlimb, placing strain on its pubio-femoral ligament and its sacroiliac and sacrolumbar junctions.

Transport stressors frequently result in horses being stiff, exhausted, dehydrated, and ill on arrival at destination, even after a comparatively short journey (Abbott 1979; Messer 1976; Owen et al. 1983). Blood tests on transported horses reveal an increase of packed cell volume (PCV), dehydration, a drop in calcium levels, a rise in cortisol and glucose levels, and a rise in serum creatinine and creatine phosphokinase (CPK) levels associated with muscular exhaustion (Caola et al. 1984).

Two types of trailers have been designed conducive to the horse's behavioral and anatomical requirements for security and balance. The original, known as the Kiwi Safety Trailer, developed in the late 1960s, was designed by a New Zealand horseman and engineer. The NZ design eliminates the natural objection of horses to entering a dark chamber (the entry system used in most road transport for horses) by virtue of teaching the horse to step parallel to the entry upon a platform, turn, and then reverse into the trailer.

Throughout the loading, the handler remains at the horse's head; no coercive tactics are used; passive, familiar aids to the horse, such as haybales for guidance, are used; and no activity takes place at the horse's blind spot—its hindquarters.

To unload, the platform is lowered to a ramp and the horses are quickly led out singly or together by one person (figure 1), an important consideration aiding the rescue when tow vehicles have caught fire.

To accommodate the change in cargo weight placement in the trailer, the axles are moved about 1.2 m (approximately four feet) forward from their position in conventional transport. The axle change places the trailer's center of gravity closer to that of the tow vehicle's and directly under the horse's center of gravity when it stands at ease, leaning over its forequarters. The change facilitates emergency braking. This design is the only such in the world that can stop within 9 meters (approximately 30 feet) at 32.2 km/hr (approximately 20 miles/hr), without upsetting the horses or jackknifing the vehicle.

During transit, the horse maintains its weight over its forequarters, which is the normal manner of a sound horse; lowers its head; and travels with one or the other hind hip in a resting position, quite unlike the straddling which horses subjected to conventional transport must do. (Trailers meeting these design criteria are available from Stratford Motor Body Builders, Stratford, New Zealand.)

Recently a North American-designed rear-face trailer has been developed. Unlike the NZ design, it requires the horse to load into the conveyance from a ramp at the bulkhead. Thus the bulkhead is not featureless, like the NZ

Figure 1. Reversed into the New Zealand design trailer from a platform, horses avoid facing a dark hole for loading. Fleshy buttocks are presented to a featureless bulkhead. Heads may be lowered for keeping nasal and throat passages drained. The horse's vulnerable head, throat and chest are no longer pitched forward on braking, nor is there any threatening activity in its rearward blind spot. (Photo: Courtesy of Rice Trailers.)

design, but utilizes a butt-bar. This device is traditional in conventional transport systems to prevent the horse, with varying degrees of success, from bursting backward toward a possible escape.

As in the NZ design, the horse may be led out by one person down a ramp at the rear of the trailer. Other features—such as the location of the trailer's center of gravity beneath the horse's balance center, and a tethering system that allows the horse to balance effortlessly with its head and thoracic sling (the latter acting much like a gimbal)—incorporate features of the NZ design. (Trailers meeting these design criteria are available from About Face Trailers, 1975 Bee Canyon Rd., Arroyo Grande, CA 93420.)

Resisting Improvements in Animal Welfare

It would seem that the bulk of transport problems of horses—e.g., those pertaining to security from attack behind, blows to the head and chest, injuries to fetlocks and pasterns, exhaustion upon arrival—are well on their way to being solved. But work in this area has demonstrated something else. It has shown that reasoned explanation of the rear-face concept has not, by itself, been enough to convince horsemen that they should do something about the abuses of transporting horses in the conventional way.

Humane workers must deal regularly with many people who, while not themselves inhumane or indifferent to suffering, will argue that animals have always suffered and always will. It has been noted that the human eye very quickly becomes accustomed to the grotesque and to deformities. Examples in the equine world are the built-up shoes on Tennessee Walking horses, forcing the horse to travel unbalanced and requiring caustic agents to sustain the gait; the contortions of rodeo bucking horses accompanied by their bawls and screams (contortions not occurring in their unhandled states); and the contortionistic efforts of a horse to protect and balance itself in conventional transport. Each are considered "normal" by many people who work with horses.

Others will contend that while some improvements are possible, only excessive suffering must be alleviated. Contributing to such conservatism are economic and religious motives. The person who is financially committed to conventional transport will find excuses to not accept improved transport for horses. One veterinarian who had just purchased a conventional trailer, on hearing about rear-face transport, spent much of a tour of her farm defending her recent acquisition on such technicalities as its stoutness and construction, its braking system, and the amount of room it allowed the equine passengers to assume their "normal" traveling position of upthrown head, base-wide hindquarters, and weight thrust to the rear!

Many sincere Christians believe that animals are a lower form of life than man, do not have souls, and cannot suffer either physically or mentally to the same degree as humans. Common to each resistance strategy—religious, economic, and "pragmatic"—is the time-honored argument of "tradition." "What was good for Grandad is excellent beyond improvement."

Even scientifically trained equine practitioners are not immune to the "Grandad syndrome." Firing as a counter-irritant in the treatment of equine lameness was inveighed against by Vegetius as early as 450 AD. This peculiar practice ostensibly "strengthens" sub-structures, such as tendons, damaged by overexertion. The warning against creating further damage to already injured tissue was repeated in the first English veterinary text (1565), and in subsequent years.

It has since been demonstrated that firing—which causes long term, excruciating pain to the equine patient—not only risks incurring laminitis, tetanus, suppurative arthritis, synovitis, and septicemia, but it does absolutely nothing to promote healing and in fact weakens the functions of skin and tendons. Recovery is solely due to the rest enforced by the procedure (Eley 1982; McCullagh et al. 1979).[2]

Despite the evidence, more than a few veterinarians will "fire" a horse at the owner's request rather than lose a fee, attempt to educate the owner, or lose the return calls to treat the disastrous side effects (Adamson 1982; Blakeborough 1983; Donaldson 1979). Some veterinarians even persist in the belief that firing is "scientific," not unlike the horsemen of the eighteenth century who cropped their horse's ears and docked the tail at the root on the theory that the horse is like a rose bush: if pruned, it would grow stronger and bigger (Dent 1983).

While conservatism, inertia, and ignorance may account for a proportion of the abuses, both active and passive to animals, there is, I believe, a deeper reason for the willingness, even eagerness, of many horsemen to use coercive tactics and painful caretaking procedures: even when these have been proved patently harmful to both man and beast.

The Psychological Basis of Resistance

It would seem that many, perhaps most, horsemen fear their horses and feel that they must dominate them. Any procedure requiring the use of force with a horse or subjection of the animals to discomfort and indignity quickly gathers a crowd of onlookers, whether at a rodeo, a show, or in a veterinary clinic. It would seem that the use of coercive tactics, even where these have been shown to be unnecessary, often contributes to the handler's psychological well-being.

For the most part, domestic animals except the horse appear to be regarded as innocuous and unthreatening. The dog has become a symbol of servility and sycophancy. It is literally a bootlicker. Men detest and try to exterminate the dog's wild and free cousins, the wolf, the coyote, the hyena, and the dingo, but love the dog because it is thought to be servile and has been subordinated to man's emotional needs. The same is true, in different ways, of most of the domestic animals.

But the horse, in most cultures including our own, has since the dim epochs of pre-history, been a symbol of freedom and the uncontrollable elements of nature. In the ancient Celtic cultures of Western Europe, which profoundly influenced the culture of British people and their progeny in

North America and Australasia, the horse was worshipped as a god, and was identified with feminine attributes and values (Gelling and Davidson 1969; Phillips 1976; Ross 1967). These were seen to be in conflict with, and threatening to, the masculine need for order, control, and power over nature and the weak. Even in pagan times the horse came to represent a dark spirit that was feared and that had to be subordinated and subjugated. The rise of the Christian and Islamic religions—with their exaltation of masculinity and their determination to destroy pagan beliefs in which the horse had a powerful, if ambiguous role—intensified the human anxiety toward the horse and the felt need to put the horse in its "proper" place, subordinate to man.

Carl Gustav Jung, the renowned Swiss psychiatrist of the early twentieth century, studied the symbol of the horse in man's unconscious as it affected man's actions. Jung was convinced that the eternal conflict, as he saw it, between masculine and feminine values—between authority, control, and power on the one hand, and freedom, spontaneity, and creativity on the other—was symbolized primarily by the horse (Cregier 1981; Jones 1983).

If this is true—and like Jung, I am convinced that it is—we need look no further for the explanation of man's abuse and mistreatment of the horse, not only in transport, but in other situations as well.

Although there is a greater empathy in our culture between women and horses, we also find many women who seem to derive pleasure from coercing their horses. In my experience, these women tend to extol masculine values and to view their own sex as subordinate and inferior. In one instance, a young woman (who, incidentally, was active in humane society work) had almost been killed while forcing her horse to load into a conventional trailer. This woman refused to consider rear-face transport for her horses, even when it was demonstrated to her that her horses could be loaded and transported easily and comfortably in a rear-face conveyance. Her rationalization was that (a) you have to show the horse who is boss, and (b) if the horse is too comfortable during a trip, it will only get into trouble. There could hardly be a better illustration of the culturally implanted human fear of the horse's natural freedom and spontaneity and the felt need of some men and women to keep this dangerous animal in conditions of subjection and indignity. This woman is an articulate rodeo supporter, especially of the bucking horse performances.

In literature, as in life, the horse has commonly served to represent man's inner conflict between "uncontrolled" instinct and controlled reason. In George MacDonald's *Lilith*, the horse there symbolizes control of will and imagination, and partial reconciliation of the female abuse of sex (which is power), and the male abuse of sex (which is sensuality).

In his work, *The Great Divorce*, the English essayist and scholar C.S. Lewis uses the horse as a symbol of the divine steed, or reconciled will and imagination, which can then carry the soul so blessed to God.

Professor George Schoolfield of Yale University, writing from his nearly four decades of familiarity with Scandinavian literature, notes the use of the horse as a symbol of this conflict. In nineteenth century Scandinavian novels, a good many of their authors use the horse in this capacity. In Verner von Heidenstam's *The Charles Men*, or J.V. Jensen's *The Fall of the King*, the Nietzschean heroes "require horses and either treat them cruelly (as Axel does in the Jensen novel) or in a showily generous fashion (as does Charles in Heidenstam)." In our time, the theme continues.H.C. Branner's *The Riding Master*, a Danish book written in the 1940s, is a commentary on the Nazi type (Schoolfield, personal communication, 1981).

This inner struggle, between cerebral and visceral, rational and imaginative, mind and spiritual sensation, animus and anima, is only resolved through a triadic reconciliation wherein the two, joined, become a greater thing than either alone. Polar opposites are reconciled in such symbolism as the Trinity, or Yin and Yang, or Logos and Eros, or the marriage union, or the symbol of the centaur during the Golden Age of Athens.

Ray Hunt, reformed from a hell-bent, beat 'em up attitude toward horses, has traveled in Australia and North America demonstrating kindness in gentling and training horses. He wonders if he will ever have any success in demonstrating to the cowboy—whether wearing western boots or English jodhpurs—that the cowboy has misunderstood his own alter ego, that he abuses the very creature on whom he most depends.

As long as mankind suffers from an epidemic proportion of spiritual illiteracy—and we see the wreckage all around us: corpses of exterminated donkeys in the Grand Canyon (*New York Times* 1984a, b); announcements by Federal authority that wild horses are to be harvested with guns (Gordon n.d.; *Wild Horse and Burro Diary* n.d; McGrory 1981)—of not coming to terms with its inner polarities, Ray Hunt's crusade has no hope. Selfstyled horsemen, witnessing the amiable meeting, the first-time trust being nurtured, the saddling and riding of previously untouched horses, all within an hour, leave muttering: Kindness to the horse. It just won't work, no matter what this buckaroo says and does (Heminway 1981).

Limits to the Educational Approach

Education, frequently invoked as a front-line defense in cases involving cruelty, has also clearly failed to deflect flagrant cruelties by authorities who profess to possess an educated background. As we have seen, some veterinarians, and even the British Horse Society, remain unconvinced that firing should be abolished in practice, from their texts, and by law (Drew 1982).

Many educational establishments themselves seemingly require education on animal issues. Those that purport to cultivate a professional attitude toward animal care may consistently refuse to consider courses in applied animal ethology. The absence of such courses, including those in horse behavior, at many veterinary schools handicaps graduating veterinarians. The latter, under the constraints of time, and in their ignorance, may resort to brutal methods of controlling animals,losing their credibility in the eyes of many clients, and certainly with the horses. The chances of handling success on subsequent visits by themselves or another veterinarian are decreased.

Surprisingly, the schools may be right, if for the wrong reason. Such educational tactics may indeed be wasted on males under forty years of age. Veterinarian R. M. Miller has demonstrated gentle medication procedures, but equine practitioners have proved resistant to his ideas. Until they begin to lose their physical strength, around the age of forty, and become more aware of the damage that a frightened animal can do to itself, equipment, and the veterinarian, the male veterinarians at Miller's demonstrations remain largely unreceptive. Women practitioners, by contrast, are considerably more amenable to his work (Miller 1984).

Professional horsetamers and trainers, such as the Australians Kell Jeffery, Maurice Wright, and Des Kirk have admitted to rough riding of thousands of horses in their younger professional years. Gentler, more lasting training was not investigated—and subsequently practiced with resounding success—by these men until they were older and had been physically disabled, if only temporarily (Blackshaw et al. 1983).

Carrying the message of gentler handling to other horsemen has proved a daunting task. "Personal prejudice and disinclination to accept change inhibits advancement in attaining new skills. In some cases the riders have to be re-educated before progress can be made" (Wright 1983).

Other horsemen concerned for better treatment of their charges report similar experiences. When his films of professionals training horses were analyzed, horseman Monte Foreman discovered that their training was steeped more in tradition and ignorance of a horse's nature than in efficiency and humaneness. When he tried to popularize approaches that put less restriction on the animal's working ability, he became "the most discussed and cussed horseman" and continues to fight an uphill battle (Foreman 1980).

Where tradition does not always aid and abet reluctance to consider things from an animal's point of view, an emphasis on an "educated" jargon may do so. A laboratory animal becomes a "model" or a "tool." It does not scream, it "vocalizes." The "tools" are not killed, but "sacrificed" (*CFHS Newsletter* 1984). In much the same way, rodeo professionals may give derogatory names such as "Widowmaker," "Crazy Mary," or "Four Eyes" to their animals. The horse which refuses to load or travel is commonly termed "stubborn," a "knothead," or simply "stupid," all labels that conveniently absolve the handler of any further thought on the matter.

Where conscience has been prodded, it is usually disguised by the bottom-line approach of utilitarianism. It is now safe to acknowledge that there is a need for stockmanship to go beyond simple technical expertise in handling, as long as the need is linked not just to more contentment in herds, but also more profit (*Vet.Rec.*: Editorial, 1984). Thus, bruise loss in slaughter animals is emphasized ahead of the necessary tools to this end, improved transport and handling. Gentle, firm, considerate handling of dairy cattle is becoming of increasing concern to dairymen, where it can be shown that such handling significantly increases yields (Seabrook 1984).

The bottom-line approach, whether it is the clockwork toting of the daily body-count in war or religious converts, or tons of horsemeat exported as a contributing factor to the gross national product, risks losing an appreciation of the intrinsic nature of the animal. The animal is studied in its natural state only to subsequently corrupt it, whether it is to take advantage of dolphin's locating abilities to plant bombs in enemy harbors, or of sheep flocking behavior to raise and ship them in densely crowded conditions.

Change Strategies

Legal, political, and to some extent, instructional approaches may help to overcome the apathy that militates against improved animal care. Where legislation is enforced, for example, better care for animals can sometimes be obtained.This evidently occurred when New York State recently passed, and enforced, legislation with reference to improved slaughter horse transport on its highways. But these tools to improvement will not, by themselves, dislodge the deep-seated psycho-cultural obstacles to improved humane care. Conventional handling practices for horses cater to those who derive a satisfaction from a physical assault on their horses—whether it is the cowboy-model for a cigarette company who introduces himself to a strange horse by stunning it with a blow to the muzzle (Chriss 1976), or the performance horseman who uses the whip to draw blood.

Any change in strategy based on legal, empathic, or educational concepts is wasted on these people. The bottom-line approach, too, has limited appeal. The rodeo cowboy is unlikely to treat his own beef cattle as he does those in roping events in the arena, simply because it is not efficient or economical.

Change strategies for the particularly resistant personality might well succeed, however, by offering them a new self-concept (Seabrook, personal communication, 1984). Conversion to improved treatment for animals, and in particular horses, by persons who are already stable names would have a significant impact upon hundreds of average horse owners who follow the herd in all respects. Particularly, macho models, or those who have a very high success profile in the arena—from notable trainers to Olympic-class equestrians—would be the most effective.

Whatever approach, or combinations of approaches are utilized-education, legislation, or role-model—a full and frank assessment of aims frequently reveals the sobering fact that man is not fit to rule a hen-roost. For this reason, English scholar C.S. Lewis wrote "even the authority of man over beast has to be interfered with because it is constantly abused" (Lewis 1975).

Not long ago a humane society brought a case against a fishing bait supplier providing crabs with an inadequate transport environment (Murray 1980). Surely it is past time that we must give as much thought to hauling our horses as we now do to crustaceans!

Endnotes

[1] North American Editor, *Equine Behavior Journal*, University of Prince Edward Island, Charlottetown, Prince Edward Island, Canada C1A 4P3

[2] See also, KG McCullaugh and IA Silver. 1981. The actual cautery: Myth and reality in the act of firing. *Eq. Vet. J.* 13(2): 81-84.; Brown et al. 1983. The clinical and experimental studies of tendon injury and healing. *Eq. Vet. J.* (Suppl. 1).

References

Abbott, E. 1979. Investigations into the physiological changes associated with transport in the horse. *Annual Report to the Animal Health Trust.* United Kingdom, n.p.

Adamson, P. 1982. Letter to the Editor. *Equi.* 2(12): 3-4.

Blackshaw, J, Kirk, D and Cregier, S. 1983. A different approach to horse handling based on the Jeffery Method. *Int. J. Stud. Anim. Prob.* 4(2): 117-23.

Blakeborough, J. 1983. Letter to the Editor. *Equi.* 3(14): 3.

Branner, HC. 1951. *The Riding Master.* London: Secker and Warburg.

Caola, G, Ferlazzo, A and Panzera, M. 1984. Livelli sierici di creatinina fosfochinasi in cavalli, dopo trasporto. (Serum creatinine and creatine phosphokinase levels in the horse after transport.) *La Clinica Vet.* 107: 46-48.

CFHS. 1984. (Canadian Federation of Humane Societies). Relieving pain in lab animals. In: *Caring for Animals.* 1(2). n.p.

Chriss, N. 1976. Marlboro men spur sales. *Moneysworth.* Sept. 13.

Cregier, SE. 1981. *Alleviating Surface Transit Stress on Horses.* Ann Arbor, MI: University Microfilms.

—. 1982. Reducing equine hauling stress. *J. Equine Vet. Sci.* 2: 186-98.

Dent, A. 1983. Shocking docking. *Equi.* 3(15):4-7.

Donaldson, RS. 1979. Firing of horses (letter). *Vet. Rec.* 105: 173-74.

Drew, H. 1982. Letter to the Editor. *Equi.* 2(12): 4.

Eley, J. 1982. "What the fire will not cure, nothing will cure"?: The truth about firing and blistering. *Equi.* 2(11): 8-9.

Follmer, D. 1980. Ride, guide and agilize: The gospel according to Monte Foreman. *Equus.* 40: 26.

Fraser, AF. 1982. The nature of cruelty to animals. *Equi.* 3(13): 28-29.

—. 1984. The behaviour of suffering in animals. *Appl. Anim. Beh. Sci.* 13:33.

Gelling, P and Davidson, HE. 1969. *The Chariot of the Sun: Ritual Symbols of the Northern Bronze Age.* New York. pp. 167-72.

Gordon, A. n.d. Looser rules urged for kills of wild horses. *Wild Horse and Burro Diary.* 20(3): 21.

von Heidenstam, CG. 1920. *The Charles Men.* New York: Am. Scandinavian. trans. from Swedish by CW Stork.

Jensen, JV. 1933. *The Fall of the King.* New York: H. Holt and Co. trans. from the Danish by PT Federspiel and Patrick Kirivin.

Jones, B. 1983. Just crazy about horses: The fact behind the fiction. In: Katcher, AH and Beck, AM. eds. *New Perspectives on Our Lives with Companion Animals.* Philadelphia: Univ. of Pennsylvania Press.

Heminway, J. 1981. The cowboy who respects horses. *Quest.* 5(5): 22-26, 84-86, 88.

Lewis, CS. 1946. *The Great Divorce.* London: MacMillan.

—. 1975. *Fern-seed and Elephants.* Glasgow: Collins. p. 19.

MacDonald G. 1962. *Lilith.* London: Eerdmans.

McCullagh, KG, Goodship, AE and Silver, IA. 1979. Tendon injuries and their treatment in the horse. *Vet. Rec..* 105: 54-57.

McGrory, M. 1981. BLM wild horse plans provoke national outrage. *Washington Star.* July. (quoted in *Am. Horse Prot. Assoc. Newsletter..* Fall 1981: 3).

Messer, NT. 1976. Acute colitis syndrome: Clinical problems in private practice. *Amer. Assoc. Equine Practitioners Newsletter..* 2: 44-51.

Miller, RM. 1984. Equine Psychology and Its Applications to Veterinary Practice. Paper delivered to the Mid-Coast Veterinary Society. Oct. 4, 1984. San Luis Obispo, CA.

Murray R. 1980. Court case on "cruelty" to crabs. (London). *Observer..* Nov. 16.

New York Times. 1984a. Burro bonanza. 8 April 1984: 49.

—. *1984b. Doomed burros find haven at Texas ranch. 25 Dec. 1984: 8.*

Owen, Rh ap Rh, Fullerton, J and Barnum, DA. 1983. Effects of transportation, surgery, and antibiotic therapy on ponies infected with Salmonella. Am. J. of Vet. Res. 44: 46-50.

Phillips, GR. 1976. *Brigantia: A Mysteriography*. London: Routledge and Kegan. pp. 189-95.

Ross, A. 1967. *Pagan Celtic Britain: Studies in Iconography and Tradition*. London: Routledge. pp. 321-33.

Sambraus, HH. 1981. Horses: Behavior patterns during transit. Remarks. *Report Series. No. SC 81/09/01*. World Society for the Protection of Animals. London. n.p.

Seabrook, MF. 1984. The psychological interaction between the stockman and his animals and its influence on performance of pigs and dairy cows. *Vet. Rec.* 115: 84-87.

Veterinary Record. 1984. Comment: Close encounters of an ethological kind. 115: July 28. n.p.

Wild Horse and Burro Diary. n.d. Wildlife federation challenged. 19(1): 5.

Wright, M. 1983. *The Thinking Horseman*. New South Wales. Self-published. p. 10.

HORSEBREAKERS, TAMERS, AND TRAINERS: AN HISTORICAL, PSYCHOLOGICAL, AND SOCIAL REVIEW[1]

Sharon E. Cregier[2]

> If racing with mere men has wearied you,
> how can you contend with horses?
> Jeremiah 12:5

To my knowledge, there has been no organized synthesis describing the historical development of horse handling, management, lore, and training. This discussion offers, in capsule form, some of the historical, psychological, and social considerations which might be taken into account when evaluating horse-handling skills.

First, I would like to describe the natures of the emotional and psychological bonds between man and horse. I will also look at the consequences of various types of bonding on horsemanship or management. We can increase our understanding of the role of the horse in our history and lives by thus seeing how the animal figures in our fantasy, and in our practice (Rhodes 1980; Van de Castle 1983).

For 98% of the last 6000 years—up to 1839—the horse was the fastest vehicle available to man. As such, it was easily associated with virtually anything, however unrelated, that offered power and speed. In our dreams and in our lives, from time immemorial, the horse became the archetype, the ready reference, for this association (Barclay 1980).

Certainly, for most of us, no other creature combines "grace, streamlining, intelligence, and tractability" so well with power and speed. The horse is at once powerful, practical, and beautiful (Barclay 1980; Jung 1956).

Artists have been very quick to extol these aspects of the horse (Rhodes 1980). A woodcarver of children's rocking horses, Tilo Kaufmann, utilizes horses with impossibly powerful stout legs to emphasize the sure victory of his horses over the demons writhing between their hooves (Rettig 1980). Where the supernatural demons fail, practical man wins, as depicted in a similarly powerful carving outside the Federal Trade Building in Washington, DC.

Today, popular psychology treats the horse as a muscular, masculine symbol. That this identification is rather simplistic, perhaps even banal, has been challenged by more thoughtful observers. Like the artists depicting equine themes, their horses support a variety of symbolic meanings.

89

For the juvenile, the horse may represent escape from a confusing world to disciplined freedom. For others, the horse is symbolic of work, service, help, and action.

For literary men such as George MacDonald, the horse symbolizes control of will and imagination, and partial reconciliation of the female abuse of sex (which is power) and the male abuse of sex (which is sensuality).

The Jungian psychologist sees the horse as symbolic of feminine attributes— such as alleged unpredictability, capriciousness, and beauty—and a strong mother symbol (Jung 1956).

As representative of a variety of human desires, needs, and perceptions, the horse is far less consistent than the dog, wolf, or ape. These animals are frequently viewed as having such largely masculine attributes as leadership, strength, and predatory skills (Jones 1983).

The lack of consistency in horse symbolism argues against one of popular psychology's pet hypotheses: that the attraction of the horse, for females especially, is chiefly the opportunity it offers to work out sexual conflicts. But this can be argued for other activities as well, and far more convincingly.

For example, the sexual conflict theme is resoundingly portrayed in bull riding in rodeo. In fact, the conflict is not worked out or resolved in this event so much as it is reinforced. The stereotypical perversions of masculine dominance through muscular strength—brawn, not brain—remain (Lawrence 1982).

Because of its varied associations, the horse offers a broad opportunity for youngsters to work out and establish current and future roles as partner, parent, boyfriend, girlfriend, or child. Interviews with juvenile horsemen by researcher Barbara Jones of the University of Pennsylvania established many such linkages.

For the most part, the young female riders showed themselves to be very adept at empathy with their horse, subtle direction,and physical balance. It was the males who had to be taught that the horse, to be mastered, must be outwitted, not overpowered. Some males, to their cost, learn this lesson almost too late in life, or not at all, as we shall see later when we discuss types of horse management and veterinary care.

When the thoroughly active horseman is studied, there is some argument to support popular psychology's contention for the physically sexual attraction of the horse for both male and female. Daily, fresh-air interaction by a psychologically sound person with an animal as demanding as the horse cannot help but contribute to an overall feeling of well-being. This feeling carries over into other activities, including but not limited to sexual (Jones 1983).

But the same can be said for other strenuous pastimes, whether jogging or kite-flying. Working with horses involves mucking out stalls, cleaning gear, fixing fence, humoring farriers, seeing to trailer repairs; adhering to a relatively inflexible schedule of feeding, veterinary attention, and grooming; paying out really staggering sums of money throughout the year; exposing oneself to the danger of kicks and falls; and having to continually concentrate on the animal in hand. Certainly there are more pleasant ways of obtaining peak

peak physical efficiency and maintaining a rich supply of the euphoria-producing beta endorphins associated with regular exertions.

All these different equine activities attract different types of people. Depending on the needs of the person drawn to the horse, this animal—especially because of its ambiguous symbolism, from masculine to feminine, from strength to frailty—is able to facilitate the expression of various components of the human personality.

Not all of these expressions are, of course, desirable. Authors have used the less attractive outcomes of human-horse associations with telling effect. For example, in Verner von Heidenstam's *The Charles Men* (1920), or in J.V. Jensen's *The Fall of the King* (1933), the alienated heroes require horses and either treat them cruelly, or in a showily generous fashion. H.C. Branner's *The Riding Master* (1951), is a commentary on this Nazi type.

In a more positive description of the symbolic and actual association between horse and man, C.S. Lewis in *The Great Divorce* (1978) uses the horse as a symbol of the divine steed, or reconciled will and imagination, which carries the soul so blessed to God (Meilaender 1978). Rooted in the collective unconscious of mankind, the theme is universal. At the Cultural Festival of India in London (July 16–August 15, 1985), The Swaminarayan Hindu Mission used horses over the welcoming gate to symbolize the soul's journey to God. The seven spirited horses, reined by God in a chariot carrying the soul, represented the five senses plus mind and subconscious.

As in literature and spiritual quest, so in life. Both the desirable and undesirable facets of our association with horses are represented in three basic approaches to our management of the horse. I have divided these approaches into three broad categories, the mechanistic approach, the shamanistic approach, and the pragmatic approach. While there is some overlapping between the three, each retains distinctive characteristics.

The mechanistic approach usually insists that the horse is worthy only when it is useful or paying for itself. And payment should be in cash. The mechanistic approach is distinguished by the practice of physical dominance over the animal, usually by the crudest tools—such as whip, chain, strait jacket, reliance on chemical controls, quick-fixes, and physical exhaustion of the animal to bring it under control (figure 1).

There is no attempt to see the horse as an individual. The mechanist sees the horse as a device for enhancing his own ego and offers nothing to the animal in return. He may argue against developing an affection for the horse, lack of affection denoting maturity and a "civilized" approach (Midgley 1984a). The mechanist nurtures callousness by limiting emotional commitments. Because it requires identification with the animal, understanding is feared and suppressed (Midgley 1984b). The horse is thus viewed as essentially a tool, a barely sensate creature.

The mechanist is also distinguished by a tendency to put his own interests first and to overwork the animal. It is noteworthy that at the time in history when cruelties to the horse were most rampant, this view prevailed. It was partly to combat this evil that *Black Beauty* was written in 1877 (Blount 1974). This book—which led to the abolition of the bearing rein—continues to have enormous influence on the treatment of horses.

SO EASY! You can MASTER and TRAIN
a horse as easy as a cat plays with a mouse!
WHEN YOU HAVE THE BEERY OUTFIT!

*Make $75 to $100 Profit
on a Single Horse!*

Figure 1. The horse's first line of defense—head-long flight from fearful stimuli—is checked by the mechanist's running-w rig. This system is still used by horse-dependent cultures in North America. (Advertisement reprinted with permission by the Beery Correspondence School of Horsemanship.)

That the battle against the extremes of mechanism is far from over, is witnessed by the efforts to halt its cruelties in the performance horse industry. It has recently been announced, for example, that all Arabian horses entered in certain competitions must be examined for whip marks, drugs, abusive equipment, correct hoof lengths and evidences of the use of ginger in the rectum. Exhibitors found to have whipped, frightened, or abused their animals will not be allowed to enter the event (IAHA 1985).

The shamanistic approach to controlling the horse is far more subtle. Sometimes, it is too subtle for both the practitioner's, and the horse's, own good. Those thought to be too close to their horses have been burned—with their horses—as witches (Howey 1958).

The shaman's, or magician's, approach is based on the closest possible empathy with the horse, an empathy which comes from long observation, practice, modification, concentration, and questioning. This is probably the hardest route of all to the most successful horse management. It is in constant flux, very little understood or appreciated even today.

Possibly this lack of appreciation is because, when exhibited, shamanism offers little in the way of entertainment (such as does the mechanistic approach) and much in the way of education. It is never as sensational as, say, mid-nineteenth century horsetamer Prof. Silas Sample's horse-taming machine. This was a rotating platform on which the unbroken or savage horse was driven and then rotated until it fell down in a dizzy stupor (Hayes 1895).

Today, while we no longer burn our horse shamans at the stake, we do send them to Coventry. When a twentieth century exponent of kindness to the horse, Ray Hunt, promulgates his theory and practice, it is not unusual for a bow-legged, saddle-weary member of the audience to shake his head and mutter, "Kindness to the horse. It just won't work, no matter what this buckaroo says or does" (Heminway 1981).

Ray Hunt has lots of company in Coventry. Among them is probably one of the greatest horsemen of the nineteenth or any century, John Solomon Rarey (figure 2). Rarey's specialty was the savage stallion and other abused horses (figure 3).

While Rarey had a simple approach to first immobilizing and then gentling these animals, it was an approach he insisted be coupled with kindness. He was also careful to emphasize to his incredulous audiences that the horse was capable of feeling and affection.

Figure 2. Lord Dorchester's "savage stallion," Cruiser, signals his confidence in the American trainer, J.S. Rarey, with softened eye, alert ears, and relaxed foreleg. Rarey saw himself as an "educator, not a gladiator." (Photo courtesy of the trustees of the Victoria and Albert Museum.)

In the wrong hands, Rarey's initial handling of a savage or simply recalcitrant animal could only worsen the situation. For this reason, he took a typically shamanistic precaution: limiting his audience to the so-called better, or more refined, social classes—among them dukes, duchesses, barons, baronesses, princes and princesses—for a substantial fee (figure 4).

Figure 3. An Equestrian Humane Society, proposed *Punch* in May 1858, would sponsor Rarey as a "groom tamer." But, as Rarey's system worked through the affections, how could he influence brutes with no affection? The tamer's approach would end the horsebreaker's dependence on starvation, drenches, twitches, roweled bits, spurs, and—at times—blinding. (Slide credit: Dan MacKinnon. Reference: *Punch* [1985]. Illustration: *Harper's Monthly Magazine.* Vol. 22, 18611862, p. 615.)

In this Rarey was consciously or unconsciously following another attribute of shamanism: initiation of the chosen few. For centuries the blood horse, especially, was the prerogative of the rich and powerful. Its ownership, care, and training were under the direction of its wealthy owner. The first horse training books were written not for the unlettered man in the stable, but for persons of position.

We can see this association of training with the "better classes" even in such a democratic society as North America. Posters depicting horsetaming sessions frequently picture the tamer dressed in formal wear, top hat and all. He might have the cavalry stripe down a trouser leg, again another symbol of exclusiveness. He was often referred to as "Professor." The reference was not to a degree holder, in this instance, but to someone who did not have to earn a living by getting his hands dirty.

Bad trainers were invariably pictured as loutish, leering, squat ignorants whose closest association with a stable should, by rights, simply be to remove the dung (figure 5). They had their historical roots in the footmen once

Figure 4. Cavalry-striped trousers befit a royal audience. Before taming by Rarey, the Zoological Society of London's zebra would suspend itself by its teeth from its hayrack, flailing its four hooves in all directions. Or it would walk on its hind legs, tearing at the rafters (Thorpe 1861). Zoo veterinarian Oliver Graham-Jones notes (1970) that "A zebra wills its body to kill you...[it] is the only animal I know of that can leap with all four feet off the ground and bite and kick you before it touches the floor again." (Slide credit: Dan MacKinnon. References: Graham-Jones, O.[1970]; Thorpe, T.B.[1861]. Illustration: *The Review*. Vol. II, No. 29, July 17, 1858, p. 41.)

selected for their "terrible voices, cruel looks," for the specific purpose of cudgeling, beating, pushing, burning, and berating a horse into compliance with its master's wishes (Chenevix Trench 1970).

Where shamanism is concerned, exclusivity persists today in a society formed, it is said, before the time of Solomon. This is the secret Society of Horsemen. Still extant in England's East Anglia and in Scotland (Evans 1979), its initiation rituals have much in common—according to one anthropologist—with animal cults around the world, notably West Africa (MacLean n.d.).

Members were chosen for initiation into the Society on the basis of their facility to work with horses. However, these were men from the hired horsemen class and not the land-owning or upper classes. By learning the secrets of control—such as mental summoning of one's horse over a great distance—and veterinary care over the horses placed in their charge, the horsemen were able to protect their livelihood, and the horses assigned to them for breaking or working.

Like many other types of shamans, they enjoy a sense of power and sacredness, shared by ritual and pledge, and historically rooted in such traditions as Daniel in the Lion's Den, Elijah fed by the ravens, and, chief of all, St. Francis of Assisi (Savishinsky 1983).

Figure 5. Charles Dickens's journal for July 1858, *Household Words*, pleaded with Rarey to form classes for grooms. "Unless the crust of mannerism and self-conceit of these men is broken down, the horse is still in the Iron Age!" (Dixon 1858). Rarey's methods, based on the horse's "intellect and affections," gave impetus to better treatment for the insane, were adopted for the classroom, and influenced the emerging humane movement. (Slide credit: Dan MacKinnon. References: Dixon, HH. [1858]. Illustration: *Harper's Monthly Magazine*. Vol. 22, 1861-1862, p. 620.)

   Strict discipline within the order and commitment to their horses, ensures that the secrets they learn are not used to the detriment or harm of the horse.
   Other than secret signs identifying members to each other, they also share a password which is still unknown to this day. It has been guessed at, however, and is significant for its implications: Both in One.
   The secret Society's initiation ritual, for example, usually includes a hair-raising experience for the initiate which parallels the fear that a horse experiences in a novel situation. At the peak of the initiate's fright, a brother horseman appears to calm the initiate and take charge.
   This intense preparation for working with horses, the practiced concentration, leads to a feeling of unity with the horse: Both in One. Horse and man together are greater than either alone (Evans 1979). A loss of the sense of self is experienced, similar to the loss of self (unmourned) when one is composing music or participating in a ritual. In such a state, the question,

Am I doing well? does not occur (Csikszentmilhalyi 1975). Once again, he that loseth his life shall save it. Thus moral (i.e., non-coercive) control of and empathy with the horse is developed.

There is little that is openly flamboyant about moral control of a horse as opposed to physical, mechanical control. The shamanistic approach requires a person who is confident and sure of himself without arrogance or ignorance.

The third approach to the horse is that of the pragmatic. Unfortunately, on close examination, it often turns out to be self-defeating. The practitioners of this approach shun the refining of critical judgment even as the shaman seeks it.

Thus a slavish conformity to the past is one of the most prominent features of this approach. It is easier to memorize the shibboleths that pass for knowledge. Probably no other area of human endeavor is so steeped in conservatism as that of our association with the horse. The Grandfather Syndrome is entrenched. "What was good enough for Grandfather is excellent beyond all improvement." We see it in every aspect of our handling of horses: whether in perpetually darkened or needlessly artificially-lit stabling, in the persistence of the barbaric practice of firing in veterinary care (Eley 1982; McCullaugh and Silver 1981); or, in the training ring, the insistence that the horse's virtually rigid thoracic spine is bending in a continuous arc when circling under tack and rider (Crossley 1983).

Each one of these practices is excused on the basis of being "practical." It took decades before the forward seat was finally accepted for jumping, cross-country, and racing. Italian cavalry officer Federico Caprilli's (1868-1907) forward seat was a carefully thought-out, logical proposition and practice. However, it was opposed by the "pragmatists." They believed, for example, that on descending from a jump the rider must lean backwards to save the horse from the jar of the rider's weight on landing, and to save it from a fall if it stumbled.

Caprilli, however, simply brought the rider's center of gravity forward, absorbing the shock of the landing in the rider's flexed leg and thigh joints. He correctly pointed out that however one sat the horse, the weight of the horse and man must inevitably be borne by the horse's forelegs and the forward position made it less of a jar for the horse. Certainly a rider could not save his horse from falling by a tug on the reins. If he could, flying would be a simple proposition (Chenevix Trench 1970). By asking the rider to conform to the horse's center of gravity, and not the horse to the rider's, Caprilli stirred up a storm which is not yet dead.

The development of scientific tools of proof did little to dispel firmly entrenched "beliefs as facts" about the horse. When Eadweard Muybridge published his detailed photographic work, *Animals in Motion* (1887), proving that the horse at a full gallop suspends and bunches all four feet under its belly, horsemen refused to accept the evidence (Ozment 1985).

Centuries earlier, the first evidence of horses in use shows them harnessed as cattle had been, to a central draft pole beneath a yoke. It wasn't until 3000 years later that the shaft and breastcollar harness, giving an economical, efficient single draft, came into use in the 3rd century BC in Han China (Littauer and Crouwel 1979).

This, and more, forces us to the distasteful conclusion that if horsemen are not exactly a bright lot, they are certainly conservative.

Today, we are still plagued with the allegedly "practical" interpretation of equine character according to head bumps, hair whorls, and number of white legs. Much of this is contradictory and, I'm certain, self-fulfilling prophecy. The chief function of such classifications seems to be to allow the handler to blame the horse for not adapting to the trainer, and perhaps even to set up the circumstances that allow the horse to react that way.

The mechanist and the pragmatist are united in their separation from the earlier shamanistic wisdom. To paraphrase and update C.S. Lewis (1947):

> For the shaman the cardinal problem is how to conform the soul to reality, and the solution had been knowledge, self-discipline, and virtue. For the mechanist and pragmatist alike the problem is how to subdue reality to the wishes of men: the solution is a technique: and both, in the practice of this technique, are ready to do things hitherto regarded as disgusting and impious— such as the mutilation of horses' feet and hooves by the Tennessee Walking Horse "industry," or the refusal to guard life by wearing protective head gear in Western riding events (Marcovitz 1985).

What are the distinguishing characteristics of the genuine horseman, as opposed to someone with a horse? It would seems that the genuine horseman first of all realizes that one cannot compartmentalize one's study of the horse.

It is one thing to study the hormonal dysfunction in horses being transported. It is another to realize that before these dysfunctions can be mitigated or eliminated, one must go beyond physical or chemical control or restraint of the transport situation. One must look into automotive and equine dynamics: the simple physics of forward and reverse, and side to side motion, and how the balanced, stationary stance is maintained by sympathetic motion or exacerbated by contra-indicative motion.

It is essential in virtually any endeavor with the horse, and not just in the area of transport, to have a knowledge of equine behavior. For example, the fact that horses are sensitive about activity in their blind zones ("hind is blind") affects every aspect of their handling, from stabling to training and transport. The biomechanics of equine anatomy must be studied. Even the bond of trust established, or not established, between the horse and its handler must be investigated.

By willingly extending his range of interests, the genuine horseman becomes perhaps one of the most authentic of world citizens, able to appreciate and sympathize with an everbroadening range of problems and interests. He becomes a person welcome at every level of healthy society, being accustomed to developing an eye for the meaning of cross-cultural signals, both alien (read inter-species) and familiar (read intra-species) (Stein 1964).

These increased contacts invariably remind the horseman of his own deficiencies in handling skills and understanding. But the skilled, mature horseman will respond to these realizations in a positive manner, selecting those new practices which will benefit the horse. He thus increases his value to horses and humans by upgrading his own skills.

This is especially important, but not always realized, in the veterinary field. Practitioner Robert Miller of California has documented the tendency of North American veterinary schools to reinforce the mechanistic approach to animal handling.

"No other horsemen," says Miller, "must handle as many horses as does the equine practitioner in a normal working day and under such disadvantageous conditions." To compound the difficulty there was nothing in the veterinarian's training in the schools he visited in North America or Europe to utilize equine psychological principles to control the horse.

While any veterinarian can learn the psychological methods of control, said Miller, not every male veterinarian is receptive. Those veterinarians who are, are almost always over forty years of age. Miller found that female veterinarians are much less likely to elicit the fear-flight response in horses, but rarely is a man under forty years of age willing to try other than the mechanistic methods of restraint.

"Somewhere around forty years of age," says Miller, "we start to enjoy outwitting our patients rather than outfighting them.... It is safer, more humane, better technique professionally, less hazardous, more enjoyable and more effective" (Miller 1984).

Miller's preferred technique of psychological control epitomizes the chief aspect of the good horseman: He is very much aware of the life-threatening danger of a frightened horse. He is conscious of the need to maintain poise, balance, and a relaxed attitude when working with horses. And he avoids force and sudden movement (Roberts 1979).

Unconsciously, good horsemen have been imitating horse language to communicate with their animals. They may, for example, practice the herdmate's reassuring nibbling of the withers by scratching the same area in a tense situation. Trainers like Pat Parelli in Clements, California, consciously utilize eye contact and their own body language to train horses with minimal physical contact and minimal, or no, tack.

There is still further room for definitive studies on the relationship of the trainer to the animal as expressed by the trainer's facial aspect, posture, and vocalizations, together with the animal's response. How do animals respond toward trainers whose approaches to the horse and use of eye contact differ? To minimize biased reportings of such studies, it is suggested that a trained ethologist observe the interactions (Manning 1983).

Throughout history the record is clear. An accurate knowledge of horse behavior can make, or break, a campaign, whether on the field of battle

(Cregier 1982), or in diplomacy. Horsemen of all stripes can, for example, appreciate the challenge offered to King Richard during the Third Crusade by his opponent, Saladin. Saladin sent Richard a fine charger. But Richard, who knew a thing or two, mounted an expendable groom on the animal which immmediately bolted back to the Saracen camp. Saladin then sent another horse without the homing instinct (Chenevix Trench 1970).

The contrast between good and bad horsemanship can be best summed up by this homily:

> Good horsemanship is a thing of skill.
> Bad horsemanship is a thing of courage.
> Good horsemanship will last through age,
> sickness and decrepitude.
> Bad horsemanship will last only as long as
> youth and strength supply courage.

Acknowledgement

I am indebted to Professor Hamish Henderson of the School of Scottish Studies at the University of Edinburgh for so generously opening his archives on the Secret Society of Horsemen to this researcher.

Endnotes

[1] This paper was presented as part of an illustrated equine behavior seminar at Nova Scotia Agricultural College, May 11, 1985.

[2] The author is the North American editor of *Equine Behavior Journal*, University of Prince Edward Island, Charlottetown, PEI, Canada C1A 4P3.

References

Barclay, HB. 1980. *The Role of the Horse in Man's Culture*. London: JA Allen. 398 pp. p. 371.

Blount, M. 1974. *Animal Land: The Creatures of Children's Fiction*. New York: William Morrow. 336 pp. pp.249-50.

Branner, HC. 1951. *The Riding Master*. London:Becker and Worburg. 159 pp.

Chenevix Trench, C. 1970. *A History of Horsemanship*. New York: Doubleday. 320 pp. pp. 89-90, 116-20, 248-69.

Cregier, SE. 1982. The Mare of Solomon's Song. *Horseman*. 26(8): March, 68,71.

Crossley, A. 1983. Flexibility of the horse's spine: The full story. *Equi*. 16: 12-15.

Csikszenmilhalyi, M. 1975. Play and intrinsic rewards. *J. of Humanistic Psychol*. 15(3): 41-63.

Dixon, HH. 1858. Horsetaming. *Household Words*. July 10. pp. 82-84.

Eley, JL. 1982. What the fire will not cure, nothing will cure? *Equi*. 11: 8-9.

Evans, GE. 1979. *Horse Power and Magic*. London: Faber and Faber. 222 pp. pp. 97-198.

Graham-Jones, O. 1970. *First Catch Your Tiger*. New York: Taplinger. p. 96.

Hayes, MH. 1896. *Illustrated Horse Breaking*. London: Hurst and Blackett. 381 pp. p. 176.

von Heidenstam, CG. 1920. *The Charles Men*. New York: American Scandinavian Foundation. 556 pp.

Heminway, J. 1981. The cowboy who respects horses. *Quest*. 5(5): 22-26, 84-86, 88.

Howey, MO. 1958. *The Horse in Magic and Myth*. New York: Castle Books. 238 pp. pp. 217-18.

I[nternational] A[rabian] H[orse] A[ssociation] takes steps against abuse. 1985. *Equus*. August(94): 104.

Jensen, JV. 1933. *The Fall of the King*. New York: Holt. 300 pp.

Jones, B. 1983. Just crazy about horses: The fact behind the fiction. In: Katcher, AH and Beck, AM. *New Perspectives on Our Lives with Companion Animals.* Philadelphia: Univ. of Pennsylvania Press. pp. 87-111.

Jung, CG. 1956. *Symbols of Transformation.* Bollingen Series No. 20, Vol. 5. New Jersey: Princeton Univ. Press. 557 pp. pp. 207, 251, 274-75, 421-22.

Lawrence, EA. 1982. *Rodeo: An Anthropologist Looks at the Wild and the Tame.* Knoxville, TN: Univ. of Tennessee Press. 288 pp. pp. 180-98.

Lewis, CS. 1947. k*The Abolition of Man.* New York: Macmillan. 121 pp. pp. 87-88.

—. 1978. *The Great Divorce.* New York: Macmillan. 128 pp.

Littauer, MA and Crouwel, JH. 1979. *Wheeled Vehicles and Ridden Animals in the Ancient Near East.* K±ln: EJ Brill. 185 pp. pp. 9-12.

MacLean, U. n.d. Letter to Professor Hamish Henderson, School of Scottish Studies, University of Edinburgh.

Manning, A. 1983. Ethological approaches to the humancompanion animal bond. In Katcher and Beck, op. cit. pp. 7-16.

McCullaugh, KG and Silver, IA. 1981. The actual cautery: Myth and reality in the art of firing. *Equine Vet. J.* 13(2): 81-84.

Marcovitz, S. 1985. A hard fight for hard hats. *Equus.* October (96): 73, 84-85.

Midgley, M. 1984a. *Animals and Why They Matter.* Athens: Univ. of Georgia Press. 158 pp. p. 122.

—. 1984b. *Wickedness.* London: Routledge & Kegan Paul. 224 pp. p.4.

Miller, R. 1984. Equine psychology and its application to veterinary practice. Paper delivered to the Mid-Coast Veterinary Society, October 4, 1984, San Luis Obispo, California.

Muybridge, E. [a.k.a. Edward James Muggeridge]. 1887. *Animal Locomotion.* Philadelphia: Univ. of Pennsylvania. Unpaginated. 781 plates.

Ozment, J. 1985. Early equine photography. *Nat. Sport. Lib. News Lttr.* June (20): 2.

Punch.. 1858. A hint to horse tamers. Vol. 34, May 8, p. 190.

Rettig, R. 1980. The eternal horse: Evoking a joyful event. *Horse, of course!* 9(12): 43-44.

Rhodes, B. 1980. The eternal horse: Capturing the horse's strength and emotion. *Horse, of course!* 9(12): 44-45.

Roberts, T. 1979. *Horse Control: The Young Horse.* Richmond, Australia: TA and PR Roberts. 212 pp. p.205.

Savishinsky, JS. 1983. Pet ideas: The domestication of animals, human behavior, and human emotions. In: Katcher and Beck, op. cit. pp. 112-131.

Stein, E. 1964. *On the Problem of Empathy.* The Hague: Martinus Nijhoff. 113 pp. p. 105.

Thorpe, TB. 1861. Rarey the horse tamer. *Harper's Monthly Magazine.* XXII. No. 131. April 29, 1861. pp. 615-24.

Van de Castle, RL. 1983. Animal figures in fantasy and dreams. In: Katcher and Beck, op. cit. pp. 148-173.

EFFECTS OF EARLY EXPERIENCE UPON ADAPTIVENESS OF HORSES [1,2]

J.C. Heird[3], R.W. Bell[4], and S.G. Brazier[5]

Introduction

The early rearing management of horses varies greatly. Whereas most breeding farms handle horses on a practically continuous basis, many horses reared on the working ranches of the southwestern United States may have experienced minimal contact with humans prior to being trained. Indeed, rearing horses with conspecifics in holding ranges which may be several hundred, or even several thousand, acres in extent mimics the habitat of the feral horse. Extensive contact with humans frequently occurs only following two to three years of free grazing, thus the subsequent process of training as work animals entails considerable stress both on horses and their human owners. An additional concern with this type of rearing is the high level of attrition of stock due to accidental injury, predation, etc. Death losses of greater than 10% are not uncommon.

Experimentation with laboratory animals, encompassing the range of usual species (mice, rats, cats, dogs, and primates), has conclusively demonstrated that handling by humans during early life facilitates subsequent development in terms of health and viability, reduced emotional reactivity, more adaptive responses to a variety of stressors (both biological and behavioral adaptations), and increased ability to adapt to changing circumstances as exemplified by increased ability to learn and solve problems (see M. Bornstein 1985, for recent review).

Two studies (described below) (Heird et al. 1981; Whitaker 1982) conducted at Texas Tech University (TTU) have extended these findings regarding the beneficial effects of early handling from laboratory animal species to domestic horses. The findings (reported below) suggest that a moderate amount of handling of young horses increases their learning ability as measured by maze-learning tasks and estimates of trainability under saddle, decreases the stress (both for horses and human trainers) associated with confining and training the horse under saddle, and possibly, reduces attrition during the initial years of rearing. This latter result is suggested by our data. More extensive sampling is necessary before it can be stated as a conclusion.

Methods

In the first experiment, three groups of 10 horses each were obtained from three sources which differed in the amount of early handling provided to the young horses. Horses in the first group, obtained from a large Texas ranch, were virtually unhandled. Their only human contact was three days of haltering as foals and minimum contact necessary to wean them at approximately eight months of age. Otherwise they were reared under range conditions. Horses in the second group, obtained from a ranch in eastern New Mexico, were handled as sucklings and frequently associated with humans. At weaning, horses were individually stalled, haltered, and gentled for one week. Following weaning, horses were housed in small corrals from which they were led, restalled, and groomed weekly. Horses in the third group were reared in small corrals on the TTU campus. At three months of age these horses were haltered, weaned, and handled by students an average of four times weekly for a ten-month period (August through May).

In the second experiment, 40 Quarter Horses were obtained from the same ranch that provided the virtually unhandled horses of the initial study. They were selected at weaning and individuals were assigned randomly to one of five conditions. These five conditions were (1) no handling except movement from one range to another, (2) one week of handling at weaning, (3) two weeks of handling at six-month intervals, (4) three weeks of handling at six-month intervals, and (5) nearly continuous handling for 18 months while housed at the TTU Horse Center. Handling for all horses consisted of haltering, leading, grooming, tying to an overhead cable, and routine care. Groups 2, 3, and 4 were reared under range conditions between handling sessions as were horses from Group 5 following the 18 months of handling. All horses were pastured together before returning to the TTU Horse Center for testing. Horses from Group 1 were halter broken prior to transport to the Horse Center.

Behavioral testing consisted of trainability ratings obtained following 10 days of habituation to a T-maze, performance in an alternation place learning task in the T-maze, and trainability ratings of performance when tested under saddle training subsequent to completion of the maze test.

The T-maze (figure 1) was constructed indoors of plywood walls measuring 2.44 meters in height. Each goal branch contained a feed tub and an exit door. During habituation, horses individually explored the maze for 10 minutes for 10 consecutive days, following which a panel of raters, who had observed the habituation trials and had handled the horses, assigned each a rating from 1 (quick learner) to 4 (slow learner). Following habituation, the horses were tested individually each day for 20 days in an alternation place learning task. Feed was placed in alternate goal arms of the maze on successive days, requiring that the horses learn a daily alternation of place. During each trial a horse was led to the maze entrance and released. When the horse had passed the choice point the response was recorded, the handler entered the maze, caught the horse and led it through the exit door and around the outside of the maze to the entrance. The route back to the

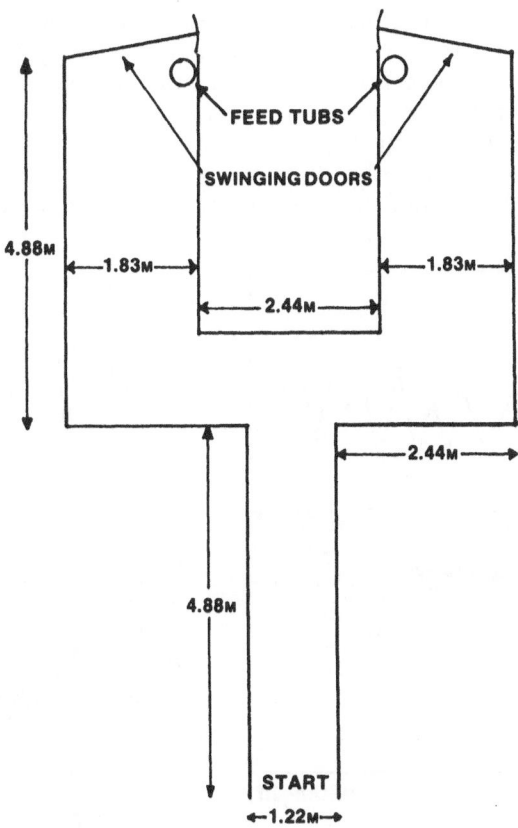

Figure 1. Diagram of maze apparatus.

entry was a random clockwise/counterclockwise sequence. Each horse received trials until it attained a criterion of 11 correct responses in a series of 12 consecutive trials with eight consecutive correct trials. A horse was retired for the day following attainment of criterion or 30 trials maximum.

Results

Figures 2 and 3 present mean maze performance for the first experiment. Group 2 animals (intermediate amount of handling) scored the highest percentage of correct responses, 92% on day 19, followed by Group 3 (most handled) with 78% on day 19. Group 1 (unhandled) persevered in entering the same goal arm on all days, thus their pattern of percentage correct reflects high and low scores on alternate days with no improvement across the 20 days. The mean trials to criterion per day reflect the same pattern as percentage correct per day. The results suggest that the unhandled animals failed to learn the place response. Both groups of handled horses exhibited learning with the intermediate amount of handling being somewhat superior

to intensive/extensive handling. All differences were statistically significant
at the $P< .05$ level. There were some indications that Group 3 (most handled)
may have simply become bored as the test did not provide the necessary
stimulus novelty for the entire 20 day period.

Results of the second experiment approximate those of the initial study.
Mean performance of all groups of horses showed no change across the last
10 days of testing, indicating that any learning already had occurred during the

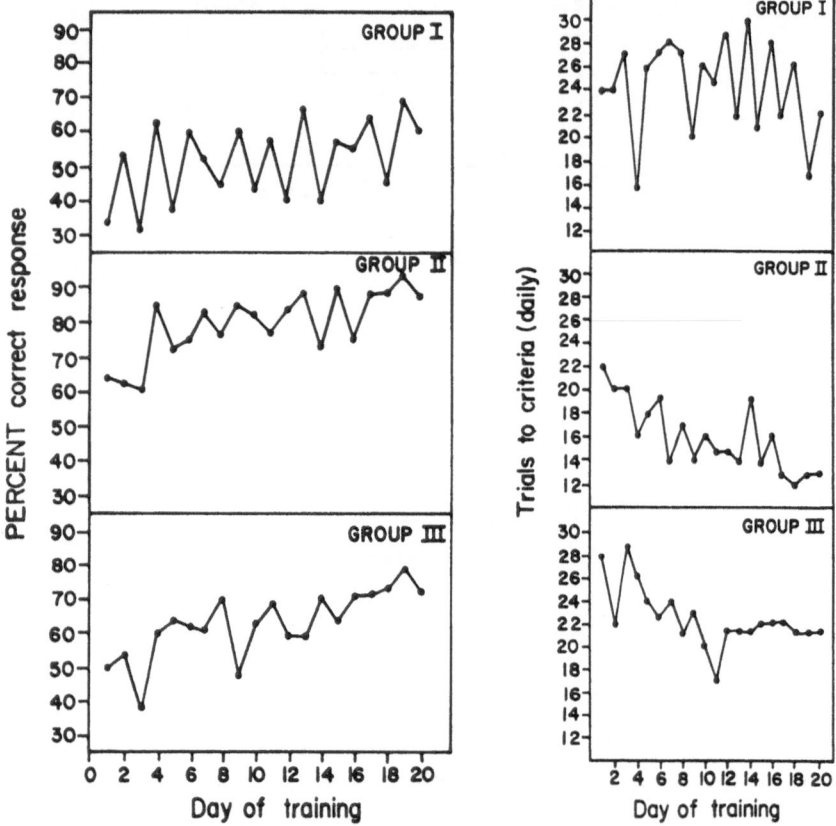

Figure 2. Percent correct responses by   Figure 3. Trials to criterion by group and
group and day of training, experiment 1.   day of training, experiment 1.

initial 10 days of testing. Performance during the first 10 days reflected the
effects of amount of handling. Group 5 (most handled) averaged 65% correct
choices while Group 1 (unhandled) averaged 56% correct choices, with the
remaining (intermediately handled) groups being intermediate ($P< .05$).
Group 5 horses also required fewer trials to criterion during daily testing
than did other groups ($P< .05$).

Trainability scores, assigned following the habituation-to-maze trials, also reflected the effects of handling. In the first experiment, animals of Group 2 (intermediate handling) received highest average ratings, followed by Group 3 with Group 1 (unhandled) receiving lowest ratings. These ratings, when assigned to animals of the second experiment, paralleled the ratings of the first study. Handled horses received higher average ratings than did unhandled, with animals of Group 4 (three weeks of handling) receiving the highest average rating.

Horses of the second experiment were trained under saddle, with four experienced horsemen rating their trainability. Animals of Group 5 received superior trainability scores when compared to other groups ($P < .05$; see figure 4).

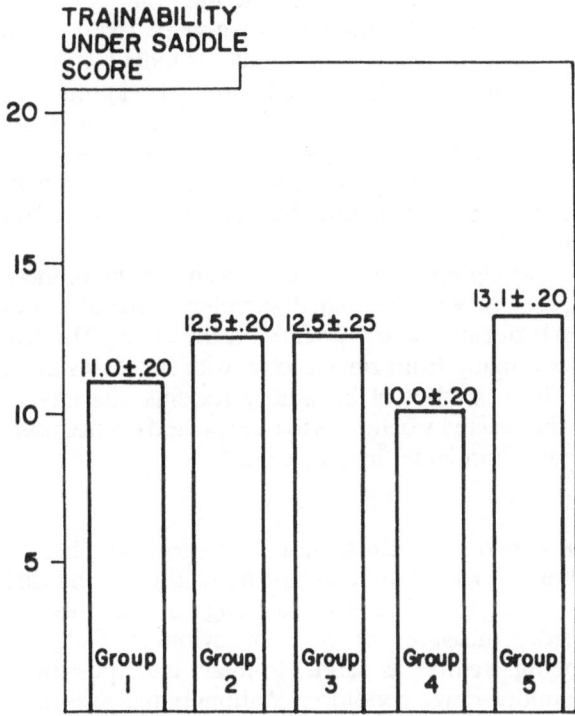

Figure 4. Trainability under saddle scores by group, experiment 2.

Discussion

The results of the experiments reported here are consistent with data on laboratory animal species which show augmented learning ability as a function of early handling. Although the systematic tests of effects focused upon learning, additional less systematic data suggest some additional and/or related effects. The ratings of trainability obtained during maze-habituation incorporate such animal behaviors as flinching, lunging, head tossing, etc., suggesting that the ratings reflect appraisals of emotionality as well as training aptitude. These same behaviors, of course, interact with the estimates of trainability under saddle. Furthermore, the student handlers reported that the unhandled horses appeared to be fearful of entering the maze and required strenuous effort on the part of the handlers to be engaged in the task. It may well be that differences in learning reflect, to a great extent, differences in emotional reactivity to a novel situation, which is consistent with the data on laboratory species.

Additional estimates of handling-induced differences were obtained from cowboys engaged in "breaking" the experimental animals to saddle. These horsemen were not trained animal scientists and indeed were skeptical of the professors and their so-called science to the point of vocal ridicule. After breaking the horses to saddle, the horsemen were asked to rate the horses' quality as potential work animals on a scale of their own devising (1 to 20 points), with no prior knowledge of the handling background of the animals. Their ratings consistently were higher for early-handled as opposed to unhandled horses with the ratings correlating substantially with scores obtained in the maze test. Many spontaneously commented that these horses (Group 5 and to a lesser extent, other handled groups) were "the best horses we've ever worked."

An effect of handling on attrition rate is another point of interest, although it is suggestive, at most, with these small samples. Unhandled groups averaged three losses each during the early years on the range. The handled groups sustained losses ranging from zero to one, with no losses in Group 5. If we can generalize from studies of laboratory rodents, the effects of handling should reduce the reactivity of horses to unexpected events, possibly reducing running panic and accidental injury or death.

Conclusions

Based on these two studies it appears that early handling by humans can significantly alter a horse's temperament/trainability in the direction of less emotional and more trainable. The data suggest that this is reflected not only in laboratory measures of these behaviors, but also in the horse's reactivity to being trained as saddle animals and, possibly, in increased survival ability under range conditions. Although these results must be inter-preted tentatively until additional data under less artificial conditions are obtained, the cowboys, foreman, and ranch manager of the cooperating ranch moved from a position of open disbelief and derision to one of endorsement. This change is reflected not only in their verbal interactions, but also in the decision (implemented) to hire a "handler" and revise their rearing management procedures in keeping with the recommendations stem-ming from the studies.

Endnotes

[1] Approved by the Dean of the College of Agricultural Science, Pub. No. T-5-184.

[2] Supported in part by funds appropriated by the Houston Livestock Show and Rodeo and the Institute of Agriculture, Texas Tech University.

[3] Assistant Professor, Dept. of Animal Science, Texas Tech University, Lubbock, TX 79409. *Current address: Dept. of Animal Science, Colorado State University, Fort Collins, CO 80523.*

[4] Professor, Dept. of Psychology, Texas Tech University.

[5] Graduate Student, Dept. of Animal Science, Texas Tech University.

References

Bornstein, M. 1985. *Sensitive Periods in Development: Interdisciplinary Perspectives.* Hillsdale, NJ: Lawrence Erlbaum Associates, Inc.

Heird, JC, Lennon, AM and Bell, RW. 1981. Effects of early experience on the learning ability of yearling horses. *J. Anim. Sci.* 53:5.

Whitaker, DD. 1982. Effects of varying amounts of early handling at weaning and later ages on the subsequent learning ability of two-year-old horses. Ph.D. Thesis. Texas Tech University.

WILDLIFE CONSERVATION AND ANIMAL RIGHTS: ARE THEY COMPATIBLE?

Michael Hutchins[1] and Christen Wemmer[2]

Introduction

In the history of Western civilization, nature was viewed primarily in utilitarian terms (Shaw 1974; White 1967). Wild animals were a source of food, clothing, and transportation, and wilderness was something to be "tamed" for the collective benefit of mankind. Indeed, the Biblical injunction from Genesis 9 was "And the fear of you and the dread of you shall be upon every beast of the earth... into your hands they are delivered." Within the last century, however, perceptions have begun to change. We have now seen the emergence of a broad-based coalition of individuals and organizations dedicated to the goal of preservation, rather than exploitation. The end result has been the formation of something completely novel to Western thought— an environmental ethic.

Aldo Leopold is universally recognized as the founding father of modern environmental ethics. His "land ethic" (Leopold 1949) is a classic statement of environmentalist philosophy, and one of the first to accord direct moral consideration to non-human animals (Callicott 1980). More recently, however, we have seen the emergence of a new ethical tradition, known as "animal rights" or "animal liberation" (Singer 1975; Regan 1983; Midgley 1983). This movement is a variation of the humane ethic, which seeks to eliminate the pain and suffering of non-human animals, especially that which results from human cruelty and indifference. Radical animal liberationists draw heavily upon the metaphors of political liberalism, claiming that animals, not unlike women and racial minorities, should be accorded equal rights, regardless of species. Extremists have demanded equal moral consideration for farm animals and other "enslaved and oppressed" non-human beings (Singer 1975; Regan 1983).

The purpose of this paper is to explore the philosophical tenets of the animal rights/humane ethic as they relate to the environmental ethic and, more specifically, as they relate to wildlife management and conservation. The two ethics will be compared in an effort to identify potential sources of conflict. Recent criticisms of the animal rights ethic, most notably by Fox (1978, 1979), Rodman (1977), Callicott (1980), Gunn (1980), and Hutchins et al. (1982) have identified several major discrepancies. The implications of these differences will be discussed.

111

Environmental Ethics and Animal Rights

Before discussing potential sources of conflict between the environmental/ conservation and animal rights/humane ethics, it will first be necessary to examine the two viewpoints in more detail. Both ethics share a concern for wild animals. Differences between the two can best be understood through a closer examination of their reasoning.

The primary goal of the environmental/conservation ethic is to preserve naturally occurring biological diversity (Ehrlich and Ehrlich 1981; Gunn 1980; Myers 1980). The term "natural" is used here to distinguish between diversity that has occurred as a result of natural evolutionary/ecological processes (i.e., speciation, colonization, "natural" extinctions), and that which has occurred because of recent human interventions (i.e., species introductions, human-caused extinctions). Aldo Leopold once said that "A thing is right if it tends to preserve the integrity, stability, and beauty of the biotic community. It is wrong when it tends to do otherwise" (Leopold 1949). Thus, the biological richness of an ecosystem, as characterized by the number and variety of species it supports, is seen as intrinsically good. Conversely, changes in naturally occurring biological diversity that occur as a result of human activities (either directly or indirectly) are perceived as intrinsic evils. Note that it is naturally occurring diversity that is important here, rather than absolute numbers of species. Tundra, for example, is relatively devoid of life when compared with tropical rain forest. However, it does support a characteristic assemblage of species, and, according to the environmental/con-servation ethic, it is this naturally occurring biotic community which should be preserved. The reader should also recognize that the aim of conservation is not to prevent change—ecosystems and species will change even in the absence of human interference. Indeed, evolution through natural selection is a dynamic, rather than a static process. According to Ehrenfeld (1972, p.7), the broad goal of conservation is "...to ensure that nothing in the existing natural order is permitted to become permanently lost as the result of man's activities except in the most unusual and carefully examined circumstances." Some exceptions have been made, for example, in the case of certain disease organisms, such as smallpox (Fenner 1980).

Species or ecosystems do not warrant moral consideration according to the animal rights/humane ethic, although they may be said to have "inherent value" (Feinberg 1978; Regan 1983). Thus, the humane moralists argue that individual organisms, rather than species or ecosystems, should be the focus of our ethical concerns. Furthermore, they argue that sentience—the capacity to experience pain—is the only relevant characteristic needed by organisms to merit full moral consideration (Singer 1975; Regan 1983). They contend that if non-human animals have the capacity to experience pain, then their suffering should be as important a matter of ethical concern as that of our fellow humans. That non-human animals may be incapable of reason, speech, forethought, or self awareness is considered irrelevant. After all, some classes of humans, (e.g., infants, and the severely mentally retarded) do not possess these abilities, and yet are accorded rights. The argument that human needs should take precedence over those of sentient non-humans is viewed as "speciesism"—a form of prejudice analogous to chauvinism, racism, or sexism (Singer 1975; Regan 1983; Midgley 1983). Thus, individual animals are seen as having a "right to life" and, except in very special cases, any attempt to kill them or to cause them to suffer pain is considered to be morally unjustifiable. The logical consequence of such an ethic is, of course, obligatory vegetarianism (Regan 1983; Singer 1975). (However, note that eating carrion does not involve the voluntary infliction of pain or death, and apparently would not be restricted by the animal rights/humane ethic.)

In this regard, it is important to recognize several degrees of rigor in the interpretation of the animal rights/welfare doctrine. Adherents to the most extreme view contend that there are few circumstances that could justify the killing of sentient animals by humans (Singer 1975; Regan 1983; Midgley 1983). However, the more liberal view is that humane and painless killing of animals is sometimes a regrettable necessity. The Humane Society of the United States, for instance, advocates the humane disposal of homeless dogs and cats. Implicit in such a policy is the assumption that death is a better alternative than the marginal quality of life these animals would experience-cone to a lack of human care and companionship (Wright 1978).

At first glance, the philosophical bases of the environmental/conservation ethic and the animal rights/humane ethic seem compatible. The welfare of animals has been a concern of both, but despite this common ground, profound differences exist. Callicott (1980) has compared the "land ethic" of Aldo Leopold (1949) with the "humane ethic" of Peter Singer (1975). While only sentient animals are afforded moral standing according to the humane ethic, the land ethic is more holistic, focusing not only on animals (both sentient and non-sentient), but also on plants, soils, waters, and other non-living things. While philosophical differences exist within various factions of both the environmental/conservation and the animal rights/welfare movements, we see their radically divergent emphasis on the individual as opposed to the population, species, or ecosystem as a crucial issue.

Ideological differences between the two ethics are evident in their contrasting view of the endangered species problem. While both ethics favor saving endangered animals, they differ in their reasons for doing so. Regan (1983, p. 360) argues that we should conserve endangered species "...not because the species is endangered, but because the individual animals have valid claims, and thus rights against those who would destroy their natural habitat, for example, or who would make a living off their dead carcasses through poaching and traffic in exotic animals, practices which unjustifiably override the rights of these animals." Thus, all sentient animals, regardless of species, rarity or other considerations, are to be given equal moral consideration according to the rights view.

In contrast, proponents of the environmental/conservation ethic argue that endangered species should be given special status solely because of their scarcity (Gunn 1980). That is, extraordinary efforts should be made to preserve rare species, especially when an organism has become rare due to some action on the part of humans (e.g., as a result of pollution, habitat alteration, or over-hunting). This follows from the underlying rationale that naturally-occurring biological diversity is intrinsically good, and that it provides a measure of the "health" of an entire ecosystem (Leopold 1949). Again, the focus is on the population, species, or ecosystem as a whole, rather than on individual organisms. There is a political element at play here as well. To the conservationist, endangered species have become representatives of the process of habitat degradation. While complexity in natural ecosystems is generally sufficient to weather the loss of some species before checks and balances are thrown out of equilibrium, it would be politically unwise to forego any loss of biological diversity without a fight. The recent controversy involving the snail darter is a case in point (Ehrlich and Ehrlich 1981).The environmental/conservation ethic is also based on the realization that the components of an ecosystem (both living and non-living) are often intricately interrelated, and that an instability in these elements can have far-reaching, and sometimes degrading consequences (Leopold 1949; Dasmann 1978; Fox 1978; Ehrlich and Ehrlich 1981).

Potential Sources of Conflict

Clearly, the animal rights/humane ethic and the environmental/conservation ethic will lead to the same decisions in many situations. For example, both ethics would consider it wrong for humans to destroy wildlife habitat, or to pollute it with chemicals and wastes. But, when the two viewpoints are compared, it is evident that disagreements will arise when the "rights" of individual organisms come into conflict with the preservation of populations, species, or ecosystems. As Regan (1983, p. 361) has pointed out, one implication of the more holistic environmental/conservation ethic is "... that the individual may be sacrificed for the greater biotic good." Regan opposes such actions, because they deny "... the propriety of deciding what should be done to individuals who have rights by appeal to aggregative considerations, including, therefore, computations about what will or will not maximally contribute to the integrity, stability, and beauty of the biotic community." Furthermore, he dismisses any attempt to subvert the rights of individual organisms to those of the species, or ecosystem as "environmental fascism."

There are many circumstances in which the "rights" of individual organisms may come into conflict with the preservation of populations, species, or ecosystems. We have chosen three such cases for more detailed consideration: (1) Problems of population regulation, (2) Incentives for wildlife conservation, and (3) Conservation-related research that harms individual animals.

Problems of Population Regulation

The culling of wild animal populations is a particularly sensitive issue for proponents of both the animal rights/humane and environmental/conservation ethics. Though their reasons may differ, proponents of both ethics are strongly opposed to the senseless killing of non-human animals. But what should be done in those situations where animals become too numerous for their own good, or for the good of the population, species, or ecosystem as a whole?

Overpopulation is a difficult concept to define. The phrase "too many animals" does not do justice to the complexity of the phenomenon. Caughley (1981) recognizes four different classes of overpopulation: (1) the animals are threatening human life or livelihood, (2) the animals are depressing the densities of favored species, (3) the animals are too numerous for their own good, and (4) the system of plants and animals is not in equilibrium, thus resulting in an alteration of the entire ecosystem. Of the four classes of overpopulation, numbers 3 and 4 appear to have the most potential for generating conflict between the two ethics and will, therefore, be discussed in greater detail.

When does a population of animals become too large for its own good? Generally, such a situation exists when individuals become so numerous that their habitat can no longer support them. An excellent example is provided by the Mule deer (*Odocoileus hemionus*) inhabiting the Kaibab Plateau of northern Arizona (Rasmussen 1941). The plateau was set aside as a game reserve in 1906, and a program of predator removal was initiated to provide more deer for recreational hunters. During a 25-year period, over 6,000 major predators were killed, including mountain lions, wolves, coyotes, and bobcats. The wolf (*Canis lupus*) was completely eliminated.

Following the eradication of predators, the deer began to multiply rapidly. The population, which had been estimated at 4,000 individuals in 1906, grew to between 60,000 and 70,000 individuals by 1923. As their numbers expanded, the animals overgrazed the vegetation, thus resulting in severe food shortages. In September, 1923, it was estimated that between 30,000 and 40,000 deer were starving. In 1925, the population "crashed" and nearly two-thirds of the herd died. By 1939, only 10,000 individuals remained.

This pattern of rapid population growth, followed by an equally rapid decrease is known as an "eruption." Caughley (1976) suggested that eruptions are characteristic of populations of large herbivores. However, all recorded instances of this phenomenon in North America have been preceded by some form of human intervention (Peek 1980). The Kaibab deer, for instance, are a classical example of poor wildlife management; elimination of the deer's predators subsequently resulted in overpopulation, habitat degradation, and widespread starvation. Unfortunately, similar situations exist to this day in many parts of the continental U.S. (Iker 1983; Klein 1981). As a result, some wildlife managers and conservationists have recommended that populations of deer and other large herbivores be controlled through culling. Animal rights advocates, on the other hand, have traditionally been anti-hunting, and recent attempts to control animal populations by killing have been vigorously opposed, whether they be in national parks or on other federally controlled lands (e.g., Grandy 1982; *The Humane Society News* 1983). Indeed,

Regan (1983, p. 361) has stated that "With regard to wild animals, the policy recommended by the rights view is: let them be!" But what are the consequences of inaction?

When a population of herbivores overshoots its food resources, individual animals can be expected to suffer an increased starvation rate (Klein 1968; Leopold et al. 1947; McCullough 1979; Taylor and Hahn 1947). Moreover, such conditions also lead to increased mortality and suffering due to disease, parasitism, and aggression-related injuries (Cheatum 1951; Christian and Davis 1964; Wilson and Hirst 1977). If a major goal of the animal rights/humane ethic is to reduce pain and suffering, then how can such situations be tolerated?

Proponents of the animal rights/humane ethic may be suspicious of this argument, since it is the same one used to justify recreational hunting (Lyons 1978; Hope 1974). However, the goals of the environmental/conservation ethic are different than those of the majority of wildlife managers and hunters. Wildlife managers are also considered among the environmentally aware. However, by definition, a manager is one who makes "judicious use of means to accomplish an end" (*Webster's Seventh New World Collegiate Dictionary*). Thus, wildlife managers make decisions which affect the environment, generally to favor the productivity of specific game animals. Furthermore, they tend to focus on disturbed rather than on natural environments, and view the cropping of overabundant animals as legitimate and necessary. Though the environmental/conservation ethic does not prohibit recreational hunting per se (*National Wildlife* 1971; Callicott 1980), it also does not condone widespread environmental manipulations that favor specific species at the expense of others. Many hunters and wildlife managers do not view such manipulations as being ecologically disruptive when, in fact, they often are (Hope 1974).

In recent years, some biologists have argued for more noninterventional wildlife management policies, especially in larger national parks, i.e., those which contain relatively complete ecosystems (Houston 1971, 1982). Noninterventional or "natural" management policies involve a "hands-off" attitude similar to that expounded by radical animal rights activists (e.g., Regan 1983). However, one result of such a policy is that natural regulatory processes are allowed to operate regardless of the consequences for individual animals. In Yellowstone National Park, for example, 200 bighorn sheep (*Ovis canadensis*) recently contracted keratoconjunctivitis, an eye disease commonly known as pinkeye (Meagher 1982; Robbins 1984). Nearly 85% of these individuals became blind, and subsequently died as a result of starvation or injuries sustained during falls. Although the animals were admittedly suffering, Park Service officials never considered treating them with antibiotics. Since the disease was considered to be naturally occurring, it would have conflicted with their policy of non-intervention. The philosophy espoused by the U.S. National Park Service is that pristine ecosystems exist in an equilibrium state in which human impact on energy flow should be minimized at all times (Houston 1971). The Park Service view is that humans should not interfere with the workings of nature, which are preceived in neutral terms, without moral judgement (see Gould 1982). Not surprisingly, this and other recent decisions by wildlife managers have been highly controversial (Blonston 1983; Chase 1986; Robbins 1984).

Figure 1. When some bighorn sheep in Yellowstone National Park contracted a blinding eye disease, Park Service personnel made no attempt to intervene. One implication of non-interventional or "natural" management policies is that normal regulatory processes, such as disease and predation, are allowed to operate regardless of the consequences for individual animals. (Photo: Gerry Ellis)

Despite Regan's (1983) plea that wild animals be left alone, it would seem that even non-interventional management policies might conflict with the rights ethic, or at least create that perception in the public mind. If we are to accept the proposition that all sentient beings have a "right to life," then the logical conclusion is that we should intervene in those cases in which sentient animals are suffering from starvation or disease (but, see Feinberg 1978 for an alternative viewpoint). In fact, animal rights/welfare organizations are generally among the first to recommend supplemental feeding for under-nourished wild animals (Grandy 1982; Iker 1983). However, such actions are often in conflict with the environmental/conservation ethic. Supplemental feeding may increase the probability that animal populations will eventually overshoot their food resources (figure 2; Robinson et al. 1980). Furthermore, by concentrating the animals at feeding stations, such practices may also increase the incidence of disease, or intensify rates of habitat alteration (Madson 1986). The obvious danger in such a policy is that short-sighted empathy can lead to much greater suffering, not only for the animals of interest, but also for the ecological community as a whole (see Hardin 1974 for a similar argument regarding human population problems).

Figure 2. Supplemental feeding of elk in Wyoming. Unless artificial feeding is combined with a regular program of population reduction, the animals may overshoot their food resources, degrade their habitat, and suffer an increased rate of starvation. (Photo: John Wilbrecht, U.S. Fish and Wildlife Service.)

The animal rights/humane ethic is clear on its position regarding human-animal relationships, but it is unclear with regard to pain and suffering inflicted upon animals by animals (Hutchins et al. 1982). In general, the doctrine aims to limit human behavior with the objective of minimizing or eliminating human-caused pain or discomfort to other creatures. At least among some animal rights advocates, what predatory animals do to their prey is beyond the realm of concern, presumably because it is done by "innocent killers," lacking in malicious intent or knowledge of the ethical consequences of their actions (Regan 1983; Feinberg 1978). This is an enlightened view compared to attitudes of the past when the conduct of animals was morally appraised (Beach 1975), and predators, such a wolves, were persecuted as "blood-thirsty killers" or "allies of the devil" (Lopez 1975). However, such arguments also open the animal rights movement to logical criticism. From the standpoint of the individual, pain is pain, regardless of the "intent" of the predator. In addition, predation is not the only way that one organism can have a detrimental effect on another. Indeed, one critical weakness of a view of nature that stresses individual rights is that it fails to recognize the interdependencies that exist within natural ecosystems.

Thus, if a population of herbivores becomes so numerous that it degrades its habitat, many other organisms may suffer as well. For example, elephants (*Loxodonta a. africana*) have become a serious conservation problem in African national parks (figure 3; Van Wyk and Fairall 1969; Buechner and Dawkins 1961; Laws 1970; Cumming 1981). The destruction of woody vegetation by elephants is causing widespread habitat alteration, including the conversion of woodlands to open grassland or semi-desert. Consequently, many other species of animals can no longer utilize these areas, and may be caused to suffer as a result (Laws 1970; Pienaar 1969). Perhaps even more importantly, such habitat alteration increases the probability of local extinctions. Those responsible for our wildlife and ecosystems are thus forced to make difficult decisions about how to promote the "greatest good."

Figure 3. Elephants have become a serious conservation problem in African national parks. With their movements restricted by park boundaries, these large herbivores overgraze the vegetation and convert woodland habitats into grasslands or semi-deserts. (Photo: Henry Klein)

The so-called "elephant problem" is due largely to human overpopulation (Laws 1970). The progressive use of land by people for living space and agriculture has restricted many large mammals exclusively to national parks. One result is that elephant populations are no longer regulated by natural processes, such as dispersal. Thus, the animals may become so numerous that they degrade their own habitat. Conservationists have generally agreed that the only solution to this problem is culling (Caughley 1976; Haigh et al. 1979; Laws 1970, 1974; Pienaar 1969; Younghusband and Myers 1986). Practical considerations make it difficult, if not impossible, to capture and translocate large numbers of adult elephants. Furthermore, most African national parks have more than enough elephants already, so there is really no place for the animals to go. (Note that immature elephants are sometimes captured and transported to underpopulated areas for release, see Haigh et al. 1979.)

How do wildlife managers decide when intervention is required? The precise answer to this question may never be known (McCullough 1979; Noy-Meir 1981; Sinclair 1981), although some estimates have been obtained through mathematical modelling. At any given point in time, the biomass of herbivores and the biomass of the vegetation are unlikely to be in perfect equilibrium, and there are frequent perturbations in the system (Noy-Meir 1981; Sinclair 1981). But, as Caughley (1970) has pointed out, unless the displacement becomes very large (more than 30% or so), little can be gained by artificial reductions of population size. If, however, the displacement from equilibrium is extreme, it can result in irreversible changes in vegetation and soils, and these are the conditions that are most likely to lead to extinctions. Unfortunately, such generalizations seldom apply in specific cases. The degree of environmental alteration caused by grazing animals can be affected by many variables, such as the type of vegetation, amounts of precipitation, and the size of the park or reserve (Laws 1970).

So far, we have limited our discussion to problems of overpopulation in indigenous animals. However, a related topic that deserves attention is that of introduced or "exotic" species. One of the many ways in which humans alter their environment is by transporting organisms across natural barriers to dispersal. Colonizing peoples have traditionally attempted to modify their new settings by releasing animals (both domestic and wild) that are native to their homelands (Laycock 1966). Many introduced forms have flourished in their new settings, and in fact there are several widespread cultivars whose origins are unknown, so vicarious was their dissemination.

The introduction of exotic mammals has often been associated directly or indirectly with pervasive changes in native fauna and flora (Courtney 1978; Courtney and Ogilvie 1971; deVos and Petrides 1967; Laycock 1966; Hutchins et al. 1982). The problems caused by introduced species are not unlike those that occur when indigenous animals become overpopulated. However, because these organisms are not native to the host ecosystem, their destabilizing effects are often greatly accentuated. It appears that island ecosystems are particularly vulnerable to disruption (Coblentz 1978; Holdgate 1967). New Zealand, for example, is an island nation in which an assortment of alien herbivores (e.g., the brush-tailed possum, red deer, sika deer, and Himalayan tahr) have virtually annihilated the native woodland communities (Howard 1964, 1967; Wardle 1974). The region's natural vegetation evolved in the absence of heavy grazing pressure, and therefore did not develop chemical or physical adaptations for protection. Plants with a history of exploitation by herbivores tend to evolve adaptations such as toxins, thorns, or rapid growth and reproductive rates to protect them from their "predators" (Janzen 1978; Pianka 1983).

Conservationists have argued that populations of destructive exotics be controlled through culling, especially in national parks, where the animals are threatening native fauna and flora (Allen et al. 1981; Laycock 1966, 1984). Such examples illustrate the potential for conflict between the environmental/ conservation and animal rights/welfare movements (Hutchins et al. 1982). Recent attempts to cull populations of feral burros in Grand Canyon National Park, for instance, have been opposed by animal rights/welfare advocates (figure 4; Laycock 1974; Stocker 1980; Reiger 1978). This occurred despite evidence that the foraging and trampling activities of burros are altering fragile desert habitats, and may be contributing to the decline of the native bighorn sheep (*Ovis canadensis*) (Carothers et al. 1976; Hanley and Brady 1977; Hansen 1980; Walters and Hansen 1978; Schectman 1978; Tennesen 1975; U.S. Dept. of the Interior 1980).

Figure 4. Burros were brought to North America by the Spanish in the sixteenth century. Thousands now roam the Southwestern United States, where they compete with native wildlife and alter fragile desert ecosystems. (Photo: Gerry Ellis)

Much of the controversy surrounding the artificial regulation of animal populations has focused on the means, rather than on the ends (Hutchins et al. 1982). Understandably, animal rights/welfare advocates are opposed to any solution which involves killing, yet current methods of population control may involve shooting, poisoning, trapping, or the introduction of disease (Anderson 1971; Fenner and Ratcliffe 1965).

In arguing against the control of certain exotic animals, some animal rights/welfare advocates have questioned whether any benefits would actually result from such actions. However, there are several instances in which the control or elimination of exotic mammals has had beneficial effects. When small exclosures were erected to study the effects of feral goats on native flora in Haleakala National Park, Hawaii, the seeds of a heretofore unknown leguminous plant began to germinate (Baker and Reeser 1972). Similarly, eradication of feral goats in the Galapagos Islands, Ecuador, resulted in the rapid regeneration of native plants (Hamann 1979). The elimination of feral rabbits from Laysan Island in the leeward Hawaiian chain saved the endemic Laysan teal (*Anas laysanensis*) from almost certain extinction (Warner 1963). At the time the rabbits were eliminated, the birds' population had been reduced to less than seven individuals. Now there is a healthy population.

Population regulation will probably continue to be a point of contention between animal rights/welfare advocates and conservationists. However, recent technological advances may help to alleviate some of this conflict. There has, for example, been increasing interest in the development of non-lethal methods of population control, such as tubal ligation, castration, chemosterilization or mechanical devices that prevent conception (Davis 1961; Johnson and Tait 1983; Matsche 1976; Singer 1975; Turner 1984).

Implanted or orally administered hormones are commonly used to inhibit reproduction in zoo animals, and this has reduced the need for euthanasia (Seal et al. 1976; Whitlock 1978). Unfortunately, the use of similar procedures for wild animals is often fraught with complications. For example, many of these methods require capturing and handling the animals, and this may lead to considerable psychological and physical trauma (figure 5; see "Conservation Research and Humane Concerns," below). Hormone implants and orally administered reproductive inhibitors often require repeated applications, sometimes on a daily basis. In addition, these methods can have many deleterious side effects (Matsche 1977a, 1977b, 1980; Remfrey 1978; Seal et al. 1976). Methods involving surgical procedures may lead to infection or death (Zwank 1981).

Even if some animals can be permanently sterilized, social factors may limit the effectiveness of control (Johnson and Tait 1983). Some rodents, for example, are poor candidates for the use of chemosterilants because of their promiscuous breeding behavior. In high density populations, a female rat may mate with as many as 20 different males. In fact, one study found that reproduction was not curtailed in a rat colony in which 85% of the males were surgically sterilized (Kenelly et al. 1970). Wild horses, on the other hand, have proven to be excellent candidates for reproductive inhibition. Horses are highly polygynous, and a single stallion may mate with from two to eleven mares. Since stallions vigorously defend their harems, the use of chemosterilants can have a significant impact on reproduction. One study obtained an 80% decrease in births by administering long-acting antifertility drugs to specific males (Turner 1984). Furthermore, the drugs can be injected

Figure 5. Biologists capture a mountain goat in Olympic National Park, Washington. To collect information essential to wildlife conservation it is often necessary to capture and handle animals or to mark them for individual identification. Despite numerous precautions, animals are sometimes injured during capture procedures. (Photo: Daryll Hebert)

by dartfiring rifle, thus precluding the need for capture and handling. Similarly, Garrett and Franklin (1983) found that prairie dog (*Cynomys ludovicianus*) populations could be controlled by feeding the animals estrogen-laced grain. These colonially living rodents are highly seasonal breeders, and hormones need only be administered during a short period of time to have a significant impact on reproduction.

It would seem that such "benign" methods of population control offer some hope for compromise between the animal rights/welfare and environmental/ conservation movements. However, reproductive inhibition is a gradual rather than a rapid method of control. Environmental alteration can therefore be expected to continue until population size eventually decreases as a result of natural mortality (Hutchins et al. 1982). Thus, from the perspective of the environmental/conservation ethic, reproductive inhibition may be a case of "too little, too late," especially when one is dealing with long-lived animals that have relatively low mortality rates. By the time population growth can be curtailed, irreversible environmental changes may have already taken place.

Live capture and translocation is another non-lethal method of population control that is becoming increasingly popular with animal rights/welfare advocates. However, it also has numerous limitations. The animals are often subjected to considerable physical and psychological stress while being captured and transported (see "Conservation Research and Humane Concerns," below). Another major consideration is the fate of the animals that are being released. There have been few studies on this topic, but existing information suggests that many, if not the majority, of translocated deer die within a year of their release, presumably as a result of intraspecific competition (D. McCullough, personal communication). One of the most thorough studies on the fate of "translocated" mammals was conducted on marsupial gliders (*Schoinobates volans*) in Australia. One thousand of these animals were displaced to adjacent areas when their forest habitats were destroyed by logging (Tyndale-Briscoe and Smith 1969). A majority were recaptured shortly thereafter, but it was discovered that they had lost up to 25% of their body weight. Furthermore, many breeding females had lost their young. The adult mortality rate also appeared to be extremely high, as less than 7% of the emigrants were recaptured one year later. It was concluded..."that displaced sugar gliders die in situ rather than emigrate to occupied forest and die there through failure to become established" (ibid, p. 658). Thus, it is questionable whether translocations constitute a completely acceptable alternative to killing in terms of humane solutions.

Cost is a major obstacle to capture and removal schemes; however, a number of animal welfare groups have offered to provide the funding. For example, the Fund for Animals reportedly spent nearly $500,000 to remove about 600 burros from Grand Canyon National Park (*Newsweek* 1981). These expenditures have been criticized as being short-sighted given the greater animal welfare interests that could be served if the money were used in other ways (Allen et al. 1981; Hutchins et al. 1982).

Some animal rights advocates might argue that most cases of animal overpopulation (whether the animals are exotic or indigenous) are ultimately caused by humans. Furthermore, they might ask why animals should be made to suffer for our mistakes. Conservationists would probably sympathize with this view but realize that wildlife does not exist in a vacuum. There are few areas left in the world that are unaffected by human activity. Indeed, most existing national parks and reserves are smaller than San Francisco's man-made Golden Gate Park. The small size of these reserves, and their isolation, make them particularly vulnerable to ecological disruption and disappearance of species (Soule et al. 1979; Miller 1979; Myers 1979; Ehrlich and Ehrlich 1981; Polunin and Eidsvik 1979).

To summarize, the rights view is characterized by a "laissez faire" attitude toward wildlife management. However, if the logic of the animal rights/humane ethic is carried to its extreme, it would seem to require intervention when wild animals are suffering from starvation or disease (but, see Feinberg 1978). By contrast, total non-intervention is expected to be favored by the environmental/conservation ethic only when remnant ecosystems are relatively pristine and large enough for natural regulatory processes to operate (Cole 1971; Peek 1980). The environmental/conservation ethic would also favor direct population reductions under certain specific circumstances. For example, if some human-precipitated ecological change were to drive a population or species close to extinction, the environmental/conservation ethic would favor intervention, regardless of the consequences for individual animals. Furthermore, direct action would be deemed justifiable when, in the absence of natural checks and balances, a population of animals becomes so large that it threatens the existence of other species, or of the ecosystem as a whole. From the perspective of the environmental/conservation ethic, direct action would be particularly justifiable in national parks, wilderness areas, and equivalent reserves (see Younghusband and Myers 1986). These few areas constitute a relatively small portion of our total land area and contain the only remaining habitats that are still relatively pristine (Houston 1971).

Incentives for Conserving Wildlife

Incentives for preserving natural ecosystems, endangered species, and wildlife in general are complex (Erhlich and Erhlich 1981). Certainly moral and aesthetic considerations lie at one end of the spectrum (Gunn 1980; Regan 1983; Stone 1975; Blackstone 1978). More recently, however, there has been an increasing recognition of the economic value of wildlife and other natural "resources." Indeed, many conservationists and wildlife managers have invoked economics in an attempt to promote conservation efforts (Bart et al. 1979; Coe 1980; Myers 1979, 1980, 1981a; Ehrlich and Ehrlich 1981; Noonan and Zagata 1982). In contrast, the rights view is vigorously opposed to human exploitation of sentient animals for economic gain. According to Regan (1983, p. 343), "A practice, institution, enterprise, or similar undertaking is unjust if it permits or requires treating individuals with inherent value as if they were renewable resources." By contrast with the animal rights/humane ethic, the environmental/conservation ethic does not seek to eliminate all human exploitation of animals, provided that it be accomplished as humanely as possible and produces minimal impact on the environment (Leopold 1949; Myers 1979; Talbot 1980).

Some conservationists might argue that the rights view fails to consider the economic and political realities in which wildlife conservation must take place (Wilson 1984). Indeed, there are many examples in which the recognition of animals as "renewable resources" may have saved species from extinction. In Papua, New Guinea, for example, people who used to kill wild salt water crocodiles (*Crocodylus porosus*) are now raising them for sale in the world market (Montague 1981). The skins of these animals are among the most valuable of all crocodilians and, although they are still abundant in New Guinea, they have been hunted to extinction throughout most of their former range. Young crocodiles, which normally experience an 80% mortality rate in the wild, are brought into captivity, raised to optimum commercial size, and then killed humanely. Profits from such farming operations go directly to the local people, thus giving them an economic incentive for conservation. Without such rewards, there would be little chance that the species could survive.

In the above example, conservation efforts were facilitated by peoples' perception of animals as renewable resources. Unlike non-renewable resources, renewable resources can be expected to provide long-term economic gains. In the absence of such attitudes, people tend to opt for short-term exploitation strategies, and this is more likely to lead to extinctions (Hardin 1978). Myers (1979) states that conservation of species must take place in a world dominated by politics and economics. In such a world, arguments based solely upon abstract philosophical notions have only limited effectiveness. This is particularly true in developing nations where impoverished people and their governments have more immediate concerns. To people engaged in a daily struggle for survival, an endangered animal is usually seen as nothing more than a source of food or income.

Economic arguments for the preservation of species are focused on human needs, and such utilitarian notions about wildlife are unlikely to sit well with animal liberationists. Indeed, Singer (1981) has gone so far as to propose that the "circle of altruism" be expanded beyond our own species to all animals that can suffer or feel pain. Similar ideals have been expressed by Stone (1975), who argues that animals and, for that matter, whole ecosystems should be given not only moral standing but legal rights as well. Most conservationists would probably applaud such ideals, but there are many who consider them to be unrealistic. According to Wilson (1984, p. 131), "...to force the argument entirely inside the flat framework of kinship and legal rights is to trivialize the case for conservation." Wilson believes that the only way to make a conservation ethic work is to "ground it in ultimately selfish reasoning." The essential component of this principle is that people will conserve land and species if they "foresee a material gain for themselves, their kin, and their tribe" (ibid, pp. 131-132).

Conservationists generally do not believe that economic arguments should be invoked in every case. Indeed, there are many species that have no immediate cash value and are therefore classified as "non-resources." Thus, the danger in relying on economic arguments exclusively is that they might be effective only for a few valuable species (Ehrenfeld 1976; Leopold 1949; Myers 1979). Alternatively, certain valuable species may be poorly managed or over-exploited, thus hastening their extinction (see Domalain 1977). Conservationists are aware of these problems, and realize that economic arguments are but one of the many potential strategies in the struggle to preserve biological diversity (Ehrlich 1980).

Conservation Research and Humane Concerns

Scientific research is one means by which humans gain an understanding of the natural world. Such an understanding is essential to wildlife conservation efforts. In fact, Poole and Trefethen (1978, p. 344) have stated that "Knowledge is the essential prerequisite to making a management decision respecting a species, population, or group of wildlife." A decision made in the absence of information about a species or population, depending on the result, is, at worst, an act of ignorance, or, at best, a stroke of good fortune." With this in mind, it is important to ask: How should we view conservation research that harms individual animals?

Animal rights/welfare advocates have been traditionally opposed to the use of animals in biomedical research, particularly when individuals are sacrificed, or caused to suffer pain (March 1984; Singer 1975; Regan 1983). Indeed, Regan (1983, p. 385) has stated that ". . . animals are not to be treated as mere receptacles or as renewable resources. Thus does the practice of scientific research on animals violate their rights. Thus it ought to cease, according to the rights view."

By contrast with the animal rights/humane ethic, the more holistic environmental/conservation ethic would not oppose the use of animals in scientific research, particularly if such research were to help ensure the survival of a population,species, or ecosystem as a whole. However, animal rights advocates would consider this view to be "utilitarian," in that "whether the harm done to individual animals in the pursuit of scientific ends is justified depends on the balance of the aggregated consequences for all those affected by the outcome" (Regan 1983, p. 392).

There are many cases in which conservation research may prove harmful to individual animals. For example, to collect essential data on population dynamics, behavior, individual growth patterns, diseases, etc., it is often necessary to capture and handle animals, or to mark them for individual identification (figure 5; Taber and Cowan 1971). Despite numerous precautions by scientists, animals are sometimes injured or killed during capture procedures (Spraker 1978). For example, some animals suffer limb fractures and lesions as a result of falls, and some may succumb to an overdose of drugs or to shock (Stelfox 1976). Others may contract capture myopathy—an often fatal muscular disorder that is induced by the trauma of capture and transport (Chalmers and Barrett 1977; Spraker 1977, 1978).

Harm that comes to individual animals during capture and handling could be considered "accidental," in that scientists are not harming animals deliberately. However, there are cases in which conservation research may involve deliberate harm. For example, thousands of animals have been sacrificed so that biologists could analyze their stomach contents (e.g., Peterson 1955), or assess their physical condition or reproductive status (e.g., Parker 1981; Sinclair 1974). Another example is provided by the work of Eaton (1972a, 1972b), who studied the development of predatory behavior in captive-reared lions (*Panthera leo*) and cheetahs (*Acinonyx jubatus*). To observe predation, he released live domestic goats, which were subsequently killed and eaten by the cats. The rights view certainly would not condone such experiments, yet, despite the unfortunate consequences for the individual goats, this work appears to be compatible with the more holistic environmental/conservation ethic. Many carnivorous species, including large predatory cats, have been forced to the brink of extinction by people (Stonehouse 1981). One method by which conservationists hope to save some of these species is through captive breeding and reintroduction (Brambell 1977). However, reintroducing captive-bred animals into their natural habitats poses many difficult problems, including the ability of the animals to obtain their own food. Although young felids come equipped with an "instinctive" sequence of predatory behaviors, practice is necessary to increase efficiency (Leyhausen 1973). Live prey can be both unpredictable and dangerous and efficiency is important. Thus, a knowledge of how captive-bred predators learn to recognize potential prey, and how predatory behavior improves with practice, will be essential to any serious reintroduction effort (Bogue and Ferrari 1976).

The rights view would place heavy restrictions on the nature of conservation research. If carried to its logical extreme, it would, in fact, eliminate all science "that violates individual rights" (Regan 1983). One implication is that information essential to wildlife conservation could not be collected, and this might increase the probability of species extinctions. Animal rights/welfare advocates might argue that scientists should work to develop more benign methods of study. In fact, scientists themselves have taken some initiative in this regard. There have, for example, been efforts to develop alternatives to tagging, branding, toe-clipping, and other types of identification techniques which involve harming animals (Ryder 1978). In addition, there has been an increased interest in the development of less invasive methods of assessing physical condition and diets. For example, physical condition can sometimes be assessed through measurements of weight, girth, blood chemistry, and horn growth rates (Bunnell 1978; Franzmann and LaResche 1978; Stevens 1983; Winters 1980). Similarly, dietary preferences can sometimes be determined by watching what animals eat (e.g., Hoeffs 1974), by analyzing feces (e.g., Owaga 1977), or by measuring the nutritional quality or abundance of the food resources themselves (e.g., Constan 1972; Miller 1974). The humane treatment of animals is therefore a continuing goal. However, it may not be possible to totally avoid suffering and pain. The problem of disappearing species is so acute and so immediate that there may not always be sufficient time to devote to such tasks.

Conclusions

Clearly, there are many cases in which the animal rights/humane ethic is in direct conflict with the environmental/conservation ethic. In fact, we consider the extreme views expressed by Singer (1975) and Regan (1983) to be largely incompatible with the goal of wildlife conservation. We agree with Wilson's (1984, p. 131) conclusion that the animal rights/welfare movement is due for "a stiffer dose of biological realism." In fact, the *only* major implication of modern biology that the humane moralists seem to have embraced is that of the evolutionary continuity between human beings and other forms of life (Callicott 1980). A recognition of our kinship with all living things is often used to argue that some nonhumans deserve "equal rights" (Singer 1981). However, other, less palatable, implications of the more holistic ecological/evolutionary perspective appear to have been conveniently ignored, especially when they conflict with the philosophic foundations of the animal rights/humane ethic (Callicott 1980; Rodman 1977).

As mentioned earlier, one of the major problems with the animal rights/ humane ethic is its focus on individual animals, as opposed to populations, species, or ecosystems. This reductionistic perspective of the natural world is biologically naive, and may itself be based on the cultural biases of its progenitors (Gunn 1980). Western cultures do tend to place more emphasis on the rights of individuals, as opposed to the welfare of society as a whole. Animal liberation is therefore an anthropomorphic philosophy (Rodman 1977), and this may explain some of its popular appeal. But what are the dangers of adopting a view of nature which is focused on the rights of individual organisms?

Myers (1979) and Ehrlich and Ehrlich (1981) have identified habitat destruction as *the* most significant threat to wild animal populations. Therefore, a concern for wild animals needs to be expressed in a willingness to protect natural ecosystems. On a superficial level animals appear to be separate entities, moving independently and freely within their environments. In fact, nothing could be further from the truth. All free-living organisms are closely tied to the habitats in which they have evolved. It is therefore difficult to separate individuals from their ecological contexts (Fox 1979). Similarly, it is equally difficult to draw strong distinctions between living and nonliving things. All living organisms, whether they be viruses or humans, are composed of non-living matter. Carnivores, for example, are as dependent on soils for their existence as they are on their prey.

In simple terms, an ecosystem must consist of a source of energy (usually sunlight), a source of raw chemical materials (rocks, soil, air, and water), "producers" capable of transforming and storing solar energy (usually green plants), "primary consumers" which feed on the producers (i.e., herbivores), "secondary consumers," which feed on the primary consumers (i.e., carnivores) and, finally, decomposers which break down the dead bodies of the producers and consumers and cycle their energy back into the system (Dasmann 1978). According to Leopold (1949), such "food chains are the living channels which conduct energy upward; death and decay return it to the soil . . . like a slowly augmented revolving fund of life." The rights view simply cannot deal with such complex interdependencies. To quote Rodman (1979, p. 89), "The moral atomism that focuses on animals and their subjective experiences does not seem well adapted to coping with ecological systems."

The humane moralists are very specific about which forms of life are to be granted full moral consideration. Indeed, many ardent defenders of animal rights have focused exclusively on the protection of sentient animals, and often their attention is concentrated only on those animals that are perceived as being appealing or "cute" (Neitschmann 1977; Rodman 1977). Fox (1979) has recognized the apparent weakness of this philosophy, noting that "The ecological imperative of responsible stewardship concerns our relationship with all of creation, both sentient and non-sentient" (p. 54). He envisions the animal welfare movement as being an important transition to a more holistic "eco-ethic." While we agree that a recognition of the "rights" of all living things is an important step toward the attainment of such a goal, we also stress that responsible stewardship may involve difficult, and sometimes painful decisions (see Fox 1978; Hutchins et al. 1982). In some cases, our actions may result in the death or suffering of other sentient beings. Of course, this does not imply that animals can be treated without care and respect. For example, when the need to control an animal population has been identified, it should be accomplished in the most humane manner possible, given the limitations of the situation. However, when the purpose of such reductions is to preserve natural ecosystems or to protect endangered animals or plants, then it should not be perceived as being "inhumane."

Regan (1983) has labelled any attempt to subvert the rights of individual organisms to the species or ecosystem as "environmental fascism." However, the more appropriate term might be "environmental socialism," in that the "rights" of individual animals are viewed as secondary to those of the species or ecosystem as a whole. From the perspective of the environmental/conservation ethic, the species and ecosystem are more important than any one organism. Indeed, without the former, there is no way that the latter could even exist! As Soule and Wilcox (1980, p. 8) have pointed out, "Death is one thing—an end to birth is something else." In fact, if animal rights/welfare advocates are unwilling to broaden their perspective to encompass the whole of nature, they will risk a total alienation of the environmental community. Moreover, "in adhering to a philosophy that emphasizes a reverence for life, but that ignores the conditions necessary for its survival, they may ultimately be unfaithful to their own ideals" (Hutchins et al. 1982, p. 333). In this respect, radical animal liberationists may have much in common with certain Hindu castes. Based on his travels in India, Sanderson (1896, p. 160) once wrote that "Many natives would not hurt the meanest insect: but though it might be merciful to put an end to suffering in many cases they cannot part from their disinclination to take life." The frustration of the colonial British with Hindu ethics illustrates an important point: A belief system that protects a well-meaning person's conscience may in fact perpetuate a greater suffering unknown to him or her (also see Rodman 1979; Callicott 1980).

Conservationists and animal liberationists have challenged our traditional perceptions of non-human animals (Wilson 1984). However, the latter view is biologically illiterate and thus ill-equipped to provide an intelligent basis for wildlife conservation. This is not to imply that our relationship with nature should never be approached in moral terms. In fact, ethical philosophy faces a severe test when it comes to the conservation problem. As Wilson (1984, p. 123) has written, "... in ecological and evolutionary time, good does not automatically flow from good or evil from evil. To choose what is best for the near future is easy. To choose what is best for the distant future is also easy. But to choose what is best for both the near future and distant future is a hard task, often internally contradictory, and requiring ethical codes yet to be formulated."

Epilogue

The objective of this paper was to compare the animal rights/humane and environmental/conservation ethics, and review potential conflicts. Though it is clear that there are significant differences in values, it is also clear that the two views are not completely antithetical. The primary difference between the two ethics lies in the scope and the primacy of their concerns. We hope, however, that this essay has not been a purely intellectual exercise, and would therefore like to explore the potential for reconciliation. Indeed, we believe that a shared appreciation of those values held in common could lead to productive compromises benefiting both wildlife and environment. As it is now, there is a strong element of evangelism among animal rights/humane proponents which views any compromise as concession, and an equally stubborn element exists among certain environmental/conservation groups.

Enlightened solutions to the problems of the humane treatment of animals and environmental concern can best be achieved through collaboration. Participation in cooperative problem solving through regular meetings, workshops, and symposia should enhance awareness of concerns vital to each group's interests. This is a significant challenge which has yet to be confronted on a useful scale. Already there have been some efforts, such as the international workshop on the problem of overabundant animals (Jewell et al. 1981). Cooperation will also permit the pooling of resources necessary to test the efficacy of humane alternatives to conventional, but disputed animal management techniques. Success in achieving mutually acceptable solutions will depend on strong and enlightened leadership on both sides, which shares the conviction that lasting solutions to complex problems cannot be found in isolation.

If we are truly concerned about the welfare of wild animals, we must also begin to question our own behavior. For example, some animal rights/welfare advocates have expressed their concern for animals by attempting to thwart recreational hunters (Relnecke 1983). Yet, in comparison with other, more "subtle" human activities, properly regulated hunting has little impact on animal welfare or wildlife conservation efforts. In fact, the long-term goals of conservationists and animal liberationists would best be served by radical changes in human life-styles (Ehrlich and Ehrlich 1981; Myers 1981b). The two greatest challenges facing preservationists today are: (1) the selfish, materialistic, and often wasteful attitudes prevalent in developed nations, like the United States, and (2) rampant population growth in the so-called Third World countries (Ehrlich 1980). Movement toward a steady state economy and zero population growth would do more for the welfare of animals than all our other efforts combined. If current trends in human population growth and habitat destruction continue, we could lose nearly one million species of animals and plants in the next two decades alone (Myers 1979, 1985; Ehrlich and Ehrlich 1981). This fact makes it even more essential that those who care about wildlife and nature, and those who care about the rights of individual animals close ranks to do battle with a common enemy.

Acknowledgements

We thank D.P. Barash, W. Conway, C. Crockett, P. Hardiston, G. Orians, and R.D. Taber for reviewing and commenting on the manuscript.

Endnotes

[1] Curatorial Intern, Department of Mammalogy, New York Zoological Society, Bronx, NY 10460.

[2] Assistant Director for Conservation and Captive Breeding, Conservation and Research Center, National Zoological Park, Smithsonian Institution, Front Royal, VA 22630.

References

Allen, DL, Erickson, L, Hall, ER and Shirra, WM. 1981. A review and recommendations on animal problems and related management needs in units of the National Park System. Report to the Secretary of the Interior. Mimeo, 18 pp.

Anderson, TE. 1971. Identifying, evaluating, and controlling wildlife damage. In: Giles, RH. ed. *Wildlife Management Techniques*. Washington, DC: The Wildlife Society.

Baker, JK and Reeser, DW. 1972. Goat management problems in Hawaii Volcanoes National Park. *U.S. Dept. Int. Nat. Res. Rep.* 2: 1-22.

Bart, J, Allee, D and Richmond, M. 1979. Using economics in defense of wildlife. *Wildl. Soc. Bull.* 7(3): 139-44.

Beach, FA. 1975. Beasts before the bar. In: Ternes, A. ed. *Ants, Indians, and Little Dinosaurs*. New York: Charles Scribner's Sons.

Blackstone, WT. 1978. Is there an environmental ethic? In: Blackstone, WT. ed. *Philosophy and Environmental Crisis*. Athens, GA: Univ. of Georgia Press.

Blonston, G. 1983. Where nature takes its course. *Science 83*. 4(9): 45-55.

Bogue, G and Ferrari, M. 1976. On the predatory "training" of captive-reared pumas. *Carnivore*. 3(1): 36-45.

Brambell, MR. 1977. Reintroduction. *Int. Zoo Ybk*. 17: 112-16.

Buechner, HK and Dawkins, HC. 1961. Vegetation change induced by elephants and fire in Murchison Falls National Park, Uganda. *Ecology*. 42(4): 752-66.

Bunnell, FL. 1978. Horn growth and population quality in Dall sheep. *J. Wildl. Manage*. 42: 764-75.

Callicott, B. 1980. Animal liberation: A triangular affair. *Env. Ethics*. 2: 311-38.

Carothers, SW, Stitt, ME and Johnson, RR. 1976. Feral asses on public lands: Analysis of biotic impact, legal considerations and management alternatives. In: *Trans. 41st North Am. Wildl. Conf.* pp. 396-406.

Caughley, G. 1970. Wildlife management and the dynamics of ungulate populations. *Appl. Biol.* 1: 183-246.

—. 1976. The elephant problem—an alternative hyphothesis. *E. Afr. Wildl. J.* 14: 265-83.

—. 1981. Overpopulation. In: Jewell, PA and Holt, S. eds. *Problems in Management of Locally Abundant Wild Mammals*. New York: Academic Press.

Chalmers, GA and Barrett, MW. 1977. Capture myopathy in pronghorns in Alberta. *J. Am. Vet. Assoc.* 171: 918-23.

Chase, A. 1986. *Playing God in Yellowstone*. Boston: The Atlantic Monthly.

Cheatum, EL. 1951. Disease in relation to winter mortality of deer in New York. *J. Wildl. Manage*. 15: 216-20.

Christian, JJ and Davis, DE. 1964. Endocrines, behavior, and population. *Science*. 146: 1550-60.

Coblentz, B. 1978. The effects of feral goats on island ecosystems. *Biol. Conserv*. 13: 279-86.

Coe, M. 1980. African wildlife resources. In: Soule, ME and Wilcox, BA. eds. *Conservation Biology: An Evolutionary Ecological Perspective*. New York: Sinauer.

Cole, GF. 1971. An ecological rationale for the natural or artificial regulation of native ungulates in parks. *Trans. North Am. Wildl. Res. Conf.* 36: 417-25.

Constan, KJ. 1972. Winter food and range use of three species of ungulates. *J. Wildl. Manage.* 36(4): 1068-76.

Courtney, WR. 1978. The introduction of exotic organisms. In: Brokaw, HP. ed. *Wildlife and America.* Washington, DC: U.S. Fish and Wildlife Service.

Courtney, WR and Ogilvie, VE. 1971. Species pollution. *Anim. Kingdom.* 74(2): 22-28.

Cumming, DH. 1981. The management of elephant and other large mammals in Zimbabwe. In: Jewell, PA and Holt, S. eds. *Problems in Management of Locally Abundant Wild Mammals.* New York: Academic Press.

Dasmann, RF. 1978. Wildlife and ecosystems. In: Brokaw, HP. ed. *Wildlife and America.* Washington, DC: Council for Environmental Quality.

Davis, DE. 1961. Principles of population control by gametocides. *Trans. North Am. Wildl. Nat. Res. Conf.* 26: 160-67.

deVos, A and Petrides, GA. 1967. Biological effects caused by terrestrial vertebrates introduced into non-native environments. In: *Towards a New Relationship of Man and Nature in Temperate Lands. Part III: Changes Due to Introduced Species.* Morges, Switzerland: International Union for Conservation of Nature and Natural Resources (IUCN).

Domalain, J. 1977. Confessions of an animal trafficker. *Nat. Hist.* 86: 54-67.

Eaton, RL. 1972a. An experimental study of predatory and feeding behavior in the cheetah (*Acinonyx jubatus*)*Zeit. Tierpsychol.* 31: 270-80.

—. 1972b. Predatory and feeding behavior in adult lions: The deprivation experiment. *Zeit. Tierpsychol.* 31: 46173.

Ehrenfeld, DW. 1972. *Conserving Life on Earth.* New York: Oxford University Press.

—. 1976. The conservation of non-resources. *Amer. Sci.* 64(6): 648-56.

Ehrlich, PR. 1980. The strategy of conservation, 1980-2000. In: Soule, ME and Wilcox, BA. eds. *Conservation Biology: An Evolutionary-Ecological Perspective.* New York: Sinauer.

Ehrlich, P and Ehrlich, A. 1981. *Extinction.* New York: Random House.

Feinberg, J. 1978. Human duties and animal rights. In: Morris, RK and Fox, MW. eds. *On the Fifth Day: Animal Rights and Human Ethics.* Washington, DC: The Humane Society of the United States.

Fenner, F and Ratcliffe, FN. 1965. *Myxomytosis.* Cambridge, England: Cambridge University Press.

Fenner, F. 1980. A welcome extinction. *Env. Conserv.* 7(3): 174-75.

Fox, MW. 1978. Man and nature: Biological perspectives. In: Morris, RK and Fox, MW. eds. *On the Fifth Day: Animal Rights and Human Ethics.* Washington, DC: The Humane Society of the United States.

—. 1979. Animal rights and nature liberation. In: Paterson, D and Ryder, RD. eds. *Animal Rights—A Symposium.* Sussex, England: Centaur Press.

Franzmann, AW and LaResche, RE. 1978. Alaskan moose blood studies with emphasis on condition evaluation. *J. Wildl. Manage.* 42: 334-51.

Garrett, MG and Franklin, WL. 1983. Diethylstilbestrol as a temporary chemosterilant to control black-tailed prairie dog populations. *J. Range Manage.* 36(6): 753-56.

Gould, SJ. 1982. Non-moral nature. *Nat. Hist.* 91(2): 19-26.

Grandy, JW. 1982. The Everglades deer massacre... and its aftermath. *The Humane Society News.* 27(4): 34-35.

Gunn, AS. 1980. Why should we care about rare species? *Env. Ethics.* 2: 17-37.

Haigh, JC, Parker, IS, Parkinson, DA and Archer, AL. 1979. An elephant extermination. *Env. Conserv.* 6(4): 305-10.

Hamann, O. 1979. Regeneration of vegetation on Santa Fe and Pinta Islands, Galapagos after the eradication of goats. *Biol. Conserv.* 15: 215-36.

Hanley, TA and Brady, WW. 1977. Feral burro impact on a Sonoran Desert range. *J. Range Manage.* 30: 374-77.

Hansen, RM. 1980. Habitat. In: Monson, G and Sumner, L. eds. *The Desert Bighorn.* Tucson, AZ: Univ. of Arizona Press.

Hardin, G. 1974. Living in a lifeboat. *Bioscience.* 24: 561-68.

—. 1978. Political requirements for preserving our common heritage. In: Brokaw, HP. ed. *Wildlife and America.* Washington, DC: Council on Environmental Quality.

Hoeffs, M. 1974. Food selection by Dall's sheep(*Ovis dalli dalli* Nelson). In: Geist, V and Walther, F. eds. *The Behaviour of Ungulates and Its Relation to Management.* Vol. II. Morges, Switzerland: IUCN.

Holdgate, MW. 1967. The influence of introduced species on the ecosystems of temperate oceanic islands. In: *Towards a New Relationship of Man and Nature in Temperate Lands. Part III: Changes Due to Introduced Species.* Morges, Switzerland: IUCN.

Hope, JE. 1974. Hunters: Useful pruners of nature or just killers? *Smithsonian.* 4(10): 78-82.

Houston, DB. 1971. Ecosystems of national parks. *Science.* 172: 648-51.

—. 1982. *The Northern Yellowstone Elk: Ecology and Management.* New York: MacMillan.

Howard, WE. 1964. Introduced browsing animals and habitat stability in New Zealand. *J. Wildl. Manage.* 28(3): 42129.

—. 1967. Ecological changes in New Zealand due to introduced animals. In: *Towards a New Relationship of Man and Nature in Temperate Lands. Part III: Changes Due to Introduced Species.* Morges, Switzerland: IUCN.

Humane Society News. 1983. The hunt that wasn't: The HSUS plays a major role in halting the National Zoo deer hunt. 28(1): 8-9.

Hutchins, M, Stevens, V and Atkins, N. 1982. Introduced species and the issue of animal welfare. *Int. J. Stud. Anim. Prob.* 3(4): 318-36.

Iker, S. 1983. Swamped with deer. *Natl. Wildl.* 21(6): 5-11.

Janzen, DH. 1978. New horizons in the biology of plant defenses. In: Rosenthal, GA and Janzen, DH. eds. *Herbivores: Their Interaction with Secondary Plant Metabolites.* New York: Academic Press.

Jewell, PA, Holt, S and Hart, D. eds. 1981. *Problems in Management of Locally Abundant Mammals.* New York: Academic Press.

Johnson, E and Tait, AJ. 1983. Prospects for the chemical control of reproduction in the Grey squirrel. *Mammal Rev.* 13: 167-72.

Kenelly, JJ, Johns, BE and Garrison, MV. 1970. Fecundity of a Norway rat colony with 85% sterile males. Ann. Meet. of the Soc. for the Study of Reproduction. Abstract.

Klein, DR. 1981. The problems of overpopulation of deer in North America. In: Jewell, PA and Holt, S. eds. *Problems in Management of Locally Abundant Wild Mammals.* New York: Academic Press.

Laws, RM. 1970. Elephants as agents of habitat and landscape change in East Africa. *Oikos.* 21: 1-15.

—. 1974. Behaviour, dynamics and management of elephant populations. In: Geist, V and Walther, F. eds. *The Behaviour of Ungulates and Its Relation to Managment.* Vol. II. Morges, Switzerland: IUCN.

Laycock, G. 1966. *The Alien Animals.* New York: Natural History Press.

—. 1974. Dilemma in the desert: Burros or bighorns? *Audubon.* 76(5): 116-77.

—. 1984. A scourge of goats. *Audubon.* 86(1): 10003.

Leopold, A, Sowls, LK and Spencer, DL. 1947. A survey of overpopulated deer ranges in the United States. *J. Wildl. Manage.* 11: 162-77.

Leopold, A. 1949. *A Sand County Almanac.* New York: Oxford University Press.

Leyhausen, P. 1973. On the function of the relative hierarchy of moods (as exemplified by the phylogenetic and ontogenetic development of prey-catching in carnivores). In: Lorenz, K and Leyhausen, P. eds. *Motivation of Human and Animal Behavior* New York: Van Nostrand Reinhold.

Lopez, B. 1978. *Of Wolves and Men.* New York: Charles Scribner's Sons.

Lyons, G. 1978. Politics in the woods. *Harper's.* 257: 27-38.

Madson, C. 1986. To feed or not to feed. *Audubon.* 88(2): 22-27.

March, BE. 1984. Bioethical problems: Animal welfare, animal rights. *Bioscience.* 34(10): 615-20.

Matsche, GH. 1976. Non-efficiency of mechanical birth control devices for white-tailed deer. *J. Wildl. Manage.* 40(4): 792-95.

—. 1977a. Micro-encapsulated diethylstilbestrol as an oral contraceptive in white-tailed deer. *J. Wildl. Manage.* 41(1): 87-91.

—. 1977b. Fertility control in white-tailed deer by steroid implants. *J. Wild. Manage.* 41(4): 731.

—. 1980. Efficacy of steroid implants in preventing pregnancy in white-tailed deer. *J. Wildl. Manage.* 44(3): 756-58.

McCullough, DR. 1979. *The George Reserve Deer Herd:Population Ecology of a K-selected Species.* Ann Arbor, MI: Univ. of Michigan Press.

Meagher, M. 1982. An outbreak of pinkeye in bighorn sheep: A preliminary report. *Bienn. Symp. North Wild Sheep and Goat Council.* 3: 198-201.

Midgley, M. 1983. *Animals and Why They Matter.* Athens, GA: Univ. of Georgia Press.

Miller, DR. 1974. Seasonal changes in the feeding behaviour of barren-ground caribou on the Taiga winter range. In: Geist, V and Walther, F. eds. *The Behaviour of Ungulates and Its Relation to Management.* Vol. II. Morges, Switzerland: IUCN.

Miller, RI. 1979. Conserving the genetic integrity of faunal populations and communities. *Env. Conserv.* 6(4): 297-304.

Montague, JJ. 1981. His "crop" is crocodiles. *Int. Wildl.* 11(2): 21-28.

Myers, N. 1979. *The Sinking Ark.* New York: Permagon Press.

—. 1980. The problem of disappearing species: What can be done? *Ambio.* 9: 229-35.

—. 1981a. A farewell to Africa. *nt. Wildl.* 11(6): 36-46.

—. 1981b. The hamburger connection: How Central America's forests become North America's hamburgers. *Ambio.* 10: 38.

—. 1985. The end of the lines. *Nat. Hist.* 94(2): 2-12.

National Wildlife. 1971. Hunters and conservationists share goals. 9(6): 18-19.

Newsweek. 1981. Wild burros still under fire.97(15): 17-18.

Nietschmann, B. 1977. The Bambi factor. *Nat. Hist.* 86(7): 84, 86, 87.

Noonan, PF and Zagata, MD. 1982. Wildlife in the market place: Using the profit motive to maintain wildlife habitat. *Wildl. Soc. Bull.* 10(1): 46-49.

Noy-Meir, I. 1981. Responses of vegetation to the abundance of mammalian herbivores. In: Jewell, PA and Holt, S. eds. *Problems in Management of Locally Abundant Wild Mammals.* New York: Academic Press.

Owaga, ML. 1977. Comparison of analysis of stomach contents and fecal samples from zebra. *E. Afr. Wildl. J.* 15: 21722.

Parker, GR. 1981. Physical and reproductive characteristics of an expanding woodland caribou population (*Rangifer tarandus tarandus*) in northern Labrador. *Can. J. Zoo.* 59: 192940.

Peek, JM. 1980. Natural regulation of ungulates (What constitutes a real wilderness?). *Wildl. Soc. Bull.* 8(3): 217-27.

Peterson, RL. 1955. *North American Moose.* Toronto: Univ. of Toronto Press.

Pianka, ER. 1983. *Evolutionary Ecology.* 2nd edition. New York: Harper and Row.

Pienaar, U. de V. 1969. Why elephant culling is necessary. *Afr. Wildl.* 23: 181-84.

Poche, RW. 1980. Elephant management in Africa. *Wildl. Soc. Bull.* 8(3): 199-207.

Polunin, N and Eidsvik, HK. 1979. Ecological principles for the establishment and management of national parks and equivalent resources. *Env. Conserv.* 6(1): 21-26.

Poole, DA and Trefethen, JB. 1978. The maintenance of wildlife populations. In: Brokaw, HP. ed. *Wildlife and America.* Washington, DC: Council on Environmental Quality.

Rasmussen, DI. 1941. Biotic communities of Kaibab Plateau, Arizona. *Ecol. Monogr.* 11: 230-75.

Regan, T. 1983. *The Case for Animal Rights.* Berkeley, CA: Univ. of California Press.

Reiger, G. 1978. Wild boars, burros, horses cause park service apoplexy. *Audubon.* 80(3): 119-22.

Relnecke, B. 1983. Hunters become the hunted: Animal lovers turn the tables. *The Seattle Times.* Wednesday, March 16.

Remfrey, J. 1978. Control of feral cat populations by long term administration of megastrol acetate. *Vet. Rec.* 28: 403-04.

Robbins, J. 1984. Do not feed the bears? *Nat. Hist.* 93(1): 12-21.

Robinson, WL, Fanter, LH, Spalding, AG and Jones, SL. 1980. Biological aspects of political mismanagement of white-tailed deer in Pictured Rocks National Seashore. *Proc. 2nd Conf. on Scientific Res. in National Parks.* pp.283-92.

Rodman, J. 1977. The liberation of nature? *Inquiry.* 20: 83-131.

Ryder, RD. 1978. Postscript: Towards humane methods of identification. In: Stonehouse, B. ed. *Animal Marking: Recognition Marking of Animals in Research.* London: MacMillan Press Ltd.

Sanderson, GP. 1896. *Thirteen Years Among the Wild Beasts of India.* London: W.H. Allen and Co.

Schectman, SM. 1978. The "Bambi syndrome": How NEPA's public participation in wildlife management is hurting the environment. *Env. Law.* 8: 611-43.

Seal, US, Barton, R, Mather, L, Olberding, K, Plotga, BD and Gray, CW. 1976. Hormonal contraception in captive female lions (*Panthera leo*). *J. Zoo Anim. Med.* 7(4): 12-20.

Shaw, WW. 1974. Meanings of wildlife for Americans: Contemporary attitudes and social trends. *Trans. North Am. Wildl. Nat. Res. Conf.* 39: 151-55.

Sinclair, AR. 1974. The natural regulation of buffalo populations in East Africa. II. Reproduction, recruitment and growth. *E. Afr. Wildl. J.* 12: 169-83.

—. 1981. Environmental carrying capacity and the evidence for overabundance. In: Jewell, PA and Holt, S. eds. *Problems in Management of Locally Abundant Wild Mammals.* New York: Academic Press.

Singer, P. 1975. *Animal Liberation.* New York: Avon Books.

—. 1981. *The Expanding Circle: Ethics and Sociobiology.* New York: Farrar, Straus, and Giroux.

Soule, ME, Wilcox, BA and Holtby, C. 1979. Benign neglect: A model of faunal collapse in the game reserves of East Africa. *Biol. Conserv.* 15: 259-72.

Soule, ME and Wilcox, BA. 1980. Conservation biology: Its scope and its challenge. In: Soule, ME and Wilcox, BA. ed. *Conservation Biology: An Evolutionary-Ecological Perspective.* Sunderland, MA: Sinauer.

Spraker, TR. 1977. Capture myopathy of Rocky Mountain bighorn sheep. In: *Trans. Desert Bighorn Sheep Council 1977.* pp. 14-16.

—. 1978. Pathophysiology associated with capture of wild animals. In: Montali, RJ and Migaki, G. eds. *The Comparative Pathology of Zoo Animals. Symp. Natl. Zool. Park.* Washington, DC: Smithsonian Institution.

Stelfox, JG. 1976. Immobilizing bighorn sheep with succinycholine chloride and phencyclidine hydrochloride. *J. Wildl. Manage.* 40(1): 174-76.

Stevens, V. 1983. The dynamics of dispersal in an introduced mountain goat population. Ph.D. Dissertation. Univ. of Washington, Seattle, WA.

Stocker, J. 1980. Battle of the burro. *Natl. Wildl.* 19(5): 14-16.Stone, CD. 1975. *Should Trees Have Standing?* New York: Discus Books.

Stonehouse, B. 1981. *Saving the Animals.* New York: MacMillan.

Taber, RD and Cowan, IMcT. 1971. Capturing and marking wild animals. In: Giles, RH. ed. *Wildlife Management Techniques.* Washington, DC: Wildlife Society.

Talbot, LM. 1980. The world's conservation strategy. *Env. Conserv.* 7(4): 259-68.

Taylor, WP and Hahn, HC. 1947. Die-offs among white-tailed deer in the Edwards Plateau of Texas. *J. Wildl. Manage.* 11(4): 317-23.

Tennesen, M. 1975. Bighorn on the run. *Natl. Wildl.* 13(6): 4-10.

Tyndale-Biscoe, CH and Smith, RF. 1969. Studies on the marsupial glider, *Schoinobates volans* (Kerr). III. Response to habitat destruction. *J. Anim. Ecol.* 38: 65159.

Turner, JW. 1984. Given a free rein, prolific mustangs gallop into trouble. *Smithsonian.* 14(11): 88-96.

U.S. Dept. of the Interior. 1980. *Feral Burro Management and Ecosystem Restoration Plan and Final Environmental Assessment.* Grand Canyon National Park: National Park Service.

Van Wyk, P and Fairall, N. 1969. The influence of the African elephant on the vegetation of the Kruger National Park with special reference to tree and shrub strata. *Koedoe.* 12: 57-89.

Walters, JE and Hansen, RM. 1978. Evidence of feral burro competition with desert bighorn sheep in Grand Canyon National Park. In: *Trans. Desert Bighorn Sheep Council 1978.* pp. 10-16.

Wardle, RE. 1974. Influence of introduced mammals on the forest and shrublands of the Grey River Headwaters. *NZJ Sci.* 4(3): 459-86.

Warner, RE. 1963. Recent history and ecology of the Laysan duck. *Condor.* 65: 3-23.

White, L. 1967. The historical roots of our ecologic crisis. *Science.* 155: 1203-07.

Whitlock, B. 1978. Seasonal use of ovaban as a means of contraception in the Kodiak bear. *Proceedings of the Annual Meeting of the American Association of Zoo Veterinarians, 1978.* pp. 178-80.

Wilson, DE and Hirst, SM. 1977. Ecology and factors limiting roan and sable antelope populations in South Africa. *Wildl. Monogr.* 54: 1-111.

Wilson, EO. 1984. *Biophilia: The Human Bond with Other Species.* Cambridge, MA: Harvard University Press.

Winters, JF. 1980. Summer habitat and food utilization by Dall's sheep and their relation to body and horn size. M.S. Thesis. Univ. of Alaska, College, AK.

Wright, P. 1978. Why must we euthanize? *The Humane Society News.* 23(11): 24-25.

Younghusband, P and Myers, N. 1986. Playing God with nature. *Intl. Wildl.* 16(4): 4-13.

Zwank, PJ. 1981. Effects of field laparotomy on survival and reproduction in mule deer. *J. Wildl. Manage.* 45(4): 972-75

WILDLIFE AND NATURE LIBERATION

Michael W. Fox[1]

Opposing World Views

Some people today with a wildlife management orientation who do not question the ethics of exploiting animals on a sustainable basis, have a world view that is the antithesis of those who oppose the killing of all wildlife. This preservationist view, which is endorsed by many animal rightists, stands in opposition to the conservationist's recognition of the need to monitor wildlife populations and at times, violate the rights of individual animals by killing or relocating them in order to maintain the integrity of their habitat-sanctuary, not for the benefit of humans, but for the benefit of all species and individual animals therein. This is very different from the management mentality where species and ecosystems are manipulated and exploited primarily for human gain. Sometimes the line between such human-centered management practices and conservation for the sake of the animals is unclear. Hence, conservationists and deep ecologists dedicated to protecting wildlife may be misjudged by humanitarians and animal rightists as being on the side of wildlife management, placing human interests before those of the animals, when their killing is encouraged. Likewise, humanitarians and animal rightists may be misjudged as being unrealistic, anthropocentric, and ignorant, especially when they fear, for example, that to condone the killing of wild animals could lead to a kind of "ecological fascism," where conservationists play God, violate the rights of animals and do not let nature take care of things. But while nature knows best, many wildlife habitats and sanctuaries are no longer natural because of human interference, ranging from adjacent farming, forestry, dam construction, acid rain, etc. Hence, the need to monitor ecosystems and all species therein is a part of responsible stewardship.

Humane ethics—animal welfare—and animal rights are not incompatible with ecologically sound wildlife stewardship. They are an integral part of it, from treating wildlife for necessary research purposes humanely, to finding humane ways to control the populations of species that are out of balance and thus threatening the viability of other species and the diversity and integrity of the ecosystem. That mistakes may be made in stewardship-management policies is inevitable. It is, for instance, difficult to know if the sudden abundance of one or more species and the dwindling of others is part of the natural process of succession and should be allowed to continue, or if these changes are abnormal and should be corrected. Perhaps the best that can be done with our present knowledge and expertise is to "freeze" many wildlife sanctuaries by endeavoring to maintain optimal species diversity and numbers. Clearly, in any of our actions, we should take the conservative, cautious approach so that if we err we can correct our errors before irreparable harm is done. Wildlife ecologist-conservationists and deep ecologists who are insensitive to legitimate animal rights and welfare concerns need to be confronted. And likewise those animal liberationists who take animal rights philosophy too far and lose sight of the ecological principles of sound steward-ship and of the rights and interests and subsistence needs of indigenous peoples.

Deep Ecology and Animal Rights

In their recent book entitled *Deep Ecology: Living as if Nature Mattered*, authors Bill Devall and George Sessions (1985) reject animal rights philosophy and vegetarianism. They state, "mutual predation is a biological fact of life" and criticize "animal liberationists who attempt to side-step this problem by advocating vegetarianism are forced to say that the entire plant kingdom including rain forests have no right to their own existence." Yet it is the rain forests that are being destroyed in part by the beef cattle industry in South and Central America, much of which is exported to the United States. It is narrow-minded for deep ecologists not to endorse vegetarianism because of its ecological significance. Devall and Sessions seem blind to the fact that raising livestock and poultry and the food for these animals propagated for human consumption, entails a massive displacement of wildlife. It has been estimated that within the next thirty to forty years, 40% of the total biomass of animal life on Earth will be comprised of people and domesticated animals, particularly cattle. Vegetarianism, or at least a drastic reduction in meat production and consumption should be an integral aspect of the deep ecology movement as it is now of the animal rights movement. Here lies one area of common ground between animal rightists and deep ecologists, both of whom are concerned about the impact of agribusiness and meat consumption upon wildlife and their habitats.

Some animal rightists contend that farm animals have a right not to be eaten. This does smack of anthropomorphism and alienates the deep ecologist who sees predation as natural and farm animals as prey species. That humani-tarian animal rightists are also concerned about farm animal welfare need not set them apart from deep ecologists unless the latter see such concern as trivial sentimentality and of lesser priority than more global ecological concerns. The abusive treatment of animals is no different from abusive treatment of Nature: both are symptoms of a lack of reverence for the sanctity and dignity of the life of the individual and of life as a whole.

Hence, I see animal rights philosophy and "ecosophy," the philosophy of deep ecology, as two sides of the same coin of a new currency: A new dialectic where the dualities of individualism and holism—specifically con-cern for the rights of the individual and for the integrity of the biospheric whole—are reconciled.

The authors uncritically cite deep ecologist John Rodman who concludes that the animal rights movement "while holding out promise of transcending the homocentric perspective of modern culture, subtly fulfills and legitimizes the basic project of modernity—the total conquest of nature by man." Then they take to task animal rights philosopher Tom Regan, who with others of like mind, "have expressed concern that a holistic ecological ethic (such as Leopold's land ethic) results in a kind of totalitarianism or ecological fascism."

It disturbs me that this otherwise excellent book has taken such a negative attitude toward animal rights philosophy. George Sessions, however, does suggest that philosophers do need to work toward nontotalitarian solutions to environmental problems and that "in all likelihood, this will require some kind of holistic ecological ethic in which the integrity of all individuals (human and non-human) is respected" (Appendix H of the book). I interpret "integrity" as rights and sanctity.

It is ironic that while the authors are so critical of the animal rights movement, they quote Arne Naess (who coined the terms ecosophy and deep ecology and is arguably the founder of the deep ecology movement) who expresses many of the views of the animal rights movement. For instance, Naess (1973) states, "The intuition of biocentric equality," [what I term trans-species democracy] "is that all things in the biosphere have an equal right to live and blossom and to reach their own forms of unfolding and self-realization...." He also observes that "with maturity, human beings will experience joy when other life forms experience joy and sorrow when other life forms experience sorrow. Not only will we feel sad when our brother or a dog or a cat feels sad, but we will grieve when living beings, including landscapes, are destroyed.... Only a very narrow range of feelings have interested most human beings until now"(Naess 1973).

The depth of feeling and empathetic awareness for other living things that Naess sees us acquiring "with maturity" is expressed in these words of Australian aborigine Bill Neidjie (1985):

> Feeling all these trees, all this country: When this wind blow you can feel it. Same for country ... you feel it. You can look, but feeling ... that make you.
>
> If you feel sore ... headache, sore body, that mean somebody killing tree or grass. You feel because your body (is) in that tree or earth. Nobody can tell you, you got to feel it yourself.

This "primitive" aboriginal's "maturity," relatively speaking, suggests a regressive or retarded condition of ego development in contemporary *Homo sapiens*.

Lack of feeling—empathy and compassion—for animals and other living things on the one hand, and for the integrity of biospheric ecosystems (Nature) on the other, are two sides of the same coin whose currency is the root and source of the holocaust of the animal kingdom and of the "progressively" destructive transformation of Nature into an industrialized wasteland.

Had Devall and Sessions a better understanding of the ethical, ecological, and wildlife conservation reasons for vegetarianism, they might have taken a step further into the domain of animal rights and considered also the issue of vivisection. They probably regard the exploitation of laboratory animals as a legitimate form of natural human predation, which is a logical and often-used extension of the unquestioned acceptance of humans exploiting animals for food, fur, and other resources. Ironically, while they recognize human overpopulation as a critical issue they are non-critical of humans living as predators, which is a relatively unnatural situation. All natural predators are far fewer in number than the prey they exploit. If the authors had been less ready to dismiss animal rights philosophy and push their own, they might have come out against the wholesale exploitation of laboratory animals because such research generally fails to address the ecological and environmental factors responsible for many human diseases. The connection between deep ecology and holistic medicine is an important one that the authors fail to make.

"Deep" ecologists, who support the philosophy of preserving the natural abundance and diversity of plants and animals in natural ecosystems, find common ground with animal rightists in opposing the "harvesting" of any living species for primarily human benefit when the natural abundance and diversity of life within an ecosystem are disrupted or threatened. The holistic philosophy of "deep" ecology and the more specific, individual and species-focused philosophy of animal rights are complementary and are opposed to the industrialized transformation of natural ecosystems into systems that primarily and most often exclusively provide limited monotypic benefit to human beings at the expense of Nature's diversity and abundance of plant and animal species. These philosophies should also recognize the absolute right of all life, human and non-human alike, to a whole and healthy environment. That industrialized society is impoverishing and poisoning the environment, destroying natural ecosystems, and forcing thousands of animal and plant species into extinction necessitates a stronger coalition between the conservation and humane, animal welfare movements. And this is happening. With the advent of animal rights and deep ecology philosophies, the ideological differences between conservationists and animal welfarists are giving way to a shared ideology and goal of Nature Liberation—which include respect for both the environment and wild plants and animals as communities and as individuals with interests, inherent value, and rights.

As Chief Seattle said over a century ago, "This we know—the earth does not belong to man, man belongs to the earth. All things are connected like the blood which unites one family. Whatever befalls the earth befalls the sons of the earth. Man did not weave the web of life; he is merely a strand in it. Whatever he does to the web, he does to himself."

More recently, the Six Nations of American Indians (Hau de no sau nee) in a proclamation made in Geneva entitled *A Basic Call to Consciousness*, stated, "The people who are living on this planet need to break with the narrow concept of human liberation, and begin to see liberation as something which needs to be extended to the whole of the Natural World. What is needed is the liberation of all the things that support life—the air, the water, the trees—all the things which support the sacred web of life." Linked with Nature Liberation philosophy is an emerging creation-centered spirituality which emphasizes man's creative participation and role as a responsible planetary steward (see Matthew Fox 1979 and 1983).

Charles Birch in an address in Nairobi before the World Council of Churches a decade ago observed that, "... It is a cockeyed view that regards ecological liberation as a distraction from the task of liberation of the poor. One cannot be done without the other. It is time to recognize that the liberation movement is finally one movement... all creatures are fellow creatures and human responsibility extends infinitely to the whole of creation... if we are to continue to inhabit the earth, there has to be a revolution in the relationship of human beings to the earth and... to each other." The accelerating rate of extinction of unique plant and animal species of diverse ecosystems and of human societies that have lived for generations in a relatively stable if not creative harmony with Nature, attests to the fact that the fate of the Earth, of the animal kingdom, and of humankind are inseparably interconnected.

The principles and goals of animal liberation, conservation, and deep ecology movements are fundamentally complementary. From different perspectives—concern for individuals, species, and whole ecosystems—they converge upon the political and socioeconomic realities of the times but are as yet not consonant with the dominant world view of industrialized technocracies. Differences aside, the supreme task of these movements is to transform the prevailing world views of all nation states to one of enlightened planetary stewardship and respect for all living things.

Endnotes

[1] Scientific Director, The Humane Society of the United States, and, Director, The Institute for the Study of Animal Problems, 2100 L St, NW, Washington, DC 20037.

References

Devall, B and Sessions, G. 1985. *Deep Ecology: Living as If Nature Mattered.* Salt Lake City, UT: Gibbs M. Smith, Inc., Peregrine Smith Books.

Fox, Matthew. 1979. *A Spirituality Named Compassion.* Minneapolis, MN: Winston Press.

—. 1983. *Original Blessings.* Sante Fe, NM: Bear and Co.

Naess, A. 1973. The shallow and the deep, long-range ecology movement: A summary. *Inquiry.* 16: 95-100.

Neidjie, B, Davis, S and Fox, A. 1985. *Kakadu Man... Bill Neidjie.* New South Wales, Australia: Mybrood. P/L., Inc.

BRINGING US TOGETHER[1]

John W. Grandy[2]

First, I thank you for the opportunity to speak before this joint meeting of the Pennsylvania Academy of Science and the Pennsylvania Chapter of the Wildlife Society. It is a pleasure to address a group with whom my professional training is so compatible. I also appreciate the opportunity to discuss a topic of far more than professional interest: bringing the scientific, management, and animal protection communities closer together.

In recent years, much has been made of the differences between animal protection/welfare/rights and conservation. In simplistic terms, the difference is said to be between a view of wild animals as individuals and as populations. Some conservationists claim to see it as a waste to devote time and energy to ensuring the survival and health of individual animals. Conversely, others seem to take the view that the health and welfare of the individual animal is of highest importance.

But like many other discussions based on philosophical differences between largely compatible philosophies, the differences are far more apparent than real—and differences are often meaningless or fleeting, particularly at the day-to-day, real world level. First, many issues arise which should be of common interest to all of our professions, and upon which philosophical differences simply have no bearing. Furthermore, philosophical views, no matter how strongly held, are hardly static. Those held by individuals, like the aggregate views of society at large, change. They pulsate; they move: sometimes slightly, sometimes perceptibly; they swing like a pendulum or they trek over long periods in a reasonably predictable direction. For example, my views on a variety of animal welfare, animal rights, and ecological/conservation issues have changed over the years. I have no doubt that many of you have experienced changes in your attitudes as well. These changes mark us as vital, healthy people, that can accept change outside and inside ourselves.

I like to refer to this process of philosophical and sociological change as a kind of evolution. My observation is that much of the debate over the differences between the emerging animal welfare/rights/protection philosophy and the traditional conservation philosophy really reflects varying speeds of evolution of thought amongst largely sympatric people or groups. (Think, perhaps, in a conservation sense, of Leopold's evolution of thought [Leopold 1949]).

But let's go back to the most basic points of apparent divergence that I identified earlier. Animal protectionists are people who care for individual animals, their humane treatment, their welfare—their rights to basic protection from preventable suffering and cruelty. Conservationists care about ecosystems, ecological processes, and the survival and vitality of populations. But are these really distinct, and separable, on a day-to-day basis?

I doubt it.

First, does anyone know true conservationists or ecologists who favor inhumane treatment or cruelty to animals? Not likely! By the same token, does anyone know animal protectionists who do not care about the vitality of wildlife populations or the health of ecosystems. Of course not! Vital wildlife populations are made up of healthy viable individual animals.

And the conservation, animal protection, and scientific communities share very important mutual goals:
—Habitat preservation
—Viable, diverse populations of wildlife non-game/game
—Endangered and threatened species protection
—Maintenance of natural habitats
—Wetland protection/wilderness protection
—Humane treatment of animals
—Maintenance of species diversity on managed habitats
—Preventing avoidable suffering
—Pollution control
—Promoting urban wildlife

But with their commonality of goals and objectives, why all the fuss, why do people write books and scientific papers which pick apart philosophical differences and raise subtle distinctions—hardly visible to the general public—to new levels of divisiveness?

First, there are writings espousing new ideas concerning human responsibilities to provide for the stewardship and well-being of individual animals, populations, and ecosystems. These writings and concurrent discussions are a form of thoughtful, nonacrimonious debate among peers and peer groups (see e.g., Regan and Singer 1976; Elliot and Gare 1983; Attfield 1983; Regan 1984). This is a largely non-public debate. It is civil in tone and marked by mutual respect and admiration among participants. It is a kind of searching for new horizons, new directions, answers, and workable concepts. Not all participants agree—sometimes none agree—but that is the nature of the process. These discussions and writings provide the grist for change and necessary reconsideration as society, thought patterns, and individual values evolve.

Yet there is more to this debate than quiet respectful dialogue. The debate between animal protectionists, wildlife managers, and scientist/conservationists often seems to erupt in a violent public debate. Indeed, some participants focus on issues such as overpopulation and starvation, permanent habitat destruction, and other infrequent problems in an apparent attempt to elucidate differences, and belittle opposing philosophical views, rather than create unity and consensus. This debate leaves the public confused (why should they fight?) and often results in divisiveness that frustrates accomplishment of our common goals. What happens in these areas? Why this public animosity and name-calling? Why do we see some of these issues elevated to such a level that they cause major schisms within the effort to achieve mutual goals?

Some of these arguments are from seemingly self-appointed idealogues who focus on extremes or the unusual (see e.g., Hutchins and Wemmer, this volume). Their writings seem intent on creating intellectual division by emphasizing extremes. Sometimes I think they are only interested in "publishing." Their writings seem to have a chip on their shoulders. They seem to be saying: "We need to cooperate and work together—so long as you agree to see everything my way." While these writers are divisive and would do better to participate in nonacrimonious quiet discussion, rather than public debate, even their papers are relatively few and far between, and contribute only slightly to visible public separateness.

Indeed, there is far more to this debate than misguided egos satisfying the publish or perish doctrine.

Let me cite a couple of examples of major controversies.

For years, we have had an ongoing predator control program in the western United States. It annually consumes millions of dollars and has killed everything from wolves to bald eagles, from black-footed ferrets to coyotes. Its supposed aim is to protect livestock. In reality, however, its major accomplishment is to kill wildlife. Predators are the target, but anything else that gets in the way is acceptable. The truth about the predator control program is that it is a not very good, simplistic political solution to a complex problem. Yet, time and again, science and wildlife management are put forth as both the philosophical underpinning and the intricate rationale for the predator control program. This is ridiculous, as anyone who has examined the program knows. And for the wildlife management and scientific community to allow themselves to be used as the rationale for the program only detracts from public confidence in the professional community. In a very real way, the wildlife and scientific communities are being used by the commercial livestock industry as justification for their own subsidy.

Next, look at the commercial trapping programs in operation throughout the United States. Why are these programs conducted? They are conducted because trappers sell animal pelts for money. Yet time and again trapping is defended as being part of an essential wildlife management or public health program: habitat protection, disease control, rabies control, starvation prevention, and animal health.

These assertions are, *at best*, well-intentioned nonsense from people either commercially motivated or motivated by an overwhelming desire to see the "tradition" of trapping continue. At the trapping industry level, these assertions are part of a carefully orchestrated campaign to preserve the commercial fur and trapping industry. Whatever the true reason, however, trappers should argue their case on the basis of whatever its merits may be and leave wildlife and science out as a "justification." Scientists and wildlife managers, for their part, would be far better off to divorce themselves from this debate since wildlife managers and scientists have no professional stake in commercial trapping. As it stands, wildlife managers and scientists are allowing themselves to be used.

In short, as these examples illustrate, a primary reason for this visible divisiveness in our communities is that the professional and scientific communities are being used as pawns in a philosophical and political fight over exploitation of wildlife for commercial purposes. As scientists or wildlife managers there may occasionally be times when we are called upon to trap or control predators or other wildlife. But, clearly, modern day predator control and trapping are not programs conducted because wildlife managers and scientists have concluded that they are good and necessary.

My strong recommendation as a professional to professionals is for us to avoid being caught up in these largely economically motivated wars. As biologists and scientists ours is not to defend either commercial trapping, or the western livestock industry. We deserve more than to be implicitly or explicitly viewed as pawns. And what, you may ask, do we do when our professional responsibilities require that we do not deal with pure science, but must take positions on nonscientific philosophical issues. My advice is to do so in a way which preserves or enhances the ability to work together with our peers on other issues of mutual concern.

Humane ethics, animal protection, and animal rights are not incompatible with ecologically sound wildlife stewardship. They are an integral part of it, from treating wildlife for necessary research humanely, to habitat preservation and protection of endangered species. We should recognize and build on our similarities. We can start by agreeing to disagree on some basic philosophical issues when our views diverge, while recognizing the common need we all have to vigorously pursue and succeed in chieving our common goals.

Endnotes

[1] Paper presented at the Joint Annual Meeting of the Pennsylvania Academy of Science and the Pennsylvania Chapter of the Wildlife Society, April 18, 1986, Seven Springs, PA.

[2] Vice President, Wildlife and Environment, The Humane Society of the United States, 2100 L St, NW, Washington, DC 20037.

References

Attfield, R. 1983. *The Ethics of Environmental Concern.* New York: Columbia University Press. 220 p.

Devall, B and Sessions, G. 1985. *Deep Ecology: Living as If Nature Mattered.* Salt Lake City, UT: Gibbs M. Smith, Inc., Peregrine Smith Books. 266 p.

Elliot, R and Gare, A. eds. 1983. *Environmental Philosophy (A Collection of Readings).* University Park, PA: Pennsylvania State University Press. 303 p.

Leopold, A. 1949. *A Sand County Almanac and Sketches Here and There.* New York: Oxford University Press. 226 p.

Regan, T and Singer, P. eds. 1976. *Animal Rights and Human Obligations.* Englewood Cliffs, NJ: Prentice Hall. 250 p.

Regan, T. ed. 1984. *Earthbound (New Introductory Essays in Environmental Ethics.)* New York· Random House. 317 p.

ASSAULT ON EDEN:
DESTRUCTION OF LATIN AMERICA'S RAIN FORESTS

Douglas R. Shane[1]

> A day will come when man will discover an
> alphabet in the eyes of the chalcedonies,
> in the markings of the moth, and will learn
> in astonishment that every spotted snail has
> always been a poem.
> —Alejo Carpentier in *The Lost Steps*

In the seemingly distant world of Latin America's rain forests, man's greed and desperation have resulted in a fire which threatens to obscure our ability to observe life's poetry. The continuing destruction of the earth's tropical rain forests is one of the most serious environmental problems confronting humanity today. Intact, these vital organisms offer an understanding of the planet's past and a key to our future; destroyed, they threaten catastrophe of global consequence.

Nature's Cornucopia

Tropical forests encircle the earth in the equatorial regions of Central Africa, Southeast Asia, and Central and South America. Comprising some 55% of the world's woodlands, tropical forests yield a cornucopia of natural resources which benefit mankind. Among this bounty are commercial timbers, resins, and gums; important pharmaceuticals derived from both plants and animals; water for drinking, transportation, and hydroelectric power; minerals and petroleum; genetic resources for maintenance and improvement of food crops; habitat for a remarkable variety of indigenous and migratory wildlife; home to sadly threatened aboriginal peoples; and hundreds—if not thousands—of as yet undiscovered biological benefits.

Yet, the planet's tropical forests are not the well-tended repositories of natural treasures that one would assume them to be. Rather, they are being destroyed at an alarming rate. In recent decades, the equatorial jungles have been viewed by government planners and private sector developers as a panacea to such urgent national problems as sluggish economic growth and overpopulation.

Although figures vary depending upon the source, the Food and Agricultural Organization of the United Nations has estimated that globally some 50 acres of tropical forest are being devastated every 60 seconds, totalling 27 million acres annually. It is believed that Latin America has lost some 37% of its original tropical forests—largely within the last 30 years—and that a substantial portion of the remainder may be gone within the next 40 years, if not sooner, due to ever-accelerating development activities.

In size, the tropical forests of Latin America are the largest in the world. Ecologically, they are models of evolutionary precision. Although the lush vegetation of the tropical forest creates the illusion of unbounded fertility, in fact, the underlying soils are impoverished, most of their nutrients leached long ago by torrential rainfalls. Only through some 60 million years of evolution have the plants and animals of the rain forest perfected methods of capturing and storing essential nutrients before they are flushed into the region's ubiquitous river systems, and finally the Atlantic Ocean.

Virtually nothing is wasted in the complex environment of the tropical forest. With most of the nutrients stored in the vegetation, the forest is a closed ecosystem, the dead matter being quickly recycled by bacteria, fungi, insects, and other organisms. The vegetation which forms the canopy about 150 feet above the forest floor cushions the impact of the rain, protecting the thin layer of soil from erosion and solar radiation. Without the protective canopy, the tropical sun would bake the soil to a brick-like crust, incapable of supporting almost any plant or animal life.

Tropical Forests Under Siege

Pursuing a centuries-old tradition, the native peoples of Latin America's rain forests practice slash-and-burn agriculture, in which small areas of jungle are partially cleared and the cut vegetation is burned gradually to release stored nutrients into the poor soil. In this manner the land can be cultivated for several harvests of food crops until the nutrient supply is depleted. When the farmer moves on to clear new forest lands, the exhausted site is left fallow for eight or more years as secondary growth reclaims the area. Thus, people can—in limited numbers—sustain themselves within the environmental limitations of the tropical forest.

While dwindling aboriginal populations and Latin colonists still employ slash-and-burn agriculture, the last several decades have witnessed the advent of ever-accelerating development pressures in the region's tropical forests. National programs of extensive highway construction have opened wilderness areas to large-scale exploitation and destruction. Among the factors contributing to the decimation of Latin America's rain forest are forestry-related industries, spontaneous and government sponsored colonization by the landless poor of the nations involved, exploration and extraction of minerals and petroleum, flooding of large tracts of forest for hydroelectric projects, and efforts to convert forested lands for the production of agricultural crops and beef cattle. Many of these activities represent national development programs

Figure 1. Slash-and-burn agriculture, practiced throughout the humid tropics, is feasible if limited to small areas. (Rondonia, Brazil; photo © Douglas R. Shane)

which are supported with financial and technical assistance from the developed countries, international lending institutions, and transnational corporations.

The reasons for the massive deforestation of the humid tropics are relatively easy to assess but stubbornly difficult to resolve. The governments of Third World tropical countries, in an effort to solve their nation's problems of explosive population growth, high unemployment, and sluggish economies have tended, during the last 30 years, to view their tropical forests as short-term solutions to predicaments that are at once economic, social, and political. Compounding this dilemma is the fact that in virtually all tropical countries, the most productive soils are controlled by a small group of landowners and are utilized for the production of cash crops for export. While these crops—chiefly coffee, bananas, sugar cane, cotton, and beef—are important foreign exchange earners for struggling domestic economies, they also create serious problems for the nations involved. In addition to being subject to the capriciousness of the weather, and to price fluctuations on the world markets, export crops grown on fertile soils mean that these nations must import food items at greater expense for domestic consumption.

This type of agrarian system in which vast tracts of arable land are owned by a small number of individuals creates a dire shortage of land ownership and employment opportunities for the burgeoning populations of each country. Thus, despite accumulating evidence that they are unproductive under "modern" farming practices, many governments continue to consider their tropical forests as essential for the settlement of colonists and as new areas for intensive agricultural projects.

Impending Apocalypse?

Extensive tropical deforestation has already proven to cause serious regional environmental consequences such as the deterioration of watersheds, floods, droughts, soil erosion, and the resultant siltation of rivers and lakes which, in several instances, has already diminished the capacities of hydroelectric facilities in various countries. In recent years, a number of hypothetical—but nevertheless frightening—postulates have been advanced by some scientists in which the widespread felling of the earth's tropical forests is related to the global environment.

Among these warnings is the possible effect that tropical deforestation may have upon the carbon dioxide levels in the earth's atmosphere. While CO_2 levels have been rising rapidly due to carbon released by combustion of fossil fuels, proponents of this postulate maintain that increased levels of CO_2 released from burned forests could further reduce the atmosphere's ability to reflect heat away from the earth's surface. This phenomenon, climatologists warn, could cause a global warming trend, which in turn could lead to a shifting of suitable agricultural areas and a decrease in presently productive lands.

Other "global effect" arguments postulate the following alarming scenarios: if the earth undergoes a warming trend, the polar ice caps would melt, thus raising sea level and inundating coastal cities; the nutrients leached from tropical forest vegetation and conveyed by river systems such as the Amazon to the oceans would seriously affect the productivity of marine ecosystems and impair fishing industries worldwide; and the large scale or total felling of the earth's tropical forests would so drastically lower global precipitation that an irreversible process of desertification would result in the grain belts of North America and other temperate regions.

Whatever the validity of the "worst case" scenarios, the existing knowledge pertaining to the biological importance of tropical rain forests makes their continuing destruction an issue of immediate concern to all of humanity. Among the myriad elements which comprise this global problem, two topics of immediate concern regarding the fate of Central and South America's rain forests are the effects of habitat destruction on indigenous and migratory wildlife and the beef cattle industry of Latin America, or, as it is more popularly known, "The Hamburger Connection."

Song of the Earth

Commuting annually between their North American breeding and nesting territories and their winter feeding grounds in the neotropics, the migratory birds of the Western hemisphere may be winging their way to oblivion. While the destruction of natural habitat threatens both indigenous and migratory wildlife worldwide, scientists warn that if the present rate of tropical deforestation continues, a "silent spring" will be only one of a variety of problems resulting from decimated bird populations.

Of the 650 bird species found in the United States, 332, or 51%, winter in Latin America and the Caribbean. Research has shown that birds migrating from North America return each year to the same habitat: 120 species live in shrub-steppes; 105 species inhabit aquatic environment; and 107 winter in tropical forests (Rappole et al. 1983).

Those birds that winter in tropical forests from Mexico through Central America to South America are considered to be the most threatened. The list, which reads like a bird watcher's dream, includes: the yellow-bellied, Acadian, and western flycatcher; the eastern and western wood peewee; the wood and Swainson's thrush; the veery; the black-billed and yellow-billed cuckoo; the yellow-throated, solitary, red-eyed, and Philadelphia vireo; the blue-gray gnatcatcher; the whip-poor-will and chuckwill's widow; and the broad-winged hawk and osprey. Among the warblers are the prothonotary, black-and-white, golden-winged, worm-eating, black-throated green, black-throated blue, Swainson's, Cape May, Tennessee, Kirtland's, Bachman's, Townsend's, Parula, bay-breasted, black-burnian, cerulean, chestnut-sided, blackpoll, Canada hooded, and Kentucky. Completing the catalogue are: the Mississippi and the swallowtailed kite, the American redstart; the ovenbird; and the scarlet, hepatic, and western tanager (Deis 1981).

Although we in North America have come to think of avian migrants as "ours," ornithologists note that many birds spend up to seven months wintering in the tropics, several weeks traveling north and south, and perhaps only eight to ten weeks breeding and rearing their young in northern climates (Pasquier and Morton 1982). Still, we recognize that "our" nesting birds are threatened due to the continuing destruction of tropical forests. According to Smithsonian Institution scientists, extensive deforestation in the neotropics has already produced observable declines of 1% to 4% annually among migratory bird populations (Myers 1985).

An eight-year study conducted on a 15 acre forest site in Veracruz state, Mexico, revealed that by the time 75% of the area was destroyed or disturbed, the Swainson's thrush, the black-and white warbler, and the worm-eating warbler were among the species that had disappeared. Some species have demonstrated a need for increased territory when an area is degraded. The hooded warbler, for example, apparently requires three times as much altered habitat as undisturbed forest. Since many birds return to the same tropical territory each year, when the habitat is destroyed or altered, they must either become wanderers or they fail to survive (Pasquier and Morton 1982).

Birds travel to the tropics for the winter months because food is scarce during the temperate zone's dormant period. Similarly, by migrating north in the spring, when insects and other sustenance are again readily available, birds alleviate the pressure of large avian populations on food resources in the tropics and increase their chances for successful breeding.

Marvels of adaptation, many migratory birds alter their feeding and social habits according to their environment. Kingbirds, the insect-eating flycatchers of open and semi-open spaces in North America, become fruit eaters when wintering as far south as the Amazon Basin of Peru and Bolivia. Swainson's and broad-winged hawks, which prey on mammals, reptiles, and other birds in northern climates, subsist on large katydids when soaring through Panama. And the Tennessee warbler, almost entirely an insectivore, includes nectar in its tropical diet.

Many birds that are solitary and aggressively territorial in North America have been observed to become members of mixed-species flocks in the tropics. The autumn molting to less colored plumage allows most wintering birds to gain acceptance to mixed flocks and reduces aggression among members of the same species. Ornithologists maintain that birds changing their diet to fruit become less aggressive and are thereby able to feed in flocks, thus providing a better chance for survival against predators. Some species, like the blue-winged and golden-winged warblers, keep their colored plumage and join mixed-species flocks, which they then regard as their "territory" and deny entry to members of their own kind (Pasquier and Morton 1982).

With deforestation activities consuming Latin America's rain forests by an estimated 4% every year, there is a possibility that by the end of this century there will be little, if any, suitable habitat for forest-seeking migrants. In fact, habitat destruction is "a larger problem with indigenous species than with migratory birds," according to Chandler Robbins of the United States Fish and Wildlife Service. Even if considerably more parks and reserves were established throughout the neotropics, migratory bird populations, already suffering from shrinking habitat, are expected to decline. The "best" that we might expect is that populations of those species preferring open habitat may increase in number.

Migratory bird populations have an important relationship to the ecology as well as to agriculture in North America. Wintering birds return to temperate climates in the spring, at a time when many insects are reappearing. Because returning birds arrive when insects are in the larvae and other vulnerable stages of their life cycles, scientists reflect that insect numbers are naturally kept in check, thus preventing serious insect predation on agricultural crops. However, if migratory bird populations continue to decline, increasing swarms of insects could mean disaster to North American agriculture (Myers 1985). The use of pesticides to control insects has already proven to be detrimental to all members of the food chain, including birds and human beings; thus the answer to this particular problem cannot be seen to lie within the realm of chemical control.

If, as expected, migratory bird populations continue to decline due to the destruction of the rain forests, scientists recognize that life forms other than birds, insects, and man will be affected. Indeed, it would be inconceivable to imagine that the total ecology of the Americas would be unaffected, should the songs of "our" migratory birds be diminished.

The Cattle Connection

Beef cattle ranching is widely considered to be the major factor in the destruction of Latin America's tropical forests. The problem is most serious throughout Central America—from southern Mexico to Panama—where, with arable lands already producing beef and other, more traditional, export crops, there is a tremendous pressure to develop "new lands"—the rain forests. While precise figures for specific uses of converted tropical forest areas are difficult to ascertain, it is known that among the countries of Central America, cattle ranching is foremost, followed by slash-and-burn agriculture. Conservative estimates reveal that some 37% of Central America's tropical forests have already been destroyed, largely within the last 30 years, and that current rates of deforestation for establishing cattle pastures continues to accelerate in countries like Panama, Costa Rica, Honduras, Guatemala, Belize, and southeastern Mexico. In South America's Amazon Basin countries—Brazil, Colombia, Peru, Bolivia, Ecuador, Venezuela, Guyana, Surinam, and French Guiana—deforestation for cattle ranching is most predominant in the first three.

Figure 2. Cattle ranching in the humid tropics of Latin America is the prime cause of the destruction of the region's vital tropical forests. (Amazon Basin of Ecuador; photo © Douglas R. Shane)

Since 1960, cattle production in the countries of tropical Latin America has increased by some 69%, while beef exports have risen by 448%. The cause for the dramatic growth of the region's cattle industry has not been due to the phenomenal increase of domestic populations; rather it can be directly attributed to the ever-increasing demand for beef by the United States, Western Europe, and Japan, and the concomitant necessity of earning much-needed foreign exchange. Conversely, the cattle industries in the once-traditional beef producing temperate countries like Argentina and Chile, have grown little or declined.

Despite the growth of the cattle industry throughout tropical Latin America, which often displaces other agricultural products such as cotton and food crops, domestic consumption of beef in the exporting nations has declined on an average of 13.5%. This is largely due to the fact that exporting countries like Costa Rica, Honduras, Guatemala, and, until recently, Nicaragua, have consistently sold 50% or more of their annual beef production to foreign markets. Statistics indicate that some 90% of Central American beef exports go to the United States, while the remaining 10% are shipped to Europe and Japan. The majority of South American beef exports are purchased by European markets. Resulting in domestic scarcity, the cost of beef in the exporting nations has been driven far above what many citizens can afford. In 1969, the United States Department of Agriculture's Foreign Agricultural Service noted that "the considerable growth in meat exports in recent years has been at the expense of domestic beef consumption." Seventeen years later the situation is only aggravated by meteoric population growth, eager export markets, and worsening domestic economies.

Cattle ranching in tropical forest areas is commonly effected by the construction of highways into wilderness areas for the benefit of timber operations and petroleum companies. Because trees of the same species grow widely dispersed in the humid tropics—a natural adaptation to prevent the transmission of numerous diseases—loggers often clear-cut forest tracts rather than take the time and expense to extract only trees of known commercial value. Using the roads as conduits to forest areas are the landless and unemployed, desperate to earn a living from the poor forest soils as subsistence farmers. Employing slash-and-burn agricultural methods, the colonists exhaust their land's potential within a few years and must then move on to a new area, deeper in the forest.

It has become common for cattle interests to buy the depleted land from colonists, even paying them to plant forage grasses, so that cattle pastures can be established. It is also not unusual for ranchers to employ colonists outright to cut down large areas of forest, burn the fallen trees, and plant pasture grasses. And, with the expansive growth of the beef cattle industry in recent years, ranchers have brought heavy equipment into the forest to speed up the destruction. An ominous procedure that is used in some areas is chemical defoliation. Largely banned from use in the United States because of their links to cancer and birth defects, herbicides such as Tordon and 2,4,5-T, commonly known as "Agent Orange," are manufactured in the United States by Dow Chemical, Hercules, Monsanto, Diamond Shamrock, North American Philips, and Thompson-Hayward Chemical. No restrictions on the use of these herbicides or pesticides are employed in tropical Latin America.

Once pastures are established, the cattle are introduced. Although European breeds (*Bos taurus*) were initially used, it was soon recognized that they could not efficiently regulate their body temperatures in the humid tropics, and tended to become feverish and unable to maintain high meat and milk production. More successful have been hybrid stocks that utilize the Zebu breeds (*Bos indicus*) of India. Also more durable are the Indu-Brasil, a cross of three Zebu breeds developed in Brazil; the Brahman, developed in the United States by mixing Zebu crosses with British stock; the Santa Gertrudis, a cross between Brahman cattle and European Shorthorn stock; and the Charolais, a French breed.

Despite the best efforts of ranchers to develop strong stock, cattle raised in the humid tropics are subject to a multitude of problems. Plagued by numerous parasitic and infectious diseases and nutritional deficiencies, cattle herds in tropical Latin America suffer high mortality rates. Among the most serious health problems are rabies; leptospirosis, which affects the kidneys; gastrointestinal parasitism and viral respiratory infections; polyarthritis, a bacterial infection of the joints; mastitis, a bacterial infection of the mammary gland; and foot rot, a common affliction of cattle in the humid tropics. The neurotoxins found in weeds in tropical pastures are estimated to claim up to 10% of herds raised in former forest areas.

Aftosa, commonly known as hoof-and-mouth disease, is mainly confined to South America and Europe. Occurring only infrequently in Central and North America, aftosa is a virus transmitted by contact or through the semen. Because aftosa is so infectious, South American beef must be cooked before it can be imported into North America. An odd alliance between environmentalists and North American cattlemen seeks to prevent the completion of the Pan-American Highway through the rain forests of Panama's Darien province—a natural barrier—for fear that the inevitable transport of cattle from South America would result in the introduction of hoof-and-mouth disease to Central and North America.

Cattle ranching in areas of tropical forest converted to pasture is both environmentally destructive and short-lived. As the soils lose their nutrients, the food value of forage grasses declines sharply. Overgrazing and compaction accelerate erosion and destroy vital successional vegetation until the pasture must be abandoned, usually within five to ten years. As the ranchers move their herds to new areas, they leave behind a veritable wasteland, unable to regenerate itself and of little value to man or animal.

As with other facets of development in rain forest areas, cattle ranching could not have become so pervasive were it not for the financial and technical assistance offered by the governments of each tropical nation, the international assistance agencies of many beef importing countries, national and international banks, regional development organizations, private sector interests, and transnational businesses.

Principal among the public sector proponents providing funding and scientific assistance to tropical Latin America's beef cattle industry are the World Bank, The Inter-American Development Bank, the United States Agency for International Development, the Organization of American States, and at least five agencies of the United Nations. Underscoring the profitable nature of cattle ranching are the investments made by individuals and companies of numerous non-Latin countries. Directly involved in various aspects of production

Figure 3. After seventeen years of intensive agriculture followed by cattle ranching, what was once tropical forest is now a "red desert." (Amazonia, Brazil; photo © Douglas R. Shane)

as well as the export and import of beef raised in the humid tropics are nationals and businesses from West Germany, Japan, Great Britain, Belgium, Italy, Austria, Canada, and foremost, the United States. These interests range from the sales of heavy machinery and herbicides to livestock feed and processing plants.

Beef from tropical Latin America is exported to a number of developed nations, including Japan and the countries of Western Europe's Common Market. But the United States is the preeminent importer of the region's beef, purchasing some 90% of Central America's exports. United States beef imports, which come from Australia, New Zealand, Canada, and several Western European countries as well as from Latin America, have increased by some 140% since 1960. By the 1980s, the United States was importing 10% of the beef consumed annually by its citizens with 17% of all imports purchased from tropical Latin American countries. Of all beef imports since the early 1970s, 13.5% was from Central America and about 3% was from Brazil. In total, beef imports from tropical Latin America account for just under 2% of the beef consumed annually in the United States.

Latin American beef enters the United States in basically two forms. If it is from Central America, it may be fresh, chilled, or frozen. South American beef, however, must be cooked prior to its importation as a safeguard against the aftosa virus.

The majority of U. S. beef imports are derived from range-fed cattle whose meat tends to be leaner, less marbled, and generally tougher than beef that is grain-fed. Because United States consumers prefer "juicy" meat, imported beef is usually mixed with meat and fat trimmings from domestically produced beef, veal, and pork.

Most of the Central American beef imported by the United States is processed as hamburger and ground beef, which accounts for about one-quarter of all beef consumed domestically in recent years. Central American beef is also sold as inexpensive cuts of meat in supermarkets and in family-style steakhouses and is used in numerous processed meat products such as frankfurters, salami, bologna, pepperoni, and other luncheon meats and sausage products. South America's cooked and canned beef imports are generally sold as corned beef and used in frozen and canned products such as stews, soups, chili, precooked dinners, pot pies, and baby and pet foods.

Although the United States is the world's largest beef producer, consumer demands continue to exceed domestic supplies. In an effort to satisfy the nation's craving for red meat while protecting the U. S. cattle industry, the federal government sets annual quotas for meat imports that are linked to U.S. production levels. Nonetheless, on-going disagreements are waged between organizations like the National Cattlemen's Association, which represents domestic interests, and the Meat Importers Council of America, which represents manufacturers who want to keep beef prices down. The fact is that the remarkable growth of the U.S. fast-food industry and increasing food costs have encouraged consumers to purchase less expensive, lean meats such as chicken, pork, and products using imported beef.

Perhaps the most controversial issue concerning the destruction of Latin America's rain forests is the link between cattle ranching in the humid tropics and U.S. imports of beef and its domestic use. Some environmentalists and scientists, anxious to stop the dissolution of the remaining rain forests in the Western hemisphere, argue that those U.S. businesses that use tropical beef imports are exacerbating the problem and should be boycotted until they cease using the commodity. Targets of a protest would be known users of imported beef.

The majority of national hamburger chains and manufacturers of products containing beef queried about the sources of their meat responded that they use only U.S. beef. Of those admitting to using "some" imported beef, the majority noted a preference for Australian and New Zealand products. A spokesman for Burger King, which admits to using "a small percentage of frozen beef from Costa Rica," said that, "any chain saying they don't use imported beef is handing you a crock."

Accusations of using imported beef from tropical Latin America have been aimed at many companies. A source of confusion for both users of beef and concerned individuals is the fact that once imported beef is cleared by U.S. customs agents, it is given a grade rating that corresponds to United States meat standards and loses its imported identity. Now officially having U.S. status, the product may go directly to a manufacturer or be held by a meat brokerage which will sell it to a processor. Although beef imports arrive in the United States in sixty-pound cartons marked with the country of origin, status, the product may go directly to a manufacturer or be held by a meat brokerage which will sell it to a processor. Although beef imports arrive in the United States in sixty-pound cartons marked with the country of origin, it is not unlikely that some buyers—particularly brokerages—repackage the product in order to hide its identity. This repackaging would occur because although Latin American beef is slightly cheaper, there remains a preference for beef from Oceania which has a lower moisture content and therefore does not require the the changing of processing formulas. So it is quite plausible that many manufacturers of hamburger and other meat products are unaware of the precise geographic origin of the beef that they use.

While Latin Americans have a traditional preference for cattle ranching, "The Cult of the Bull" has received substantial encouragement from interests representing the developed Western nations, particularly the United States. Until the factors that have stimulated the growth of the beef industry in tropical Latin America are dealt with, the fires will continue to rage throughout the hemisphere rain forest with a chilling effect (Shane 1986).

Saving Eden From Ourselves

Imagine a planet where the most intelligent inhabitants knowingly pursue a suicidal course of planetary destruction; where demand, greed, and desperation have supplanted the knowledge that extraction of natural resources and environmental management must be conducted with respect for Nature's sustainability.

In recent years, scientists representing a wide variety of disciplines and concerned environmental organizations have called for a more rational approach to utilizing the earth's tropical forests. But still the forests are being felled or flooded. It can no longer be argued that the decision makers are unaware that the poor soils of the humid tropics are unsuitable for large-scale, sustainable development projects; the hundreds of thousands of acres of degraded forest and resultant "red desert" already bear their lifeless testimony. The fate of the rain forests reflect the best and the worst of humanity's ability to grasp a serious problem—and either solve it, or ignore it, until it resolves itself in the form of mass extinctions and, quite possibly, biological failure on a global scale. Solutions are not easy, but they are possible—and essential.

Many of the factors leading to the destruction of rain forests lie with the social and economic problems of the tropical countries themselves. Land reform and the appropriate utilization of productive soils, the increased transfer of technological capability from developed nations in order to expand economic and employment potential, commitment to birth control practices; these and other important issues must be dealt with effectively.

Foreign influences must be tempered by an awareness of global mutuality. There are recent signs that such sentiments can be translated into effective action. In the spring of 1985 a coalition of environmental organizations, Indian rights advocates, and bipartisan members of the United States Congress were able to cause the World Bank to halt funding on a five million dollar development project in northwestern Brazil. The action marked the first time the bank halted a project for environmental reasons. And, in both the United States Senate and House of Representatives, legislation is pending that deals with the protection of tropical forests, the preservation of biological diversity, and the necessity of limiting foreign assistance to only appropriate and sustainable activities.

But there is more to do. The sale of dangerous herbicides and pesticides— which are banned in the United States—to developing countries must be prohibited. Evidence from the United States Department of Agriculture's Food Safety and Inspection Service and other research institutions have shown that the presence of carcinogenic chemical residues and other adult- erates in agricultural products—including beef—imported from Latin America has more than doubled in recent years. Unfortunately, this represents minimal knowledge since agricultural imports are only randomly sampled.

A boycott of one or more companies known to use beef imported from the humid tropics would be of value if it led to educating the public and the companies about the problems of rain forest destruction. The end result of such a boycott, however, should be federal legislation banning imports of beef raised in tropical forests. This action is necessary because even if the hamburger chains ceased using beef imported from the humid tropics, other manufacturers would certainly use it, knowingly or not.

By closing the lucrative U. S. market to beef raised in the humid tropics, the United States could encourage the producing nations to use both their productive lands and tropical forests more rationally. Because the United States has been the major market for the region's beef exports—accounting for just under 2% of the beef consumed domestically each year—it is questionable as to whether the tropical countries would find alternative markets for their beef. Although Japan does import a small amount of tropical Latin American beef, most of its meat imports are obtained from Australia and New Zealand. Western Europe, now a major beef producer itself, imports small quantities of meat from various countries, including the United States, Canada, Australia, New Zealand, and Latin America. (The European Economic Community, among others, is responsible for the escalating destruction of tropical rain forests in Africa done primarily to increase the production of beef for export to Europe.) However, due to the cost of transportation and high tariffs imposed by European nations, tropical Latin American producers would not find as profitable a market as they have with the United States. Latin American economies would not be expected to suffer from a beef ban since other agricultural commodities would retain their importance. It should also be noted that cattle ranching is the least productive form of land use, and it is doubtful that the beef industry, as practiced on the poor soils of converted tropical forests, will ever be anything but an environmentally destructive enterprise.

The United States, with important political and economic interests in Latin America, must accept a role of rational responsibility in the region. Foreign assistance must lead to sustainable activities that first benefit the southern nations. The trade policies of the U. S. government and its private sector must act with an understanding that the conservation of vital natural resources is essential for the well-being of not just a few nations, but for the entire planet's inhabitants. The earth exists for all—human beings, indigenous and migratory wildlife, forests—and all must share it. We, as the stewards of the planet, have the responsibility of ensuring a healthy biological future for the unborn generations.

Endnote
[1] Environmental Consultant, 1805 Shallcross Ave, Wilmington, DE 19806.

References
Deis, R. 1981. Again silent spring. *Defenders Magazine.* 56(2): 6-10.

Myers, N. 1985. How the song birds of America choked on fast food. *Manchester Guardian Wkly.* Jan. 6: 19.

Pasquier, RF and Morton, ES. 1982. For avian migrants a tropical vacation is not a bed of roses. *Smithsonian Magazine.* 10: 169-88.

Rappole, JH, Morton, ES, Lovejoy, TE and Ruos, JL. 1983. *Neoarctic Avian Migrants in the Neotropics.* Washington, DC: U.S. Dept. of the Interior, Fish and Wildlife Service.

Shane, DR. 1986. *Hoofprints on the Forest: Cattle Ranching and the Destruction of Latin America's Tropical Forests.* Philadelphia, PA: Institute for the Study of Human Issues Publications.

Recommended Reading
Caulfield, C. 1985. *In the Rainforest..* New York: Alfred A. Knopf.

Part II

Proceedings

Animals and Humans: Ethical Perspectives
Conference held at:

Moorhead State University
Moorhead, Minnesota, April 21–23, 1986

HUMANS AND OTHER ANIMALS:
A BIOLOGICAL AND ETHICAL PERSPECTIVE[1]

Ashley Montagu[2]

Dedicated to the Memory of Dian Fossey (1933-1986)
Naturalist, Protector, and Friend of the Virunga Gorillas—

> The modern science of ecology is showing us that the fate of the Earth and
> of humanity are inextricably connected. This is as much a biological fact as it
> is a spiritual condition, from which arise the ethical principles and moral
> sensibility to live in respectful harmony with the rest of creation.
> (From the *Preamble* to the Program of the International Network for Religion
> and Animals, 1986.)

The first thing to be said about our title is that all humans are animals.
Our biological kinship with the whole of animated nature, to use an old-
fashioned phrase, in one Great Chain of Being should be unequivocally
clear. In this connexion should also be included our intimate relationship
to the world of inanimate nature, precisely because it is inanimate, for it
speaks to us in a voice no less appealing and meaningful than that other, of
animate nature. In our Father's house, we are members of one family, but
in its mansions we are guests in common with all its other inhabitants,
neither superior nor inferior, but members of an extended family, a kingdom
of animate and inanimate nature comprising several millions of variegated
species, and sustaining a physical environment the most wonderful and
beautiful to behold. As an anatomist, a biological anthropologist, and social
biologist, I have learned that we are made of the same essential materials
as are all other animals, that ultimately, indeed, we are constructed of the
same stuff as is our world, and that our kinship is with the whole of nature,
that we are guests upon this earth, and should conduct ourselves accordingly,
with sensitivity, thoughtfulness, reverence, enjoyment, and gratitude.

Gradually we are beginning to understand that it is upon the clear recognition of our biological continuity and community with the whole of nature
that our very survival, that the survival of this marvelous world, depends.

Whether we choose to view nature and evolution as God's way of creation,
or whether we dispense with the idea of a supreme being as a creation
myth common to most known peoples, the fact of evolution is clear, and

certainly it is the best authenticated explanation of the origins and development of the vast variety of animate and inanimate forms on this earth (Montagu 1984). Evolution is just another word for development, or the maximization of the improbable, and development comes about largely as a consequence of the adaptive "responses" as it were, which the organism makes to the challenges of the environment. Such responses do not imply any conscious decision on the part of the organism, but rather the "selection," for the most part, by the environment of those organisms possessing the variations that most successfully, that is, adaptively, fit the organism to the challenges of the environment. Of all the improbable species that have come into being in this manner, humankind is in many ways the most improbable, *Homo sapiens*. Oscar Wilde, has not altogether unjustly described the name that Linnaeus, in 1758, bestowed upon our species, as the most oafishly arrogant, prematurely self-serving description ever perpetrated. Certainly it is true that the sapience, the wisdom, is there as a potentiality, but there can be no doubt that by the measure of his performance in human relations, and to the remainder of nature over the last twelve thousand years, in civilized societies, at least, he deserves no better appellation than *Homo sap*. A confused mindedness which has come about largely as a result of the unique evolutionary history of our species. Not only are we the only species that is able to weep, that is, to cry with tears, but we are also the only species able to laugh vocally. It is true that humankind has more to weep and also to laugh about than any other species. Perhaps that is why we are able to do so (Montagu 1960a, b). Be that as it may, it turns out that these two human traits constitute a saving grace in a rather cubistically dilapidated human landscape. To weep and to laugh are adaptive traits of supreme value, for without them sympathy, compassion, and self-criticism would scarcely be possible. A species so endowed cannot ever be without hope.

The trait, however, which beyond all others distinguishes humans from all other creatures is *educability*. It is our defining and outstanding characteristic as a species.

We have few, if any, remnants of instincts, and whatever we do as human beings, *as human beings*, we have to learn from other human beings. We share, of course, innumerable traits, both physical and behavioral, with other animals, but everything we come to know and do *as human beings*, those things which render us distinctively unlike any other animal, we must learn under the tutelage of other human beings. For example, we are all born with the capacity, that is, the potentiality for speech, a trait which is unique to our species, but no one of us would ever speak were it not for the teaching, the training, we receive from others, that capacity would never develop into an ability. A capacity is a potentiality, an ability is a trained capacity. Thinking as potentiality is a capacity, but how we come to think will depend largely upon the conditioning to which we are exposed. It is because of those capacities that to be born human is to be in danger, for by virtue of our unparalleled educability we are capable of learning more unsound things as well as more sound ones than any other creature: and when one puts the two together, one doesn't get intelligence, what one gets is confusion, a state in which most of our species now more than ever dangerously flounders.

In our technological age we have very nearly perfected the knack of arranging the world so that we can muddle through it on the basis of pseudological rationalizations built on unexamined and unanalyzed habits of stereotyping. The greater our progress in the development of technological substitutes for thought, the more spiritually illiterate we become. For quite some time now we have been flattering ourselves that we can make machines that think like human beings, without ever having grasped the fact that for a much longer time we have been turning out human beings who think like machines, that is, who don't think at all.

What we teach in our schools is *what* to think, not *how* to think. Jane Taylor (1783-1824), the gifted young English author, early in the nineteenth century, put it very well:

> Though man a thinking being is defined,
> Few use the grand prerogative of mind,
> How think justly of the thinking few;
> How many never think, who think they do!

Alas, the proportion of people who think today is probably much less than in Jane Taylor's day, for in our own time we are besieged with such institutions as commercial television, which has appropriately been called "the lobotomy box," or "the chewing gum of the mind," which, as is well known, is designed to provide formula fodder, accompanied by a mindless excess of shockwaves, for what is generally known in the industry as "the lowest common denominator." Children now spend more time watching the programs and commercials than they spend in school, with the result that they have been largely robbed of their childhood. The few programs that could be described as good make it abundantly clear what a marvelous medium television could be for a genuinely humane education and enlightenment without slighting entertainment one bit.

Our schools and colleges have become institutions, to a far larger extent than we realize, for the training in an incapacity to think, and in the ability, with the aid of ritual incantation, to compartmentalize incompatible ideas without the slightest discomfort, and, indeed, to grow quite comfortable with the disparity between what one solemnly declares one believes, and the very contrary of what one does in practice.

Early on we learn to engorge large quantities of rote remembered facts, and to disgorge them at ceremonial occasions dubbed "examinations" when after those who have displayed the greatest disgorgative capacities are graded the brightest and the best, and are the most highly rewarded. We go on from schools to colleges and universities, which have mostly come to be regarded as job-qualifying institutions, where we take A.B. degrees, M.A. degrees, and doctor's degrees, dying in the process both spiritually and intellectually by degrees.

In such a world, the mark of a truly educated person is one who has overcome the deficiencies of the educational system, one who has become an independent thinker, a questioner of the obvious, and who is not only able to use his mind as a fine instrument of precision, but also who is able to feel for others that sense of involvement and sympathy which would make the whole world kin. I speak here of love, for knowledge is not enough. It is loving kindness that must be joined to knowledge and to cleverness.

I am reminded here of those charmingly apposite verses by Dame Elizabeth Wordsworth (1840-1932), the first Principal of Lady Margaret Hall, at Oxford, entitled "The Good and the Clever":

> If all the good people were clever,
> And all the clever people were good,
> The World would be better than ever
> We thought that it possibly could.
> But somehow 'tis seldom or never,
> The two hit it off as they should,
> The good are so harsh to the clever,
> The clever are so rude to the good!
> So friends let it be our endeavour
> To make each by each understood;
> For few can be so good like the clever,
> Or clever so well as the good.

Goodness, I shall hold, in its own way is as rational as cleverness, if not more so. Goodness, love, lovingkindness, are all words for the same thing, best resumed in the one word, "love." But what is love? It is a question like that other which Pontius Pilate did not stay to answer, that most people have asked, and equally have not done well for an answer. Corinthians 13 is very good, indeed, on the matter, but even better is George Chapman (1559-1634), poet, playwright, translator of Homer into English, and friend of Ben Johnson. In his play, *All Fooles*, acted in 1599, in the first scene of the first act, Chapman makes his hero Valerio break forth into the following paean:

> I tell thee Love is Nature's second sun,
> Causing a spring of virtues where he shines;
> And as without the sun, the world's great eye,
> All colours, beauties, both of Art and Nature,
> Are given in vain to men, so without love
> All beauties bred in women are in vain,
> All virtues born in men lies buried,
> For love informs them as the sun doth colours,

And as the sun, reflecting his warm beams
Against the earth, begets all fruits and flowers;
So love, fair shining in the inward man,
Brings forth in him the honourable fruits
Of valour, wit, virtue and haughty thoughts,
Brave resolution, and divine discourse:
Oh, 'tis the Paradise, the heaven of earth.

That is not bad for 1599 or any other time. It is, in fact, the soundest description of the role of love in human development with which I, as a student of human development am acquainted, and the most memorable.

From the biological point of view, love can be described as behavior designed to confer survival benefits in a creatively enlarging manner upon the other. Spelled out, this means that by your behavior you not only enable the other to live, but to live more fully fulfilled than they would otherwise have been. It means that love is the demonstrative communication to the other of your profound involvement in their welfare, such that you provide them with all the encouragements, stimulation, support, and succor that they require for healthy growth and fulfillment of their potentialities, the communication that you will never commit the supreme treason of letting them down when they are in need of you, that they can depend upon you always standing by ministering to their needs. For becoming what? What you are being to them. That is love, and it has never been better said than by George Chapman.

The important thing for us to understand is that our love must extend to all and every part of nature, and not be limited to our own kind. We have seen the effect of such limitations in the cruel denial of their human rights to whole peoples, as in the case of "The Final Solution" of the Nazis, resulting in coldblooded systematic murder of six million Jews, not to mention the millions of members of other miscalled "races," while the civilized world stood by and deliberately closed its eyes (Gilbert 1981, 1986; Ross 1980; Wyman 1984, 1985). The whole concept of "race" is a myth and a fraud, nevertheless it is subscribed to and acted upon by millions of people throughout the civilized world, not least to this day, in the United States of America (Montagu 1974).

As for the civilized world, I am reminded of Mr. Gandhi who, when asked by an American reporter what he thought of Western civilization, mused for a bit, and then replied, "You know, I don't think it would be such a bad idea."

And speaking of racism, it should be pointed out that the attitudes of many people towards animals is a kind of racism, that is, in the belief that human beings are superior to animals, that animals are a lower order of being, subordinate to and subject to the whim and will of man. The most powerful influence in spreading and sustaining this view of man's relationship to the animal kingdom has been the Old Testament teaching set out in the Book of Moses in Genesis 26, in which it is written, "And God said, let us make man in our own image, after our likeness; and let them have dominion over the fish of the sea, and over the fowl of the air, and over the cattle, and over all of the earth, and over every creeping thing that creepeth upon the earth." And as if to rub the message in, the injunction is repeated in the next but one paragraph 28, even more strongly than in 26, "And God blessed them, and God said unto them. Be fruitful and multiply, and replenish the earth, and subdue it: and have dominion over the fish of the sea, and over the fowl of the air, and over every living thing that moveth upon the earth."

These injunctions from God himself, perhaps more than anything else, have contributed to the appalling view that it is man's God-given right to prevail over all other creatures. This homocentric view has given him license, not only in relation to other animals but also to himself, to execute the divine will first by following the philosophy of the cancer cell in multiplying by thoughtless and uninhibited reproduction, second by subduing and devastating the earth, and third by inflicting upon other living creatures the cruelest pains, unspeakable tortures, and death, in the name of husbandry, food, trade, sport, hunting, science, medicine, yes, and even in the name of humanity. Civilized man has already exterminated hundreds of species. In justification of such practices the most transparent insupportable rationalizations have been adduced: animals have no feelings, or at least they are much less sensitive to pain and fatigue than humans, lobsters may be dropped into a cauldron of boiling water because they are only lobsters, and one wouldn't want to eat a live lobster, for that would be uncivilized, not to mention unpalatable. I have heard women and fishermen claim that fish have no feelings, and I have heard all sorts of explanations of this kind offered in support of all the depredations that man has visited upon the animals. And when I say man, I mean socalled "civilized" man, for from the blanket indictment must be exempted virtually all so-called "primitive" peoples, for among them there is reverence and affection for animals and for the environment, of which they consider themselves an intimate part. The creation myths of these people contain no "racist" reflection upon the plant and animal life, or upon the earth which they inhabit. Indeed, quite the contrary. Through their creation myths, these peoples are aware that all things in this natural world are related, and that the various forms they exhibit are merely different expressions of the same original creative materials. Such myths are often extremely beautiful and intelligent, providing a complete account of the origin and even evolution of all things. This is clearly to be observed in the totemic relationships with which different clan members are associated, not only with animal forms but with the sky, the clouds, water holes, rocks, plants, and the like. Among some indigenous peoples, the totemic object is not merely regarded as kin, but is identified with oneself, or else that a vital part of oneself is also an intrinsic part of the revered object, whether plant, animal, or other object (Frazer 1935).

Listen to Jamake Highwater, America Indian and distinguished scholar, who was born and raised on the Blackfeet Blood Reserve in northern Montana. He begins by saying that as he learned English and compared it with the language in which he had been raised, he came to realize that "languages are not just different words for the same things but totally different concepts, totally different ways of experiencing and looking at the world.

"As artists have always known, reality depends entirely on how you see things. I grew up in a place that was called a wilderness, but I could never understand how that amazing ecological park could be called 'wilderness,' something wild that needs to be harnessed. Nature is some sort of foe, some sort of adversary in the dominant culture's mentality. We are not part of nature in this society: we are created above it, outside of it, and feel that we must dominate and change it before we can be comfortable and safe within it. I grew up in a culture that considers us literally a part of the entire process that is called nature, to such an extent that when Black Elk called himself brother of the bear, he was quite serious. In other words, Indians did not need Darwin to find out that they were part of nature.

"I saw my first wilderness, as I recall, one August day when I got off a Greyhound bus in a city called New York. Now that struck me as being fairly wild and pretty much out of hand. But I did not understand how the term could be applied to the place where I was from" (Highwater 1982).

The care bestowed upon animals in indigenous cultures, is among the most engaging of their qualities. Pets sleep together with the family, and small animals are often breast fed by the women. The first animal to have been domesticated appears to have been the dog, during the Mesolithic, some 12,000 years ago. The indigenous peoples of the earth appear to have lived by a view of their relationship with it, which is, except for the first paragraph, encapsulated in Job 12: 6-9:

> The tabernacles of the robbers prosper, and they that provoke God are secure; into whose hand God bringeth abundantly.
> But ask now the beasts, and they shall teach thee; and the fowls of the air, and they shall tell thee:
> Or speak to the earth, and it shall teach thee: and the fishes of the sea shall declare unto thee.
> Who knoweth not that the hand of the Lord hath wrought this?
> In whose hand is the soul of every living thing, and the breath of all mankind.

Note how Job here contrasts the corrupt ways of man with the lessons we have to learn from both animate and inanimate nature.

A similar passage occurs in the Koran, the sacred scripture of Islam, compiled during the seventh century A.D., in which it is written that "There is no beast on earth nor fowl that flieth, but the same are a people like unto you, and to God they shall return."

Is not that a beautiful sentiment? Both the passage from Job and that from the Koran recognize the profound kinship of animals with humans, an appreciation which is not achieved again until the appearance of St. Francis of Assisi in the twelfth century. The combined austerity and poetic gentleness of St. Francis, his vow of poverty, and his love for all living creatures, addressing them as brothers and sisters, and so regarding them, introduced a new dimension into the perception of man's relation to other creatures. Through the religious Order which he founded, the Franciscan example and teaching was carried throughout Europe. That St. Francis saw no essential difference between himself and other creatures is illustrated by his words to the wild turtledoves, whom he liberated from the complaisant hands of the youth who had captured them. "O my sisters, simple-minded turtledoves, innocent and chaste, why have ye let yourselves be caught. Now would I fain deliver you from death and make you nests that you may be fruitful and multiply, according to the commandments of your Creator." Here St. Francis views the commandment to increase and multiply as applying to all living creatures, and sees them only as different people unto himself.

It is of interest, in passing, to note, that when I consulted several encyclopedias in order to refresh myself on the life and work of St. Francis, I could find adequate biographical accounts of him but no mention whatever of his profound sense of brotherhood with beast and bird. A somewhat inadequate exception is Julien Green's book (1985) on St. Francis. This, surely, tells us something concerning the interest of contemporary man in our relation to animals. Even the 53 pages devoted to animals in Hastings' great *Encyclopaedia of Religion and Ethics* omits any reference to man's ethical relationship to animals. Indeed, the only article in that eminent work which discusses the subject is by Henry Salt, the author of the famous *Animal Rights*, which was published in 1905. Salt's article was devoted to "Humanitarianism," and being the pioneering humanitarian that he was, pays handsome tribute to St. Francis (Salt 1923).

Christianity played a dominant role, especially early Christianity, in the perpetuation of the callous indifference to the fate of animals, for since animals were without souls, they were beyond the pale of hope, and at the same time beyond the pale of sympathy. The one outstanding exception to this was the extraordinary St. Chrysostom (A.D. 347-407), who wrote that we should show animals "great kindness and gentleness for many reasons, but above all because they are of the same origins as ourselves." During the Middle Ages, the indifference of the Catholic Church to the claims of animals was broken only by the eccentric example of St. Francis. It is not for nothing that to this day the gentle Buddhists speak of Christendom as "the hell of animals." And it is an object-lesson in humanitarianism to watch a Buddhist gently remove an ant from the hem of his robe, and carefully place it out of harm's way (Pallis 1940).

It is with the Renaissance and the revival of humane learning that we find humane sentiments making their appearance in the writings of such men as Thomas More, Erasmus, Montaigne, Bacon,and others, thus opening the way for the fuller development of humane sentiments which occurred in the eighteenth century in the writings of such authors as Voltaire, Thomson, Pope, Goldsmith, Goethe, Herder, Blake, Burns, and many others.

It was not until the beginning of the second half of the nineteenth century with the development of experimental studies on animals in the name of physiology, medicine, and what was called vivisection, that the outcry against the mistreatment of animals began to find its voice. This movement acquired increasingly greater momentum during the twentieth century. With the rise of modern experimental medicine and industrial testing chiefly for cosmetics, involving the sacrifice of millions of animals each year, the public and legislatures have become involved as never before. There are today more organizations devoted to the rights and protection of animals than there have ever been, and with the power of these agencies, the future for humanitarianism, even in this conservative age, would seem to be assured. But this would be a dangerous posture to adopt, for in the matter of the rights of animals, like liberty, eternal vigilance is necessary.

What I have been hoping to do in this talk is to provide the scientific basis for the biological kinship of humans with other animals in particular and the whole of nature in general, and to show that the ethical perspective to which such a demonstration leads is inherent in the very nature of nature, that cooperation, love, *not* conflict and aggression, as we have long been led to believe, is the dominant principle by which living creatures are designed to live with each other. It was not Darwin, but the muscular Darwinists, like Herbert Spencer, who wasn't a biologist at all, but a desk philosopher, who coined the term, "the survival of the fittest," a misnomer which Darwin unfortunately adopted, but later regretted. The term, as we have better come to understand the facts, was a blunder, for it is the "fit" who are most likely to survive, not the "fittest," for the fittest are likely to be overspecialized, where flexibility, adaptability, is required.

In spite of what many of us in the Western world believe, it is not "Nature, red in tooth and claw," as Tennyson sung, that is the law of nature, but cooperation. In spite of some superficial appearances compounded by their misinterpretation, animals live in a state of cooperation with each other, in the miscalled "wild," yes, even the lion with the gazelle. Of course, some animals eat others, but not because they are in conflict with them or because they dislike them, but rather because they like them. When a lion (usually the lioness) hunts and brings her kill back to her family, it is precisely as when a woman goes shopping, and has nothing to do whatever with aggression or hostility. Lions live in perfect balance and harmony with their environment, serving a very real purpose in the balanced maintenance of the very populations from which they cull their food (Montagu 1949, 1955, 1976).

The facts have never been more soundly stated than by J. Arthur Thomson and Patrick Geddes in their monumental work, *Life: Outlines of General Biology*, published in 1931, and which did for biology what H.G. Wells' 1922 *Outline of History* did for the story of civilization. This is what they wrote:

> What has got into circulation is a caricature of Nature—an exaggeration of part of the truth. For while there is in wild Nature much stern shifting, great infantile

and juvenile mortality, much redness of tooth and claw, and—even outside of parasitism—a general condemnation of the unlit lamp and the ungirt loin, there is much more. In face of limitations and difficulties, one organism intensifies competition, but another increases paternal care; one sharpens its weapons, but another makes some experiments in mutual aid; one thickens its armour, but another triumphs by kin-sympathy. It is realized by few how much of the time and energy of living creatures is devoted to activities which are not to the advantage of the individual, but only to that of the race. Not that this is deliberate altruistic foresight, it is rather that in the course of Nature's tactics survival and success have rewarded not only the strong and self-assertive, but also—and yet more—the loving and self-forgetful. Especially among the finer types, part of the fitness of the survivors has been their capacity for self-sacrifice....The fact is that the struggle for existence need not be competitive at all; it is illustrated not only by all the endeavours of parents for offspring, of mate for mate, of kin for kin. The world is not only the abode of the strong, it is also the home of the loving.

(Thomson and Geddes 1931)

The myths we have created concerning the violence of life in the "wild," "the jungle," and "innate depravity," become the rationalizations for the justification of our attitudes toward animals, and the maintenance of our indifference to their suffering. Our attitudes toward animals are as unnatural as our cities, which are, in fact, the only real jungles. The most shocking thing is that not only do we have the most erroneous views about animals, but in our unwillingness to face the conditions of life for so many in our cities, we project the unsavory image of them upon the screen of Nature, and call it and its inhabitants, the animals, by names which far more accurately apply to us than to them. The fact is that we do not descend from something we wouldn't like to meet in a forest at night, but rather that something descends from us we would not care to meet in broad daylight on a crowded city street.

The Order of animals to which we belong, the Primates, has spent some 60 million years living in forests. Forest-dwelling animals are predominantly vegetarian, and vegetarians have long intestinal tracts—all 22 feet of them. This strongly suggests that we are designed to be vegetarian. It also suggests that many of our ills maybe due to our meat-eating habits.

Our biological ancestry indicates, in brief, that we are programmed to live in peace and harmony with the whole of Nature, and most especially, with all living creatures, that it is more than an ethical necessity—it is a biological imperative if we are to survive. As Albert Schweitzer put it, "Until man extends the circle of his compassion to all living things, man will not himself find peace" (Schweitzer 1934).

The discovery of agriculture some 12,000 years ago lead to the first village settlements and also to the control of animal reproduction. This was followed by the development of cities and theocratic government, with all their attendant evils, and the dehumanization and exploitation of both man and animals (Childe 1962). Looking back, it would seem that civilization has a natural resistance against improving itself. But let us not despair. I believe that the rehumanization of man may follow upon the rehumanization of his attitudes toward animals, and from that, I believe, will follow the repudiation of his xenophobia and the rehumanization of his attitudes toward his fellow man.

I am old enough to recall the days when, some three-quarters of a century ago, antivivisectionists, members of groups for the protection and against cruelty to animals, protesters against fox hunting, and poets like William Blake, who wrote, "A robin redbreast in a cage/ Puts all Heaven in a rage," and "A dog starv'd at his master's gate/ Predicts the ruin of the state," led to nothing but ridicule and derision. Such persons and their supporters were bracketed, in the open scorn for whom it was the fashion to indulge, with suffragettes, conscientious objectors, and other "oddballs." As always, fashion is what people follow who have no taste or minds of their own. But behold, in Switzerland, that most conservative of countries, women have just been granted the vote, so there is hope for humanity yet, even though in the United States it is not uncommon to hear such epithets as "bleeding hearts," "secular humanists," and "liberals," applied with contempt to those who plead the cause of humanity, of humanitarianism. But that, too, will pass, though it will only do so when genuine education, a humane education embracing the whole of nature, will have become the birthright of every citizen.

I see no other possibility in a free society of ensuring the growth and development of personalities who are able to think for themselves and to feel that compassion for all living creatures, for the whole of nature, that will ensure equal freedom for growth and development. I would therefore like to devote my final remarks to what I consider education to be.

Let me say at once that I do not think that genuine education exists anywhere in the Western world. What we call education is nothing but instruction, a training in techniques and skills. Whatever we teach we tend to teach as a kind of technology, in which we emphasize the structure of things in order to be able to put what we learn to practical use. Because our schools, colleges, and universities are locked into an outmoded form of thinking about what human beings are for, and about their place in nature, and because what they produce are for the most part stereotypers, the lovers of ritual incantation, who have never been taught that a word should ask itself what it means, and that the meaning of a word is the action it produces, that the world is in the sorry state it is, and that it is principally for these reasons that the world and that the attitudes of people toward animals are what they are.

In spite of the age-old dissociation between thinking and feeling, thinking and feeling are reciprocally indissolubly interrelated, and that, indeed, is what they are designed to be. Without for a moment slighting the prenatal period (Montagu 1962, 1978), what we need to understand is that education begins at birth, and that it is from that critical period on that we must revise our approaches to the child who, at the present time we continue to deform in such wrongheaded ways, and continue to do so for the rest of its life, so that we may truly describe most adults as nothing more or less than deteriorated infants, for most individuals have been failed in the education, the satisfaction and fulfillment of their potentialities as healthy human beings (Montagu 1981).

As a consequence of the confusion into which we have fallen as a result of our miseducation, we stand very near the brink of self-annihilation. It was no less a thinker than H.G. Wells who, in 1922, twenty-four years before his death, wrote in his widely read *Outline of History*, what is today even more than ever ominously true, that "Human history becomes more and more a race between education and catastrophe."

The hour is, indeed, late—it is therefore critically necessary for us to recognize that because we are human beings,and because we are capable of shedding our errors, that we *are* educable, that we need to be educated in humanity beyond everything else, to learn to live as fulfilled human beings, to learn to live as if to live and love were one.

The word "education" is derived from the Latin, *educare*, which means to nourish and to cause to grow. To nourish and to cause to grow what? The answer to that question,by the measure of all the verifiable evidences is: to nourish and to cause to grow those basic behavioral needs or potentialities with which every newborn is endowed, the needs for love, sensitivity, friendship, speech, thinking, to know, to learn, to organize, to work, for curiosity, wonder, play, creativity, imagination, open mindedness, experimental-mindedness, explorativeness, resiliency, humor, joy, laughter, tears, optimism, honesty, trust, compassionate intelligence, dance, song, enthusiasm, touch, and others.

These are the basic behavioral needs the development of which must be encouraged by caregivers throughout the life of the young, if they are to grow into healthy human beings. By a healthy human being I mean one who is able to love, to work, to play, and to use one's mind critically. It is not possible to define all these terms here. I have done so elsewhere (Montagu 1981). It is imperative for us to always bear these basic behavioral needs in mind, and to implement them by sound strategies.

Finally, I would like to conclude with the words of Henry G. Maurice, from a lecture he delivered in Brussels in 1946, and published in English in 1948, in the summer of which I read it,and which has since inspired my thinking on the subjects of humanity's relation to animals.

"To understand man," wrote Maurice, "one must study him in the light of the ecology within which he has been developed and of which he is an integral part. I give no pledge that this study will reveal to us a remedy for the ills that man has brought upon himself, but I am not afraid to assert that we can never understand man and the problems which afflict him until we have learned to appreciate the ties which link man to his fellow creatures. As we pursue the study of this relationship we cannot fail to admire the wonders and beauties of Nature, and, as we compare the behavior of men with that of the beasts we may begin to understand that those whom, in our insolence, we call the lower' animals can teach us many lessons of tolerance and good will. And, for my part, I can envisage the possibility that, stifled as we are by the atmosphere of despair that has invaded the world, and crushed by the misfortunes engendered thereby, the contemplation of the harmonious equilibrium and the regularly renewed beauties of Nature may renew and strengthen in our hearts the hope that is well nigh dead, and that with renewed hope may come that humility which is the beginning and the hallmark of true wisdom" (Maurice 1948).

Endnotes

[1] *Keynote Address:* Presented at the national conference, "Animals and Humans: Ethical Perspectives," Moorhead State University, Moorhead, MN, April 21-23, 1986.

[2] 321 Cherry Hill Rd., Princeton, NJ 08540.

References

Childe, VG. 1962. *Man Makes Himself.* New York: New American University.

Frazer, JG. 1935. *Creation and Evolution in Primitive Cosmogonies.* London: Macmillan.

Gilbert, M. 1981. *Auschwitz and the Allies.* New York: Holt, Rinehart and Winston.

—. 1986. *The Holocaust.* New York: Rinehart and Winston.

Green, J. 1985. *God's Fool: The Life and Times of Francis of Assisi.* New York: Harper and Row.

Highwater, J. 1982. As native Americans perceive the world. In: Schwartz, M. ed. *TV and Teens.* Reading, MA: AddisonWesley.

Maurice, HG. 1948. *Ask Now the Beasts.* London: Society for the Preservation of Fauna of the Empire.

Montagu, A. 1949. The origin and nature of social life and the biological basis of cooperation. *J. Soc. Psychol.* 29: 267-83.

—. 1955. *The Direction of Human Development.* New York: Harper Bros.

—. 1960a. Why man weeps. *Think.* 26: 7-9.

—. 1960b. Why man laughs. *Think.* 26: 30-32.

—. 1962. *Prenatal Influences.* Springfield, IL: C.C. Thomas.

—. 1974. *Man's Most Dangerous Myth: The Fallacy of Race.* Fifth ed. New York: Oxford University Press.

—. 1976. *The Nature of Human Aggression.* New York: Oxford University Press.

—. 1978. *Life Before Birth.* New York: New American University.

—. 1981. *Growing Young.* New York: McGraw-Hill.

—. 1984. (editor). *Science and Creationism.* New York: Oxford University Press.

Pallis, M. 1940. *Peaks and Lamas.* New York: Knopf.

Ross, RW. 1980. *So It Was True: The American Prostestant Press and the Nazi Persecution of the Jews.* Minneapolis, MN: Univ. of Minnesota Press.

Salt, HS. 1923. Humanitarianism. In: Hastings, J. ed. *Encyclopaedia of Religion and Ethics.* 6: 836-40.

Schweitzer, A. 1934. *The Philosophy of Civilization.* London: A. and C. Black.

Thomson, JA and Geddes, P. 1931. *Life: Outlines of General Biology.* New York: Harper and Bros.

Wells, HG. 1922. *Outline of History.* New York: Macmillan.

Wyman, DS. 1984. *The Abandonment of the Jews.* New York: Pantheon Books.

—. 1985. *Paper Walls.* New York: Pantheon Books.

THE CASE FOR ANIMAL RIGHTS[1]

Tom Regan[2]

I regard myself as an advocate of animal rights—as part of the animal rights movement. That movement, as I conceive it, is committed to a number of goals, including:

—the total abolition of the use of animals in science;
—the total dissolution of commercial animal agriculture;
—the total elimination of commercial and sport hunting and trapping.

There are, I know, people who profess to believe in animal rights but do not avow these goals. Factory farming, they say, is wrong—it violates animals' rights—but traditional animal agriculture is all right. Toxicity tests of cosmetics on animals violates their rights, but important medical research-cancer research, for example—does not. The clubbing of baby seals is abhorrent, but not the harvesting of adult seals. I used to think I understood this reasoning. Not any more. You don't change unjust institutions by tidying them up.

What's wrong—fundamentally wrong—with the way animals are treated isn't the details that vary from case to case. It's the whole system. The forlornness of the veal calf is pathetic, heart wrenching; the pulsing pain of the chimp with electrodes planted deep in her brain is repulsive; the slow, tortuous death of the raccoon caught in the leg-hold trap is agonizing. But what is wrong isn't the pain, isn't the suffering, isn't the deprivation. These compound what's wrong. Sometimes—often—they make it much, much worse. But they are not the fundamental wrong.

The fundamental wrong is the system that allows us to view animals as *our resources*, here for *us*—to be eaten, or surgically manipulated, or exploited for sport or money. Once we accept this view of animals—as our resources—the rest is as predictable as it is regrettable. Why worry about their loneliness, their pain, their death? Since animals exist for us, to benefit us in one way or another, what harms them really doesn't matter—or matters only if it starts to bother us, makes us feel a trifle uneasy when we eat our veal escalope, for example. So, yes, let us get veal calves out of solitary confinement, give them more space, a little straw, a few companions. But let us keep our veal escalope.

179

But a little straw, more space and a few companions won't eliminate—won't even touch—the basic wrong that attaches to our viewing and treating these animals as our resources. A veal calf killed to be eaten after living in close confinement is viewed and treated in this way: but so, too, is another who is raised (as they say) "more humanely." To right the wrong of our treatment of farm animals requires more than making rearing methods "more humane;" it requires the total dissolution of commercial animal agriculture.

How we do this, whether we do it or, as in the case of animals in science, whether and how we abolish their use—these are to a large extent political questions. People must change their beliefs before they change their habits. Enough people, especially those elected to public office, must believe in change—must want it—before we will have laws that protect the rights of animals. This process of change is very complicated, very demanding, very exhausting, calling for the efforts of many hands in education, publicity, political organization and activity, down to the licking of envelopes and stamps. As a trained and practicing philosopher, the sort of contribution I can make is limited, but, I like to think, important. The currency of philosophy is ideas—their meaning and rational foundation—not the nuts and bolts of the legislative process, say, or the mechanics of community organization. That's why I have been exploring over the past 10 years or so in my essays and talks and, most recently, in my book, *The Case for Animal Rights*. I believe the major conclusions I reach in the book are true because they are supported by the weight of the best arguments. I believe the idea of animal rights has reason, not just emotion, on its side.

In the space I have at my disposal here I can only sketch, in the barest outline, some of the main features of the book. Its main themes—and we should not be surprised by this—involve asking and answering deep, foundational moral questions about what morality is, how it should be understood, and what is the best moral theory, all considered. I hope I can convey something of the shape I think this theory takes. The attempt to do this will be (to use a word a friendly critic once used to describe my work) cerebral, perhaps too cerebral. But this is misleading. My feelings about how animals are sometimes treated run just as deep and just as strong as those of my more volatile compatriots. Philosophers do—to use the jargon of the day— have a right side to their brains. If it's the left side we contribute (or mainly should), that's because what talents we have reside there.

How to proceed? We begin by asking how the moral status of animals has been understood by thinkers who deny that animals have rights. Then we test the mettle of their ideas by seeing how well they stand up under the heat of fair criticism. If we start our thinking in this way, we soon find that some people believe that we have no duties directly to animals, that we owe nothing to them, that we can do nothing that wrongs them. Rather, we can do wrong acts that involve animals, and so we have duties regarding them, though none to them. Such views may be called indirect duty views. By way of illustration: suppose your neighbor kicks your dog. Then your neighbor has done something wrong. But not to your dog. The wrong that has been done is a wrong to you. After all, it is wrong to upset people, and your neighbor's kicking your dog upsets you. So you are the one who is wronged, not your dog. Or again: by kicking your dog your neighbor damages your property. And since it is wrong to damage another person's property, your neighbor has done something wrong—to you, of course, not to your dog. Your neighbor no more wrongs your dog than your car would be wronged if the windshield were smashed. Your neighbor's duties involving your dog are indirect duties to you. More generally, all of our duties regarding animals are indirect duties to one another—to humanity.

How could someone try to justify such a view? Someone might say that your dog doesn't feel anything and so isn't hurt by your neighbor's kick, doesn't care about the pain since none is felt, is as unaware of anything as is your windshield. Someone might say this, but no rational person will, since, among other considerations, such a view will commit anyone who holds it to the position that no human beings feel pain either—that human beings also don't care about what happens to them. A second possibility is that though both humans and your dog are hurt when kicked, it is only human pain that matters. But, again, no rational person can believe this. Pain is pain wherever it occurs. If your neighbor's causing you pain is wrong because of the pain that is caused, we cannot rationally ignore or dismiss the moral relevance of the pain that your dog feels.

Philosophers who hold indirect duty views—and many still do—have come to understand that they must avoid the two defects just noted: that is, both the view that animals don't feel anything as well as the idea that only human pain can be morally relevant. Among such thinkers the sort of view now favored is one or other form of what is called *contractarianism*.

Here, very crudely, is the root idea: morality consists of a set of rules that individuals voluntarily agree to abide by, as we do when we sign a contract (hence the name contractarianism). Those who understand and accept the terms of the contract are covered directly; they have rights created and recognized by, and protected in, the contract. And these contractors can also have protection spelled out for others who, though they lack the ability to understand morality and so cannot sign the contract themselves, are loved or cherished by those who can. Thus young children, for example, are unable to sign contracts and lack rights. But they are protected by the contract none the less because of the sentimental interests of others, most notably their parents. So we have, then, duties involving these children, duties regarding them, but no duties to them. Our duties in their case are indirect duties to other human beings, usually their parents.

As for animals, since they cannot understand contracts, they obviously cannot sign; and since they cannot sign, they have no rights. Like children, however, some animals are the object of the sentimental interest of others. You, for example, love your dog or cat. So those animals that enough people care about (companion animals, whales, baby seals, the American bald eagle), though they lack rights themselves, will be protected because of the sentimental interests of people. I have, then, according to contractarianism, no duty directly to your dog or any other animal, not even the duty not to cause them pain or suffering; my duty not to hurt them is a duty I have to those people who care about what happens to them. As for other animals, where no or little sentimental interest is present—in the case of farm animals, for example, or laboratory rats—what duties we have grow weaker and weaker, perhaps to vanishing point. The pain and death they endure, though real, are not wrong if no one cares about them.

When it comes to the moral status of animals, contractarianism could be a hard view to refute if it were an adequate theoretical approach to the moral status of human beings. It is not adequate in this latter respect, however, which makes the question of its adequacy in the former case, regarding animals, utterly moot. For consider: morality, according to the (crude) contractarian position before us, consists of rules that people agree to abide by. What people? Well, enough to make a difference—enough, that is, *collectively* to have the power to enforce the rules that are drawn up in the contract. That is very well and good for the signatories but not so good for anyone who is not asked to sign. And there is nothing in contractarianism of the sort we are discussing that guarantees or requires that everyone will have a chance to participate equally in framing rules of morality. The result is that this approach to ethics could sanction the most blatant forms of social, economic, moral, and political injustice, ranging from a repressive caste system to systematic racial or sexual discrimination. Might, according to this theory, does make right. Let those who are the victims of injustice suffer as they will. It matters not so long as no one else—no contractor, or too few of them—cares about it. Such a theory takes one's moral breath away... as if, for example, there would be nothing wrong with apartheid in South Africa if few white South Africans were upset by it. A theory with so little to recommend it at the level of the ethics of our treatment of our fellow humans cannot have anything more to recommend it when it comes to the ethics of how we treat our fellow animals.

The version of contractarianism just examined is, as I have noted, a crude variety, and in fairness to those of a contractarian persuasion, it must be noted that much more refined, subtle, and ingenious varieties are possible. For example, John Rawls, in his *A Theory of Justice,* sets forth a version of contractarianism that forces contractors to ignore the accidental features of being a human being—for example, whether one is white or black, male or female, a genius or of modest intellect. Only by ignoring such features, Rawls believes, can we ensure that the principles of justice that contractors would agree upon are not based on bias or prejudice. Despite the improvement a view such as Rawls's represents over the cruder forms of contractarianism, it remains deficient: it systematically denies that we have direct duties to those human beings who do not have a sense of justice—young children, for instance, and many mentally retarded humans. And yet it seems reasonably certain that, were we to torture a young child or a retarded elder, we would be doing something that wronged him or her, not something that would be wrong if (and only if) other humans with a sense of justice were upset. And since this is true in the case of these humans, we cannot rationally deny the same in the case of animals.

Indirect duty views, then, including the best among them, fail to command our rational assent. Whatever ethical theory we should accept rationally, therefore, it must at least recognize that we have some duties directly to animals, just as we have some duties directly to each other. The next two theories I'll sketch attempt to meet this requirement.

The first I call the cruelty-kindness view. Simply stated, this says that we have a direct duty to be kind to animals and a direct duty not to be cruel to them. Despite the familiar, reassuring ring of these ideas, I do not believe that this view offers an adequate theory. To make this clearer, consider kindness. A kind person acts from a certain kind of motive-compassion or concern, for example. And that is a virtue. But there is no guarantee that a kind act is a right act. If I am a generous racist, for example, I will be inclined to act kindly towards members of my own race, favoring their interests above those of others. My kindness would be real and, so far as it goes, good. But I trust it is too obvious to require argument that my kind acts may not be above moral reproach—may, in fact, be positively wrong because rooted in injustice. So kindness, notwithstanding its status as a virtue to be encouraged, simply will not carry the weight of a theory of right action.

Cruelty fares no better. People or their acts are cruel if they display either a lack of sympathy for or, worse, the presence of enjoyment in another's suffering. Cruelty in all its guises is a bad thing, a tragic human failing. But just as a person's being motivated by kindness does not guarantee that he or she does what is right, so the absence of cruelty does not ensure that he or she avoids doing what is wrong. Many people who perform abortions, for example, are not cruel, sadistic people. But that fact alone does not settle the terribly difficult question of the morality of abortion. The case is no different when we examine the ethics of our treatment of animals. So, yes, let us be for kindness and against cruelty. But let us not suppose that being for the one and against the other answers questions about moral right and wrong.

Some people think that the theory we are looking for is utilitarianism. A utilitarian accepts two moral principles. The first is that of equality: everyone's interests count, and similar interests must be counted as having similar weight or importance. White or black, American or Iranian, human or animal—everyone's pain or frustration matter, and matter just as much as the equivalent pain or frustration of anyone else. The second principle a utilitarian accepts is that of utility: do the act that will bring about the best balance between satisfaction and frustration for everyone affected by the outcome.

As a utilitarian, then, here is how I am to approach the task of deciding what I morally ought to do: I must ask who will be affected if I choose to do one thing rather than another, how much each individual will be affected, and where the best results are most likely to lie—which option, in other words, is most likely to bring about the best results, the best balance between satisfaction and frustration. That option,whatever it may be, is the one I ought to choose. That is where my moral duty lies.

The great appeal of utilitarianism rests with its uncompromising *egalitarianism*: everyone's interests count and count as much as the like interests of everyone else. The kind of odious discrimination that some forms of contractarianism can justify—discrimination based on race or sex, for example-seems disallowed in principle by utilitarianism, as is speciesism, systematic discrimination based on species membership.

The equality we find in utilitarianism, however, is not the sort an advocate of animal or human rights should have in mind. Utilitarianism has no room for the equal rights of different individuals because it has no room for their equal inherent value or worth. What has value for the utilitarian is the satisfaction of an individual's interests, not the individual whose interests they are. A universe in which you satisfy your desire for water, food, and warmth is, other things being equal, better than a universe in which these desires are frustrated. And the same is true in the case of an animal with similar desires. But neither you nor the animal have any value in your own right. Only your feelings do.

Here is an analogy to help make the philosophical point clearer: a cup contains different liquids, sometimes sweet, sometimes bitter, sometimes a mixture of the two. What has value are the liquids: the sweeter the better, the bitterer the worse. The cup, the container, has no value. It is what goes into it, not what they go into, that has value. For the utilitarian, you and I are like the cup; we have no value as individuals and thus no equal value. What has value is what goes into us, what we serve as receptacles for; our feelings of satisfaction have positive value, our feelings of frustration negative value.

Serious problems arise for utilitarianism when we remind ourselves that it enjoins us to bring about the best consequences. What does this mean? It doesn't mean the best consequences for me alone, or for my family or friends, or any other person taken individually. No, what we must do is, roughly, as follows: we must add up (somehow!) the separate satisfactions and frustrations of everyone likely to be affected by our choice, the satisfactions in one column, the frustrations in the other. We must total each column for each of the options before us. That is what it means to say the theory is aggregative. And then we must choose that option which is most likely to bring about the best balance of totalled satisfactions over totalled frustrations. Whatever act would lead to this outcome is the one we ought morally to perform—it is where our moral duty lies. And that act quite clearly might not be the same one that would bring about the best results for me personally, or for my family or friends, or for a lab animal. The best aggregated consequences for everyone concerned are not necessarily the best for each individual.

That utilitarianism is an aggregative theory—different individuals' satisfactions or frustrations are added, or summed, or totalled—is the key objection to this theory. My Aunt Bea is old, inactive, a cranky, sour person, though not physically ill. She prefers to go on living. She is also rather rich. I could make a fortune if I could get my hands on her money, money she intends to give me in any event, after she dies, but which she refuses to give me now. In order to avoid a huge tax bite, I plan to donate a handsome sum of my profits to a local children's hospital. Many, many children will benefit from my generosity, and much joy will be brought to their parents, relatives, and friends. If I don't get the money rather soon, all these ambitions will come to naught. The once-in-a-lifetime opportunity to make a real killing will be gone. Why, then, not kill my Aunt Bea? Oh, of course I *might* get caught. But I'm no fool and, besides, her doctor can be counted on to cooperate (he has an eye for the same investment and I happen to know a good deal about his shady past). The deed can be done...professionally, shall we say. There is *very* little chance of getting caught. And as for my conscience being guiltridden, I am a resourceful sort of fellow and will take more than sufficient comfort—as I lie on the beach at Acapulco—in contemplating the joy and health I have brought to so many others.

Suppose Aunt Bea is killed and the rest of the story comes out as told. Would I have done anything wrong? Anything immoral? One would have thought that I had. Not according to utilitarianism. Since what I have done has brought about the best balance between totalled satisfaction and frustration for all those affected by the outcome, my action is not wrong. Indeed, in killing Aunt Bea the physician and I did what duty required.

This same kind of argument can be repeated in all sorts of cases, illustrating, time after time, how the utilitarian's position leads to results that impartial people find morally callous. It *is* wrong to kill my Aunt Bea in the name of bringing about the best results for others. A good end does not justify an evil means. Any adequate moral theory will have to explain why this is so. Utilitarianism fails in this respect and so cannot be the theory we seek.

What to do? Where to begin anew? The place to begin, I think, is with the utilitarian's view of the value of the individual—or, rather, lack of value. In its place, suppose we consider that you and I, for example, do have value as individuals—what we'll call *inherent value*. To say we have such value is to say that we are something more than, something different from, mere receptacles. Moreover, to ensure that we do not pave the way for such injustices as slavery or sexual discrimination, we must believe that all who have inherent value have it equally, regardless of their sex, race, religion, birthplace, and so on. Similarly to be discarded as irrelevant are one's talents or skills, intelligence and wealth, personality or pathology, whether one is loved and admired or despised and loathed. The genius and the retarded child, the prince and the pauper, the brain surgeon and the fruit vendor, Mother Teresa and the most unscrupulous used-car salesman—all have inherent value, all possess it equally, and all have an equal right to be treated with respect, to be treated in ways that do not reduce them to the status of things, as if they existed as resources for others. My value as an individual is independent of my usefulness to you. Yours is not dependent on your usefulness to me. For either of us to treat the other in ways that fail to show respect for the other's independent value is to act immorally, to violate the individual's rights.

Some of the rational virtues of this view—what I call the rights view—should be evident. Unlike (crude) contractarianism, for example, the rights view *in principle* denies the moral tolerability of any and all forms of racial, sexual, or social discrimination; and unlike utilitarianism, this view *in principle* denies that we can justify good results by using evil means that violate an individual's rights—denies, for example, that it could be moral to kill my Aunt Bea to harvest beneficial consequences for others. That would be to sanction the disrespectful treatment of the individual in the name of the social good, something the rights view will not—categorically will not—ever allow.

The rights view, I believe, is rationally the most satisfactory moral theory. It surpasses all other theories in the degree to which it illuminates and explains the foundation of our duties to one another—the domain of human morality. On this score it has the best reasons, the best arguments, on its side. Of course, if it were possible to show that only human beings are included within its scope, then a person like myself, who believes in animal rights, would be obliged to look elsewhere.

But attempts to limit its scope to humans only can be shown to be rationally defective. Animals, it is true, lack many of the abilities humans possess. They can't read, do higher mathematics, build a bookcase or make *baba ghanoush*. Neither can many human beings, however, and yet we don't (and shouldn't) say that they (these humans) therefore have less inherent value, less of a right to be treated with respect, than do others. It is the *similarities* between those human beings who most clearly, most non-controversially have such value (the people reading this, for example), not our differences, that matter most. And the really crucial, the basic similarity is simply this: we are each of us the experiencing subject of a life, a conscious creature having an individual welfare that has importance to us whatever our usefulness to others. We want and prefer things, believe and feel things, recall and expect things. And all these dimensions of our life, including our pleasure and pain, our enjoyment and suffering, our satisfaction and frustration, our continued existence or our untimely death-all make a difference to the quality of our life as lived, as experienced, by us as individuals. As the same is true of those animals that concern us (the ones that are eaten and trapped, for example), they too must be viewed as the experiencing subjects of a life, with inherent value of their own.

Some there are who resist the idea that animals have inherent value. "Only humans have such value," they profess. How might this narrow view be defended? Shall we say that only humans have the requisite intelligence, or autonomy, or reason? But there are many, many humans who fail to meet these standards and yet are reasonably viewed as having value above and beyond their usefulness to others. Shall we claim that only humans belong to the right species, the species *Homo sapiens*? But this is blatant speciesism. Will it be said, then, that all—and only—humans have immortal souls? Then our opponents have their work cut out for them. I am myself not ill-disposed to the proposition that there are immortal souls. Personally, I profoundly hope I have one. But I would not want to rest my position on a controversial ethical issue on the even more controversial question about who or what has an immortal soul. That is to dig one's hole deeper, not to climb out. Rationally, it is better to resolve moral issues without making more controversial assumptions than are needed. The question of who has inherent value is such a question, one that is resolved more rationally without the introduction of the idea of immortal souls than by its use.

Well, perhaps some will say that animals have some inherent value, only less than we have. Once again, however, attempts to defend this view can be shown to lack rational justification. What could be the basis of our having more inherent value than animals? Their lack of reason, or autonomy, or intellect? Only if we are willing to make the same judgment in the case of humans who are similarly deficient. But it is not true that such humans—the retarded child, for example, or the mentally deranged—have less inherent value than you or I. Neither, then, can we rationally sustain the view that animals like them in being the experiencing subjects of a life have less inherent value. *All* who have inherent value have it *equally*, whether they be human animals or not.

Inherent value, then, belongs equally to those who are the experiencing subjects of a life. Whether it belongs to others—to rocks and rivers, trees and glaciers, for example—we do not know and may never know. But neither do we need to know, if we are to make the case for animal rights. We do not need to know, for example, how many people are eligible to vote in the next presidential election before we can know whether I am. Similarly, we do not need to know how many individuals have inherent value before we can know that some do. When it comes to the case for animal rights, then, what we need to know is whether the animals that, in our culture, are routinely eaten, hunted, and used in our laboratories, for example, are like us in being subjects of a life. And we do know this. We do know that many—literally, billions and billions—of these animals are the subjects of a life in the sense explained and so have inherent value if we do. And since, in order to arrive at the best theory of our duties to one another, we must recognize our equal inherent value as individuals, reason— not sentiment, not emotion—reason compels us to recognize the equal inherent value of these animals and, with this, their equal right to be treated with respect.

That, *very* roughly, is the shape and feel of the case for animal rights. Most of the details of the supporting argument are missing. They are to be found in the book to which I alluded earlier. Here, the details go begging, and I must, in closing, limit myself to four final points.

The first is how the theory that underlies the case for animal rights shows that the animal rights movement is a part of, not antagonistic to, the human rights movement. The theory that rationally grounds the rights of animals also grounds the rights of humans. Thus those involved in the animal rights movement are partners in the struggle to secure respect for human rights—the rights of women, for example, or minorities, or workers. The animal rights movement is cut from the same moral cloth as these.

Second, having set out the broad outlines of the rights view, I can now say why its implications for farming and science, among other fields, are both clear and uncompromising. In the case of the use of animals in science, the rights view is categorically abolitionist. Lab animals are not our tasters; we are not their kings. Because these animals are treated routinely, systematically as if their value were reducible to their usefulness to others, they are routinely, systematically treated with a lack of respect, and thus are their rights routinely, systematically violated. This is just as true when they are used in trivial, duplicative, unnecessary or unwise research as it is when they are used in studies that hold out real promise of human benefits. We can't justify harming or killing a human being (my Aunt Bea, for example) just for these sorts of reason. Neither can we do so even in the case of so lowly a creature as a laboratory rat. It is not just refinement or reduction that is called for, not just larger, cleaner cages, not just more generous use of anesthesia or the elimination of multiple surgery, not just tidying up the system. It is complete replacement. The best we can do when it comes to using animals in science is—not to use them. That is where our duty lies, according to the rights view.

As for commercial animal agriculture, the rights view takes a similar abolitionist position. The fundamental moral wrong here is not that animals are kept in stressful close confinement or in isolation, or that their pain and suffering, their needs and preferences are ignored or discounted. All these *are* wrong, of course, but they are not the fundamental wrong. They are symptoms and effects of the deeper, systematic wrong that allows these animals to be viewed and treated as lacking independent value, as resources for us—as, indeed, a renewable resource. Giving farm animals more space, more natural environments, more companions does not right the fundamental wrong, any more than giving lab animals more anesthesia or bigger, cleaner cages would right the fundamental wrong in their case. Nothing less than the total dissolution of commercial animal agriculture will do this, just as, for similar reasons I won't develop at length here, morality requires nothing less than the total elimination of hunting and trapping for commercial and sporting ends. The rights view's implications, then, as I have said, are clear and uncompromising.

My last two points are about philosophy, my profession. It is, most obviously, no substitute for political action. The words I have written here and in other places by themselves don't change a thing. It is what we do with the thoughts that the words express—our acts, our deeds—that changes things. All that philosophy can do, and all I have attempted, is to offer a vision of what our deeds should aim at. And the why. But not the how.

Finally, I am reminded of my thoughtful critic, the one I mentioned earlier, who chastised me for being too cerebral. Well, cerebral I have been: indirect duty views, utilitarianism, contractarianism—hardly the stuff deep passions are made of. I am also reminded, however, of the image another friend once set before me—the image of the ballerina as expressive of disciplined passion. Long hours of sweat and toil, of loneliness and practice, of doubt and fatigue: those are the discipline of her craft. But the passion is there too, the fierce drive to excel, to speak through her body, to do it right, to pierce our minds. That is the image of philosophy I would leave with you, not "too cerebral" but *disciplined passion.* Of the discipline enough has been seen. As for the passion: there are times, and these not infrequent, when tears come to my eyes when I see, or read, or hear of the wretched plight of animals in the hands of humans. Their pain, their suffering, their loneliness, their innocence, their death. Anger. Rage. Pity. Sorrow. Disgust. The whole creation groans under the weight of the evil we humans visit upon these mute, powerless creatures. It *is* our hearts, not just our heads, that call for an end to it all, that demand of us that we overcome, for them, the habits and forces behind their systematic oppression. All great movements, it is written, go through three stages: ridicule, discussion, adoption. It is the realization of this third stage, adoption, that requires both our passion and our discipline, our hearts and our heads. The fate of animals is in our hands. God grant we are equal to the task.

Endnotes

[1] Reprinted by permission from *In Defense of Animals,* Basil Blackwell, Oxford, England. Paper presented at the national conference, "Animals and Humans: Ethical Perspectives," Moorhead State University, Moorhead, MN, April 2123, 1986.

[2] Professor, Department of Philosophy and Religion, North Carolina State University, Raleigh, NC 27695-8103.

A CASE AGAINST ANIMAL RIGHTS[1]

Jan Narveson[2]

Introduction

Down through the past decade and more, no philosophical writer has taken a greater interest in the issues of how we ought to act in relation to animals, nor pressed more strongly the case for according them rights, than has Tom Regan, in many articles, reviews, and exchanges at scholarly conferences and in print. It is a pleasure to join him on this symposium, to explore this interesting and important set of issues.

The importance of the issues is plain enough when we consider the recommendations to which Regan's important book, *The Case for Animal Rights* (1983) leads up: no animal food in our diets, virtually no use at all of animals for experimental purposes, and more. Few of us would be unaffected by so sweeping a set of strictures. And I know of no better, more philosophically powerful defense of animal rights than Tom Regan's work. It certainly behooves us, then, to follow his arguments with care, to appreciate the challenge of his position, and to provide a satisfactory alternative that gives us, if I may put it that way, a better shake. But there is another reason, as I have said before (Narveson 1977, 1980) : The issue of animal rights is one of the watersheds of moral philosophy. If we are indeed to leave ourselves in anything like the position we presently are inclined assume in relation to animals, an alternation in our understanding of the subject of a rather radical kind, may be in order.

To present all this to you in the time available for this talk is no easy matter. I shall have to be rather skeletal at points, and beg your indulgence if I have gone too far—in either direction!

I shall begin by outlining, as fairly as I can, Regan's view of the matter, and then sketch my alternative. Regan has in fact criticized certain aspects of my position at some length in his book. My replies to those criticisms will be largely implicit here, and I will not dwell on them when outlining his view.

Regan's Case for Animal Rights

The first three chapters of *The Case for Animal Rights* argue for the intermediate conclusion, that considerations of welfare, of well- or ill-being, do literally apply to animals. Animals do actually have a welfare or illfare that we can either cater to or ignore. It is interesting that some philosophers in the past have actually denied this claim. Ren ± Descartes comes foremost to mind, of course. So it is perhaps in order for me to point out that I accept this thesis without reservation. Animals surely do feel pain, have desires, perhaps emotions, and in some rudimentary sense, thoughts.

Regan, however, goes further than most of us would be inclined to on this important point. For example, he insists that animals have "concepts," and indeed that they have "perception, memory, desire, belief, self-consciousness, intention, a sense of the future—these are among the leading attributes of the mental life of normal mammalian animals aged one or more" (p. 81—all otherwise unattributed references are page references to *The Case for Animal Rights*). Obviously, we must wonder whether this hasn't been overstated. Most of us have had pets at one time or another and are familiar with the range of behavior on the basis of which Regan makes these strong attributions. To most of us, it will seem not plausible to assert that animals have all of these capacities without severe qualification. In particular, we would be inclined to point to the evidence concerning *linguistic* behavior as an indication that the mental life of animals is pretty thin stuff compared to that of normal humans. As Regan is aware, recent work on the linguistic capabilities of chimpanzees, the most promising of land-based mammals for these purposes, points to the conclusion that chimpanzees "do not have the ability for language acquisition equal to that of young children" (14). Regan was only concerned to explore this aspect of animal attainments as a test for their possession of *consciousness*. I accept the view that you can have consciousness without knowing any language at all. But linguistic ability is surely evidence of mental complexity, while the inability to acquire language strongly suggests lack of intellectual ability. And it seems that even very retarded human beings, evidently, are very far in advance of even very bright animals in this regard.

The problem with chimpanzees is not that they don't happen to have acquired such fluency despite their cognitive capacity for acquiring it: It is that they *lack that capacity*, so far as all current evidence indicates. Most of us would think that this is not a fact to be sneezed at. Of course we might be all wrong about this. But Regan's assessment of the facts here suggests that in his view, higher mental attainment is really *irrelevant* to the question of whether a being deserves moral consideration. And this seems to me, and I am sure to most people, very disputable indeed. Animals are conscious, indeed: but maybe being conscious isn't all that counts.

There are, of course, other extremely impressive animal skills one can point to. The singular homing abilities of some species, for instance, the incredible mechanisms for self-defense or attack, the astonishing (in human terms) capacity for self-sacrifice displayed by ants and bees, and so on. All these are certainly enough to intrigue even the most minimally curious among us. But do they provide support for the view that these beings have *moral rights*?

Interestingly enough, such attributes of animals have no obvious role to play in Regan's case at all. He is very concerned to show that animals have welfare, that they suffer and they enjoy, that they do indeed have *lives* to live and lose. But why, we may ask, should we be impressed by that? What we want to know is why these things *matter*: specifically, why they matter enough to give us good reason to refrain from the many activities in which we make use of animals to their detriment, e.g., by eating them or experimenting with them or wearing their fur.

In a crucial chapter (Ch. 4, "Ethical Thinking and Theory"), Regan articulates his methodology in thinking about these issues. A number of criteria are laid down for making an "ideal moral judgment," and more for appraising proposed moral principles, most of them uncontroversial. But I want to have a closer look at two which loom very large in his arguments: (1) impartiality and (2) "conformity to our intuitions."

Impartiality is something moral philosophers regard as uncontroversially essential to basic moral principles. We can't have a basic moral principle that says "such-and-such is right for John Jones but wrong for everyone else." Can we have one that says, "such-and-such is right for people, but wrong for everyone else"? Recent writers on this matter have coined the ugly word, "speciesism," to stand for precisely such views. They wish to hold that "speciesism" is as untenable as racism, or John Jonesism. (Many of you will be unfamiliar with this unusual view on moral philosophy. It says, in brief, that everyone must do whatever John Jones says.) I think they are right about this, actually; but not, I think, in just the way they think. We shall return to this.

Now, consider this argument of Regan's: "If to cause suffering is wrong, then it is wrong no matter who is made to suffer" (129). Not many would disagree. Indeed, how could they? No doubt *if* suffering, just as such, is wrong, it *follows* logically that it is wrong (prima facie) in all cases. But *is* it thus wrong? *Does* suffering matter, just all by itself? Or does there have to be a *reason* why the sufferings of a given being matter? That is the question; and it is one which simple logic will not decide for us.

This brings us to the other criterion: conformity to our moral "intuitions." Once upon a time, philosophers were inclined to argue that we have a special "faculty" of moral truth, a little black box in the soul that tells you when something is right and when not. Regan doesn't buy this view, of course—it's easily shown to be untenable. Another thing he doesn't mean is that we can test moral principles just by asking any Tom, Dick, or Henrietta whether they happen to agree with them. When he talks about intuition, he means of course, *reflective* intuition, as appealed to by such eminent philosophers as Henry Sidgwick, W. D. Ross, and John Rawls.

Nevertheless, there is an evident and classic objection to the use of intuition as a *criterion of correctness in ethical theory*: What do we do when intuitions conflict—when your considered intuition says that a certain kind of thing is wrong, mine that it is right? Regan has a reply to this, arguing that it "fails to recognize the difference between a principle's *being* valid for all and our *knowing* that it is," and he goes on to say that "When we have subjected an ethical principle to the tests of consistency, scope, precision, and conformity with our reflective intuitions, and when we have done this while making a conscientious effort to make an ideal moral judgment, we have done *all we can reasonably be required to do* to be in a position to justify that the principle is binding on all moral agents" (139-emphasis added). But this mistakes the issue, and indeed, sweeps it under a theoretical carpet. The problem is that if appeal to intuition is to be a *test* for the truth (= acceptability) of a proposed moral principle, then it is a test such that mutually contradictory proposed moral principles could *each* pass it. Such a test cannot, by definition, be *sufficient*. Once we allow that men of good will can both be wrong, we have as good as admitted that appeals to good will aren't enough, if what we are looking for is the truth.

The appeal to intuition, in short, is theoretically bankrupt. It cannot be a *source* of moral knowledge. Nor will it do to offer the pious hope that all men of good will *would* agree when they have gone through all the steps. For as long as an appeal to intuition is one of the admissible steps, there will be the same problem: The steps simply don't give us any good reason to expect agreement, for they offer us nothing to explain what we are agreeing or disagreeing *about* when we have moral arguments. So if people did agree, with sufficient use of intuition, it would have to be an accident, not a fact of theoretical significance.

We could partly solve problems of this kind by adopting a stance of tolerance and liberalism: You do it your way (in accordance with your intuition) and I'll do it mine (in accordance with my intuition). That, in fact, is part of the solution I recommend, and indeed substantially the solution currently accepted. But of course it is not a solution available to Regan, for in his view, my practice is *wrong*. That is not compatible with saying, "to each his own." And indeed, as is evident from his concluding chapter, if Regan had his way about it, I would not even be legally allowed to do things my way (e.g., eating hamburgers). The liberal idea is itself, of course, a moral stance, and not, for instance, another component of a neutral idea of "moral reasoning." But in its application to the present subject, it is not a stance Regan can take.

But let us return to the exposition of Regan's basic argument. On the one hand, basic moral principles must be impartial, universal, general. That is the major premise. The minor premise is an observation about current intuitions. There are two important things which, he thinks, we all believe: (1) that human infants, feeble-mindeds, and other of what I have elsewhere called "marginal cases" *have rights*, and (2) that they have those rights "in themselves" or inherently.

The cases I have just mentioned belong to a class of what Regan calls "moral patients," a term needing some explanation. A moral agent is, of course, one which acts, and in particular acts (or at least can act) *in the light of* moral requirements and other moral considerations. A "moral patient," by contrast, is one which we can affect by our actions, for better or worse, but which cannot itself act in response to moral considerations at all. Such, of course, would be animals. And one way of putting the basic question of this controversy is this: What are our duties, if we have any at all, to moral patients? (And, of course, why?)

Now the view which, as I say, Regan attributes to all of us concerning *some* moral patients, (namely human ones) is that we do indeed have duties to them, and moreover to them "in themselves," or "inherently." The contrast would be with what Regan calls "indirect" duties. Regan discusses these views extensively in his book[3], where they are characterized as holding that such duties are really owed to other beings, rather than to the animals, infants, etc., *themselves*.

Now as we will see, I do want to advocate a view of some such kind myself. But I am uncomfortable with this way of characterizing the distinction. Suppose that I have promised you that I will take care of your child if you die suddenly. You do so, and now here is the child on my hands. Do I not owe *that child* my love, concern, and attention, even though the reason I provide them is that I made a promise to you, rather than to it? (And even though I definitely do not think I have such a duty to all the other children in the world, or any of them for that matter?) The distinction, in other words, is really a distinction in the fundamental *grounds* of our duties rather than their *objects*, and it is somewhat misleading to characterize them in a way that invites confusion on this head.

We should also have a look at the distinction between "moral agents" and "moral patients," as Regan draws it. The former, as usual, are those who act and whose actions are liable for moral appraisal. The latter term, however, Regan defines in a narrower way: as referring to those who are patients and *not* agents. This is unfortunate, since all of us agents are capable of being affected for better or worse. I should like to put this by saying that we are *both* agents *and* patients. If we use the term "patient" in Regan's "exclusive" way, however, then we may have at least a terminological problem. For that may leave us bereft of suitable vocabulary for expressing what seems to me an important possibility: namely, that the passive *aspects* of agents may matter morally, even though they nevertheless matter only by virtue of belonging to agents—despite the fact that they are not those aspects of agents in virtue of which they *are* agents.

To explain this perhaps somewhat complicated idea a little more: Suppose that a certain moral agent—Sally, we'll call her—has, in her wisdom, taken out an insurance policy in virtue of which certain doctors and others have the duty to cater to her headaches. Now, headaches are passive conditions. Nevertheless, the reason why the doctors involved in this insurance plan have the duty to minister to Sally's headaches is that they've signed up for it with her insurance company. Those doctors would have the duty to cater to Sally even though it might not be true that those doctors had duties to other individuals who had headaches but no insurance. Nowadays, to be sure, many people do think that all doctors have a duty to fix all headaches, regardless. But even if that is true—which I doubt—it would remain that Sally's doctor has the duty to cater to *her* headaches over and above any obligation that doctor might have to "patients" in general.

The point is that we must not let ourselves be steamrollered into the assumption that patients are to be catered to *as such*. The example just given shows how the duty to cater to a certain need or want, even though that is a patient's aspect of the person in question, may nevertheless be founded on some agent's aspect of that individual. Another interesting example has to do with our dealings with loved ones. Regan agrees that we have the duty to loved persons to prefer a greater harm to someone else over a lesser harm to them, on occasion. And of course we shouldn't confine it to harms: we surely are also morally justified in preferring the lesser good of loved ones to the greater good of others. Given Regan's universalistic proclivities, there is a problem of how this can be justified. Regan, dipping once again into his fairly deep satchel of intuitions, simply puts these down as "*special considerations*" (316—his italics). But putting something in italics is not equivalent to giving an account of it. The fact is that love relations, though obviously directed toward others, are relations of self-interest—very deep ones though those interests may be.It is important to *you* that you chose that man or that woman, as friend, lover, spouse; that *you* wanted to have that child; and so for many other such cases. All that is accomplished by preferring that loved one's lesser needs or wants to the greater needs or wants of some stranger is that your and their lives are better as a result-not, for all you know, better than that of the stranger (by hypothesis, the marginal improvements to the respective lives in question are supposed to favor the stranger in these cases). If moral relations are generated essentially by rational agents promoting their own well-considered, long-run interests, these cases make sense. On the Reganian dispensation, they have to be listed as "special cases," no more said.

This brings us to the central issue: What, if any, duties do we have to patients *qua* patients, and why? And here, I fear, Regan's meta-ethical proclivities work ill. In his discussion of "Direct Duty Views," Regan has gotten to the point where he feels justified in rejecting theories because they don't support *his* intuitions about animal rights. Thus, for example, he rejects the "cruelty-kindness" view (i.e., that the reason we shouldn't inflict harm on animals is that to do so is to display a vice, cruelty) on the ground that "kindness is not something we *owe* to anybody, is not *anyone's* due" (199)—i.e., because this position would not imply that animals have the right not to be treated in those ways! But of course those of us who don't think that animals *do* have rights and who have not been persuaded by his other arguments, are not going to be impressed at this.

Nor, similarly, will we be impressed at his rejection of utilitarianism for similar reasons. His discussions of utilitarianism are of considerable interest in their own right, to be sure; and he is quite right to point out that utilitarianism does not provide a firm foundation for the things he is convinced of: vegetarianism, the almost complete prohibition on the use of animals for experimental purposes, and so on. But is this a reason for disavowing utilitarianism—or is it instead a reason for disavowing those particular Reganian intuitions?

Then Regan introduces his heaviest theoretical weapon, the notion of "inherent value" (Ch. 7, "Justice and Equality"). He denounces utilitarianism, among other theories, on the ground that in their view, individuals are "*mere receptacles*" of what has intrinsic value (viz., pleasure or pain, for instance). "They have no value of their own; what has value is what they contain..." (205). Opposed to this, we are told, is the idea that individuals are valuable in their *own* right; that they have "inherent value," this being a sort of value which is (1) *incommensurate* with other values, and in particular, (2) "not reducible to the intrinsic values of an individual's experiences" (235). Inherent value is a property such that possession of it can only be equal in that respect. It is not, apparently, capable of variation in degree. Any given thing either has it or doesn't. And if it does, it has (for instance) rights; if not, not. If we think of individuals as "receptacles" for what has value, then the idea is that "It's the cup, not just what goes into it, that is valuable."

One would like to have a clearer idea of just what this postulate is supposed to imply, and why. For instance: One might think, given this idea that it is the "cup, and not what goes into it," that is "valuable," this being a value grounding a sort of respect that is not commensurate with any other considerations of value, that, for instance, we should have an absolute duty not to commit suicide, since the person contemplating suicide does so merely because the life awaiting him or her looks intolerable-extremely painful, or boring, or whatever. These, however, are mere intrinsic values of contents, and presumably aren't supposed to be capable of overturning the treatment that is due an individual in virtue of that individual's "inherent value." Regan doesn't bring out this implication, and presumably doesn't think it follows. But if not, why not? We are not told.

Regan is obviously very taken with this notion, and we soon find him treating it rather as if it were a sort of *fact* about the individuals who, he thinks, possess it that they do so. Thus in searching for the ground of inherent value, we find him looking for the "relevant similarities" among the individuals who have this sort of value—rather than demonstrating to those of us whose moral intuitions don't seem to be quite up to the job of spotting this interesting property that we have overlooked it, or that we even need to keep an eye out for it!

Nevertheless, there is an argument here. Having laid down, or "noted" (as we should probably put it) that moral agents have inherent value, the question then is whether moral patients do. And Regan claims that "the attempt to restrict inherent value to moral agents is arbitrary" (239). The argument seems to be this:

1) Moral agents have inherent value,
2) it is in virtue of that value that we have the duty not to harm them,
3) but (some of) the harms we can do to them (and which are morally prohibited by [2]) are just like the harms we can do to animals, and
4) we have a direct duty not to inflict those harms on those animals.

Therefore,

5) those animals also have inherent value.

(My summary of his argument, 239-40.)

But the conclusion doesn't follow from its premises. There is no *logical* reason why the ground for prohibiting harm to individuals with inherent value couldn't be *different* from the grounds for prohibiting the same harm to other individuals who don't have it. The claim that it is "arbitrary" to do so is either another appeal to intuition, or, if meant as a stronger claim, is just false. Note again the case of the health-insured patient, Sally, to whom Dr. Crowne *owes* an operation, even though Sally's need for it is identical to that of Albert who, however, lacks the relevant insurance. It obviously *can* be the case that one person can have a right to the very same patient-affecting thing that another person, identical *qua* patient, does not have a right to. And there are various other possibilities. The point is that Regan's argument doesn't rule them out—even if we accepted his premises, which I do not.

Now, I am inclined to agree that we should reject the suggestion that it is the "intrinsic value of the experiences" of a given subject that constitute the whole and only relevant fundamental ground of treatment of that being. (Again, my case of the insurance policy will illustrate this well enough). But if we go on to ask why it should be plausible to make any such distinction, Regan's idea of "inherent value" doesn't help. For if we ask how we are to know whether a given organism *has* this property or not, then the only answer is that it's a matter of perception. We don't need, and can hardly have, an *argument* for the "view" that grass is green. But then, from the fact that grass is green nothing seems to me to follow directly regarding what we should do about it. We would need, in addition to that information, some such information as that we *like* green things, or that being green promotes some other states of affairs that we value, before we can get any conclusions about action. Similarly, why should the fact that a certain being would feel pain if I were to perform action X constitute a reason why I should refrain from X? Now, is it obvious that the "fact" that it has "inherent value" should have that implication? It is, I think, not obvious unless its having "inherent value" is just another way of stating that implication; in which case, we are back to the original question: Why should the fact that it would hurt organism X constitute a good reason for me to refrain from hurting it? Here, then, we turn to positive theory.

The "Why" of Rights for Marginal Persons and Animals

One important possibility in answer to my last question would be that I *dislike* pain—not only my own, but that of others. That would explain why I should refrain from inflicting it on them as well as on myself. But unfortunately, it seems that not everyone is constituted that way. And what do we say to *those* people? What Regan will say is clear enough: They are morally below par.

I don't expect that those people would be impressed by that answer. They'd like to know *why* it's wrong to inflict pain on other creatures, apart from the fact that Tom Regan and maybe some other people have a unique and apparently inexplicable intuition, *that* it is wrong, or the fact—if indeed it's a different fact!—that lots of other people, such as myself, have a considerable aversion to doing so. I take it that neither of those facts would be sufficient, in and of itself, to give these skeptics a reason for accepting that they morally ought to refrain from those practices.

Perhaps it will be asked why there needs to be a reason of the kind I am looking for: that is, one that would be acknowledged by the skeptics in question. For note that the right word is "would," and not "should," in any irreducible or, of course, any moral sense. That is: in coming on as I do, I am implying that if action X is morally wrong, then there has to be a reason from the point of view of *everyone* for accepting the claim that it is wrong. For a moral argument to be any good, it cannot merely preach to the converted, who in the present case are those who already accept the view that it is wrong to kill or inflict pain on animals, even if doing so serves our ends. If we say to them that they "should" acknowledge such a reason, then they will raise the question why they should do so. And if in saying that they should, the advocate of animal rights was merely reiterating his own view about animal rights, then his argument is a nonargument, an exercise in circular reasoning. It is for this reason that I think we must reject appeals to intuition in ethics: where they are really needed, they are useless and merely irritating.

It is satisfactory to show that these others "should" acknowledge a certain reason for accepting a moral principle if we mean, in so saying, one of the following two things: (1) that it would follow from other things that they already accept; or (2) that there is good reason to think that they would accept it if they would consider points (a-n), these being either demonstrable facts or items which, from what we know of their psychological makeups, would appeal to them. The proof of this latter pudding, of course, is in the eating: Show them (a-n) and see what it does. In the absence of either of these two ways of establishing to a given individual, B, that B should accept some moral view, it seem to me that we have not succeeded in establishing that view with him.

This explication of individual reasons amounts, I should note, to the following: A has a reason for doing something, X (of for accepting some practical claim, some claim upon A to do X) if it would on the whole conduce to the *realization of A's values* that A do X. A's values are the things that A values, that A holds to be valuable, attaches value to. (A less fancy—or is it fancier?—way of saying this is that it would "maximize A's utility.") Obviously I do not here mean by "values" A's *moral* values.

Now "he," in the preceding argument, refers to just *anybody who doesn't already accept* whatever moral principle we are trying to establish. This does raise a question just who "anybody" is here. For present purposes, I think, it is any member of the class of moral agents whose actions we think are properly subject to the principle being advocated. Let us call this group the relevant "moral community." We can then formulate my thesis about moral arguments as follows: That in order for an act X to be morally wrong, it has to be true that there is a reason for everyone in the relevant "moral community" to acknowledge its wrongness:a reason, that is, that *would move* that agent into action.

But this requires further explanation. For I do not think that we must show that every member of the community has sufficient reason, just like that, for refraining from doing X.It is not luminously clear to me, nor I think to most people,that we do always have sufficient reasons for refraining from doing X even when we agree that X is immoral. But this should not daunt us from insisting, nevertheless, that there must be good individually considered reasons for morality. What we need to look for good reasons for, namely, is the *imposition* of a uniformity of conduct on the community in question. This imposition is done in all the familiar ways: of criticism, especially, and of subtle rewards and punishments. And it is done by everyone, and done with respect to everyone. This, I think, is the essence of morality: It is a set of rules for everyone in the moral community in question, and to be imposed by everyone. We have a satisfactory rational moral rule when we can see how everyone in the community benefits from there being a rule of that kind, including those who might nevertheless consider themselves on occasion to be better off for disobeying that rule. The rationality of the rule may not prevent them from such disobedience. What it will do is to prevent them from having any sort of *case* against those who attempt to impose it.

Note that the set of individuals who are members of the community in question are the moral *agents* of that community. The thesis that an acceptable moral rule must also be acceptable, by way of being of benefit, to the non-agents in the community—infants, the very feeble-minded, and animals—is not part of the view I am arguing for.

"Why not?", it will be asked. The reason is fairly simple and straightforward, but pretty devastating to many views of morality. In order for morality to have a point, it must have point for any individual subject to it. Now, the set of organisms in my environment with whom I must deal in the course of life may be divided into two classes: those who are capable of engaging in voluntarily self-regulated behavior with respect to me, of a type that I can have some influence on, and those not thus capable.

This distinction, I believe, is radical and fundamental. With regard to those in the latter category, transactions from me to them are in a fundamental sense *one-way*: I can decide what to do about them, but they can't react by making adjustments in their principles of conduct, for they have none. But with regard to those in the former class, things are very different. When dealing with such entities, it is a fact I ordinarily would not be able to ignore, and certainly not one I could afford to ignore, not only that my actions make a difference to their behavior, and vice versa, but also that both of us can take a view in advance of what we are up to and communicate with the other what those lines of conduct might be and how we are going to interact with each other. Moreover, each of us knows that actions of the other are capable of being beneficial or detrimental to us, and this matters to us. Put this all together, and we have a case—an *individually considered case*—for considering the establishment of the sort of mutual principles known as "moral" ones. Each of us would prefer that others perform actions that are beneficial to us, or at very least not detrimental. And each of us knows that the other may well have options beneficial to him or her that are detrimental to ourselves. Obviously we have an interest in seeing to it that others by and large choose their actions from the former class and not from the latter.

In order to do this, however, we must make it a good thing from their point of view to adopt this preference. How do we do that? In part by threatening the other with undesirable consequences if he or she doesn't adopt that preference, and partly by promising similar behavior on our own part. In short, we shall adopt principles of mutual advantage.

But with regard to animals and other non-agents, this procedure isn't available. In the first place, communication with them is not efficient or extensive enough to enable any such mutuality of principle to be adopted. And in the second, they may not have enough to offer us to make it a good deal from our point of view to modify our antecedent line of behavior with respect to them—which is, roughly, to do whatever seems to serve our advantage best, regardless of its effect on them. This, indeed, is our "antecedent line of behavior" with regard to everyone, I take it; but we learn rapidly enough that as a principle of conduct for dealing with like-minded other beings, it just won't do.

It's not obvious that it won't do for animals and the like, however. We would seem to require some reason for abandoning that tendency that can hardly be of the same kind as it is when we consider our relations with our fellow rational beings. In their case, there's no difficulty, for anyone of remotely normal psychology, in seeing the reasons for limiting our tendency to consider each other only as useful or useless in relation to our own ends. But there certainly is such difficulty in regard to animals and the like.

A bit more can and should be said about the "and the like" part of this last sentence. All of the members of this class, that I know of, anyway, are what I have elsewhere referred to as "marginal cases" of human beings—to put it somewhat misleadingly. There is no intention to derogate entities so described, I hasten to add. Some of the humans for whom I have the greatest affection are, or until recently were, members of that set. My relations with some of those members, in short, are determined by sentiment. Which does not mean that they are not determined by reason. Nothing can be more reasonable than taking care of and giving pleasure to those we love and who love us. The point about sentiment, though, is that as far as any particular other being is concerned, we may *or may not* have it toward that individual. And where we do not, it is no good saying we "should." Either we do or we don't, so far as sheer, native sentiment is concerned.

Now it happens, though, that humans all come from other humans and that in the normal course of events, those other humans do have such relations toward us, and we toward them—not invariably, to be sure, but normally, and indeed overwhelmingly often. In general, in other words, so far as any particular "marginal human" is concerned, we may be virtually certain that there are a few people who have a sentimental interest in the well-being of that individual. Nor is that all. For it is also true that we have, ordinarily, little if anything to gain by treating those individuals badly;whereas those who love them have much to lose if we do. If we turn from the special cases of paraplegics, imbeciles, and the like, to infants, there is a further and extremely important consideration: These are the organisms which will become full-fledged persons by and by. How we treat them now has a great influence on how they will treat us later on. It has a great influence on what the future community will be like, and *that* matters to us. The fact that they are not, as they stand, original members of the moral community means that it is up to us to decide what their status is to be; but the reason for giving them substantial rights is very strong indeed. It is also true enough, I think, that there is a general and positive sentiment toward fellow organisms of our species, as claimed by David Hume. It is not, I think, of sufficient strength to do the job Hume wanted it to do, but it is strong enough, when supplemented by the other point I have just been developing, to support the claim that there is a general public interest in having humans treat other humans well or at least not badly.

But if we turn to the various species of animals, things are different. We do not have much in the way of sentiment toward animals in general, for one thing. And, more important still, it is not true that for each animal there is some particular human who has a great interest in its welfare. Nor is it the case that for each particular human, there are particular animals in whom that human has an enormous interest, so as to generate a public interest in the well-being of animals in general. Whereas there *is* a great interest in animals which gives us strong incentive to kill them prior to the point in time at which they might have died anyway: viz., the interest in using them for food. This is not a *vital* interest, so far as I know. My information is that we could do without animal products in our diets and get on well enough, nutritionally. It is rather in the nature of an aesthetic interest: we *like* hamburgers, chicken a la king, swordfish steak, and so on and so on. And in the case of animal experimentation, we have interests of a major kind in the sort of information we can get from such experiments. Some of this information is certainly essential for the greater well-being of the human race, though a good deal of it is not.

Now if, as Peter Singer and most other enthusiasts for animal rights would have it, we had to justify these experiments or this slaughter by showing that our interest is *greater* than that of the animals who lose their lives or endure discomfort in the course of our pursuit of those interests, then it seems to me it would be game over, for the most part. From the point of view of the cow, I suppose, your and my interest in steak and the like may be pretty trivial compared to its interest in remaining alive. I may, no doubt, be giving too much benefit-of-doubt to the cow; I suspect that your typical cow doesn't really care all that much about its future, and indeed I am sure that your typical cow has essentially no notion of its future. But for the sake of argument, let us grant the cow more than the facts strictly support. My question remains: Well, so what?

What we have here, it seems to me, is a sort of "existential" question, if I read the existentialists rightly (and I take no responsibility for doing so!). As individuals, we simply have to make an ultimate choice: How much do we care about the cow? Or the pig? Or the swordfish? And so on for the rest of the animal kingdom. As for my part, I am quite clear about the answer: Not very much! Moreover, I am sure that is true for most people as well.

But if this is true, then there simply is no public interest that would support the view that animals literally have rights. I do not deny that they logically *could* have rights. If infants can have them, so can animals. But it is we who bestow these rights, and we must bestow them on the basis of considerations of collective interest. It is not enough that some few, or even quite a few people, are inclined to grant rights to all animals of various species. The rest of us are involved here, and we are not so inclined. Nor, I think, is this a matter of majority vote. Whether you or I have the right to life against people in general is not a matter of majority vote, and neither is the question of animal rights. It does seem to me that once a majority got overwhelmingly large on such a question, the view that there is a public interest in animals sufficient to support the granting of rights to them would become plausible. But nothing of that sort obtains at present in this society.

Can we even make a case against wanton cruelty to animals on the premises I am operating on? Here I am inclined to think that the overwhelming majority I refer to does exist. Nearly everyone sympathizes with the pains of other beings sufficiently to have an aversion against sheer sadism with respect to them, and against total and utter disregard of those feelings. This may in part, I think, be due to a suspicion of psychological transference as well: We find it hard to believe that someone who could utterly disregard the pains of any being in particular might not do likewise with regard to at least some humans, and the fundamental interest of all of us in not being the victim of any such persons is strong indeed. But I don't think this is a very strong consideration.

On the whole, then, I see no good case for a general extension of rights to animals, at least at this level of generality. I do not rule out the possibility that some public-interest consideration might eventually be found. But I know of none at present. And until that time, I think we can eat meat and perform animal experimentation in good conscience.

Endnotes

[1] Paper presented at the national conference, "Animals and Humans: Ethical Perspectives," Moorhead State University, Moorhead, MN, April 21-23, 1986.

[2] Professor, Dept. of Philosophy, University of Waterloo, Waterloo, Ontario, Canada N2L 3G1.

[3] Regan's discussion, unfortunately, is confined to the rather primitive exposition of my views found in my "Animal Rights" (Narveson 1977). A much better locus, I think, is the succeeding "Animal Rights Revisited" (Narveson 1980); as this was available to him at the time of writing, it is puzzling that he did not use it. I hope that the present effort will strike him as an improvement.

References

Narveson, J. 1977. Animal rights. *Canadian J. Philosophy.* 7(1): 161-78.

—. 1980. Animal rights revisited. *Anim. Reg. Stud. 2.* Amsterdam: Elsevier. pp. 223-36. Also published in: Miller, H and Williams, W. eds. *Ethics and Animals.* Clifton, NJ: Humana Press. pp. 45-60.

Regan, T. 1983. *The Case for Animal Rights.* Berkeley, Los Angeles: Univ. of California Press.

THE CASE FOR THE USE OF ANIMALS IN SCIENCE[1]

James A. Will[2]

Animals are now used extensively in research and teaching, and the appropriateness of their use appears to be questioned. Some people believe that we are in a new era where the animal activists have become much more influential, and that the antagonism between the scientists and these groups is worse than it ever has been. This does not appear to be the case. The preeminence of various influences seems rather cyclic, even perhaps influenced by such things as economic conditions or wars. At present, the question is often asked, "Should we continue to use animals in science?" The real question should be, "How do we use animals in research and teaching responsibly?" Anyone asking the first question begs credibility, while the second question implies that the questioner is realistic and responsible, with a concern for humanity as a whole.

By way of presenting the arguments for the continued use of animals in research and teaching, I will discuss responsibility of developed society as a whole, the responsibility of individuals, societal and individual priorities, the process of discovery in science, the safeguards for animals in science, and so-called alternative methods.

The European, some Asian, and the American societies (and especially the latter), are very privileged societies that constitute a small minority of the world's population. At present, this small minority controls the destiny of most of the world; rightly or wrongly. Thus, in many ways we must be held accountable not only for our own destiny, but also for the destiny of the world as a whole. We are privileged in that most of us know full well where our next meal is coming from, we have shelter available even though it may be our choice not to use it, and we have the opportunity to be instructed or even educated. It is in this state of being that we can enjoy the luxury of thinking about our responsibility toward other human beings, and other creatures in the world that may be as sentient as we but unable to control their own destinies.

We have a responsibility to these thousands of millions of people that is as demanding as any priority we establish. A look at the statistics in figure 1 proves the magnitude of some of these disease entities. Figure 2 shows that the distribution area of malaria is not small, and that in the past or in the future much more of the world has been or may yet be threatened.

Diseases like AIDS have seemingly mutated in man from a nonpathogenic and therefore non-threatening organism in non-human primates, to a very virulent killer. Only prior research on this class of virus has allowed us to move as quickly as we have to identify the virus and develop a rationale for intervention.

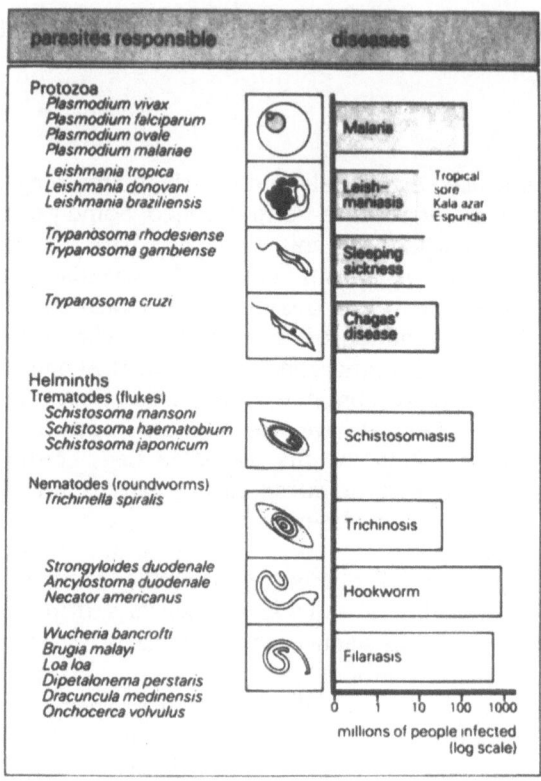

Figure 1. The most important parasitic infections of man and the numbers of people affected. The precise figures for leishmaniasis and sleeping sickness are not known. (Figure used with permission from *Immunology*, 1985, by Roitt, Brostoff and Male. Glower Medical Publishing, London.)

If we are to concede that there is some societal value in the use of animals in science, how do we assign responsibility for the appropriateness of animal use? In our present society, it is easy to put the blame for inappropriate use of animals on someone else. There is no escape from the fundamental responsibility for our own actions. It is the responsibility of each person involved in the sciences which use animals to assume responsibility for their own actions. There can be no shifting of this responsibility to the unaccountable "they."

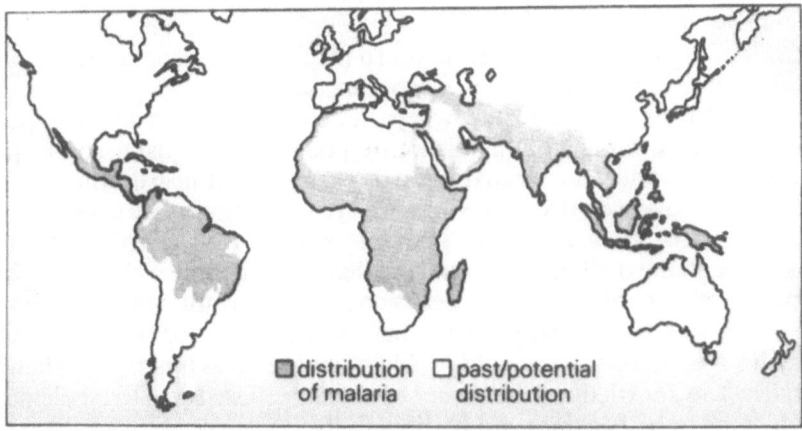

Figure 2. The geographical distribution of malaria is shown for 1981. The potential distribution is also shown and this represents a "time bomb" we live with. Diligence and enormous research efforts are the only protection society has today. (Figure used with permission from *Immunology*, 1985, by Roitt, Brostoff and Male.Glower Medical Publishing, London.)

To my way of thinking, each person must decide for him- or herself what their decision will be regarding this whole question. The decision to eat meat, to hunt, or to use the products of research which required animals are all fractions of a central question concerning our relationship with animals. Because these are personal decisions, there is no reason why we should not develop a pattern that is consistent. Furthermore, since these are personal decisions, we should respect the right of each person to live by the decisions they make. I believe it is also consistent that we put priorities on our concerns about how animals are cared for and used because at any given time the resources of humanity cannot mitigate, much less satisfy, all of the needs perceived as being necessary and good for the world. I believe that these priorities should be exercised in the most responsible way to alleviate the suffering of all animals, including man, as equitably as possible. With our present state of knowledge, we cannot accomplish the task of improving or even maintaining the health of humans and animals without the use of experimental animals in research and teaching.

As a practical person who knows little, if anything, about theoretical philosophy, it seems unimaginable to me that we could exercise our responsibility toward other peoples and higher animals in this world without the use of animals in research and teaching. Whether we like it or not, we live in an evolutionary era characterized by the necessity for animals in science. This era may pass, but not in our lifetimes and most likely not during the lifetimes of our children and grandchildren. How then shall we conduct ourselves in this era? Responsibly. There is no room in good science for any other standard. This does not mean that there will not be waste in the use of animals, nor will we abolish abuse. People are not infallible, and no plethora of regulations or laws will govern man's moral sense; society cannot afford one policeman per animal experiment, nor can society afford one policeman for any single situation where there is potential for abuse. Appreciation and understanding of the relationship between the quality and responsibilities of life will only change through example and instruction.

How then have we arrived at where we are today, how does science really work, and what does the future portend? Science has arrived where it is today through the dedicated lives and intelligence of people with a more-than-average level of curiosity and the persistence to pursue this curiosity. Without them, it is safe to say that probably half of us would not be on the earth today. We are still fighting age-old problems and seem to find new challenges each day. The knowledge that a better diet and exercise seem to be elements that help us to be healthier and perhaps to live longer is not something that suddenly arose from nowhere; nor is it knowledge based on evidence from a single experiment. The truth behind information such as this is the result of hundreds, and more likely, thousands or millions of experiments testing many hypotheses. One might wonder about the relevance of many of these hypotheses at the time they are tested by experimentation. Sometimes these experiments were very basic in nature and seemingly totally unconnected to the problem they are used to solve even generations later. Because of the ethical considerations, the problem of establishing the "need" to do an experiment "at this time" is not one that is easily solved. Many instances can be demonstrated in which an observation made in an experiment 10 or even 70 years earlier provided the key or clue to a mechanism that now explains how something works, and, knowing how this system works allows therapy or preventive treatment to be developed and implemented (Comroe 1977). It is easy to look through our retrospectroscope and proclaim that this mechanism was very obvious and that the animal experiments done were unneeded. An example that is frequently cited is the recent remarkable decrease in heart disease attributable to community education programs. The community programs were important and even essential to effect this great change in the incidence of heart disease, but no one should forget that the knowledge allowing us to develop these programs came from very basic research, much of it in animal models. The world continues to be challenged with new threats to a healthful existence, with Legionnaires' disease, with AIDS, and with the major causes of morbidity and mortality (which are not such civilized diseases of affluence like heart disease or cancer), as well as the parasitic diseases such as malaria. These parasitic diseases have an obligate requirement for a whole animal host to complete their cycle. Epidemiological evaluation of these diseases in affected populations has provided valuable insight into the disease syndromes, but such evaluation is ineffective in answering critical and basic questions. Tissue culture is widely used to study portions of the cycle and to examine particular mechanisms, but in the end, only the whole animal can confirm or reject the hypotheses proposed.

How does science work in the typical situation? Usually this process begins with an observation that something occurs in a specific situation. Questions begin to proliferate. Someone comes up with a theory as to how the phenomenon might be explained and possible mechanisms for the explanation. From these discussions, a theory develops and then the hypothesis or hypotheses to test are formalized. At this point it is most appropriate to do some pilot studies to decide if there is a chance that the hypothesis might be plausible and go further, or to reexamine the problem and develop new hypotheses. Up to this stage of the process, the numbers of animals used are very small. We assume that before using any animals, the investigator has searched the literature to find clues that either support or dispute the hypothesis. At this point, the investigator reviews the literature, critically asking such questions as was the animal model appropriate and will age make a difference? The investigator also considers the implications of the genetic strain used, looks for clues in the articles that would indicate that the animals used were not disease free (a factor which might influence the results), evaluates the analytical techniques used to determine if they are adequate in light of today's technology (and therefore if the conclusions would likely be the same), and finally, determines if the statistics used were adequate to discriminate between possible type one and type two errors. These are just general questions that the investigator must ask; for each specific area of science the questions are more specific and appropriate to the particular hypothesis to be tested.

I hope from this example to have demonstrated that the problems of duplication in research are not simple. Most so-called "duplication" is research that is taking place almost simultaneously at several places in the world, and this type of duplication is virtually impossible to prevent. This is just as much a facet of progress as the competition between two manufacturers of automobiles or household appliances.

This is where we are today—in an era of fantastic growth in knowledge and yet as naive as newborns about the mysteries of life and how biology works. Although we have made great strides, it seems that each answer we get is accompanied by a hundred additional questions, each a little more difficult to answer. What are the prospects for the future? We will never be a disease-free society; new challenges to our intellect and resources will come as fast or faster than our solutions to the existing problems. If we slacken our efforts to meet these challenges, we will be worse off as a society than we are today or were perhaps even decades ago. The plagues we are presently experiencing are just new names added to the lists of the old.

There are safeguards already in place for the animals used in research and teaching. The radical groups of animal activists are not satisfied that these safeguards are sufficient because the stated goal of these groups is to abolish the use of animals in research and teaching. I do not believe that this is a reasonable stance, nor do I believe that the present system of regulations is cost effective. By this I mean that the pendulum has swung too far in the direction of over-regulation of animal protection for very little increase in abuse prevention and is therefore not a good use of the world's finite financial and natural resources. I would make the point that a large majority of investigators are very concerned and compassionate individuals who abhor animal abuse as much or more than the most concerned animal activist. Guidelines for the responsible use of animals such as the ones we all adhere to now are not new either. Dr. H. Newell Martin, the pioneer physiologist who became the first head of biology at Johns Hopkins University in 1876, drew up guidelines for animal experimentation in March 1885 (Fye 1985). The Animal Welfare Act is certainly not the first national or state effort to protect animals from abuse, nor will it be the last.

The institution I represent has undergone a change in the last years, not in the level of responsibility, but in awareness of what this responsibility means and the public expression of this responsibility. This change probably occurred as a result of the efforts of the animal activist movement. Although I firmly believe that there was, in fact, little abuse of animals in teaching and research, there have been changes within science itself that have tended to improve the care and use of animals. One of the greatest influences has been the progress in cellular and subcellular research that allows us to do more definitive experiments capable of elucidating mechanisms at this level.

Another reason for change is, purely and simply, money. As research becomes more sophisticated, the cost of doing each experiment becomes greater. This means that investigators can no longer afford to use animals that are not best suited to test an hypothesis. Special strains and animals that are disease free are much more in demand than they were 10 years ago. This does not mean that animals of random breeding are no longer useful. It is by using such animals that idiosyncratic responses were uncovered. This makes sense if you think about it. We know that there is a great deal of genetic variation in man many times manifested in overt syndromes resulting in an easily visible deformity. Many more genetically different characters in man are not apparent through casual observation; they are not expressed as phenotypic but as genotypic characteristics. It seems logical then to suppose that this same degree of genetic variation occurs in the random populations of other animals. This is the way things happen in nature. The most important finding of an experiment using random source animals may be that one animal responds much differently than the majority. An inquisitive investigator will ask "why?" and this inquiry may lead to a heretofore undiscovered mechanism. The ability to recognize and utilize serendipity and the role of the importance of the differences between experimental results is a sign of maturity in an investigator. This does not mean that every unusual result should be followed and the original direction of inquiry dropped. This maturity requires a trained mind and training always requires some waste; some mistakes. In this way, science is not a mysterious society, but, in fact, little different from any other occupation.

What are we doing at The University of Wisconsin to increase awareness and to help ensure that abuses will not take place? We believe very strongly in education, as opposed to oversight which smacks of policing. We have a program of mandatory certification. This program is oriented to making each person who uses, cares for, or supervises the use of animals in research and teaching aware of his or her responsibilities under the current laws and regulations. The program also deals with the historical perspective of animal regulation, development of ethical considerations, and the zoonotic disease potential of contact with research animals. There is a test associated with the document entitled "The Responsible Care and Use of Laboratory Animals." This test is not intended to be a test of anyone's intelligence, but rather acts as a form of certification demonstrating that the people who will be using animals have taken the time to read the document and are aware of their responsibilities under the current laws and the University policy. Thus far, we have certified more than 1600 people on the Madison campus alone. Variations of this program are in place at each of the 26 components of the University of Wisconsin System. At present, this certification is the only mandatory requirement. We have offered seminars and have now erected 21 poster boards in the largest animal units on campus where we are attempting passive teaching through attractive displays. Our goal is to raise the level of awareness and to increase the level of expertise of all those who require animals in research and teaching.

Another question that arises is, "Why not use alternatives, which are in most cases, really adjunct methods?" I and other authors have addressed this issue at length in previous publications (Smyth 1978; Will 1985; Fox 1986). The first argument is *for* the use of adjunct methods. Examples of adjunct methods that can be very useful are tissue culture and computer models. These provide different kinds of information. The computer models usually provide more general kinds of information than does tissue culture. For example, at our present state of knowledge, a computer can tell us if it is probable that an enzyme system is present, what the system appears to do, and at what metabolic site it may act. The computer can be right or it can be entirely wrong, depending upon our state of knowledge. By this I mean that a computer can only make decisions based on the data we are able to supply. These data usually come from experiments that have been performed in the whole animal first to make the original observation that a certain phenomenon occurs, then from more definitive experiments that let us test hypotheses in experimental animals. At this point, perhaps, critical experiments using an adjunct method are used to define a mechanism, and finally the hypothesis is most probably tested in the heterogeneous environment of the whole animal once again to see if it really works in the way we have surmised. Many times when whole animals are used, the experiments do not result in severe pathology or mortality and, as in the case of large domesticated animals, they are returned to the herd or flock.

As I have indicated, the adjunct method may play a very important role in defining the mechanism; however, the fact that the mechanism works in a certain way in tissue culture, for instance, does not always mean that it works this way in the whole animal. In the whole animal, many other systems may modify the action of the mechanism. These mechanisms may only work under certain disease conditions and not in the normal laboratory animal. It is as important to understand when the mechanism will not work as it is to know when the mechanism is operative. Tissue culture has its strong and weak points. The term "tissue culture" is a broad one, and encompasses several types of organ and cell culture techniques. If organ culture is used, it means that an animal is killed, and the organs removed and studied, either whole or in parts in the isolated state. Presumably organ culture may decrease the numbers of animals used, but this is not always the case.

Primary cell cultures are a second type of tissue culture, and these cultures may be quite different from established cell lines. Primary cell cultures are established by killing an animal, harvesting the cells required, and growing the cells in a culture medium. As long as these cells stay alive, they usually respond as they would in vivo. However, when the cells divide or reproduce, they may revert to a more primitive cell type with properties that are no longer the same as those of the cells when present in the animals or tissue (or organ). Cell lines established in this manner are available as frozen cells that researchers can order from a cell bank or collection. These banked cells are useful in that they are predictable in response to various environmental and other exogenous influences, but these cell lines may bear no resemblance to primary cell cultures or to the cells as they are in vivo.

There are some diseases in which the causative organisms will not grow in tissue culture. Examples are the organisms that cause leprosy, foot and mouth disease, herpes virus infections of man and animals, and many respiratory viruses. It seems amazing that we still haven't conquered leprosy, a disease well described in biblical times. Peculiarly enough, the only good model to work with in this disease is the armadillo. The results of a diagnostic test for leprosy in the armadillo are directly transferable to man.

In other situations the fact that the response is not the same as in man is equally important. Foot and mouth disease is a scourge of cattle and other ruminants in many parts of the world. In the research with this disease, there are many instances of a complete lack of correlation between the in vitro and in vivo responses to antigens or vaccines. Schistosomiasis (see figure 1) is a major disease of the world and in this particular disease, no correlation is demonstrated between the presence of the causative organism, the intensity of the disease, and immune status of the victim. For example, one of the most frequent maladies of schistosomiasis is a kidney disorder, and it is difficult to associate the severity of kidney disorder with disease intensity or immune status.

What should we conclude about whether the whole animal or an adjunct method is most appropriate to use in a specific instance? The most appropriate method to answer the question is that method which offers the best results, regardless of the cost or time involved. In the practical situation, it does often not work this way because equipment, expertise, and money to change technologies to use an adjunct method may not be available. In this instance, adjunct methodology may not be cheaper. Furthermore, granting agencies are reluctant to fund an individual who proposes to change technology because agencies believe that the investigator may not have the requisite expertise to make the grant worthwhile and productive using this new technology. All of these factors tend to impede the use of adjunct methods. In virtually all documented cases of development of new technology, the development has occurred because the new method improves research capability, i.e., improves resolution by having greater specificity, increased sensitivity, or lower cost. Most investigators would be eager to use any method that would allow them to use fewer or no animals at all, but somewhere along the line the drug or mechanism proposed must be tested in the whole animal. In cases where the organs must be collected, this obviously must be done in an experimental animal.

I have attempted to provide, in an unemotional way, a perspective that justifies the prudent and responsible use of animals in research and teaching. I would remind the reader of what I perceive to be the responsibility of those of us in the world who are more fortunate than others. I believe the era for the need to use animals in science is with us now and will not end in the foreseeable future, but as our knowledge increases, we will have less dependence on animals. In the meantime, it is imperative that every investigator follow the principle of the three Rs in animal use—reduction, refinement, and replacement—wherever possible.

Endnotes

[1] Paper presented at the national conference, "Animals and Humans: Ethical Perspectives," Moorhead State University, Moorhead, MN, April 21-23, 1986.

[2] Professor, Veterinary Science, College of Agricultural and Life Sciences, Anesthesiology, Medical School, and Director, for the Graduate School, Research Animal Resources Center, University of Wisconsin, Madison, WI 53706.

References

Comroe, JH, Jr. 1977. *Retrospectroscope: Insights into Medical Discovery.* Menlo Park, CA: Von Gehr Press.

Fox, MH. 1986. *The Case for Animal Experimentation.* Berkeley, Los Angeles, London: The Univ. of California Press.

Fye, WB. 1985. H. Newell Martin—A remarkable career destroyed by neurasthenia and alcoholism. *J. of the Hist. of Med. and Allied Sci.* 40: 133-66.

Smyth, DH. 1978. *Alternatives to Animal Experiments.* London: Scolar Press.

Will, JA. 1985. An overview of alternatives. *Lab Animal.* Jan/Feb: 37-39.

THE CASE AGAINST THE USE OF ANIMALS IN SCIENCE[1]

Donald J. Barnes[2]

I was very good at my job as an experimental psychologist, Principle Investigator; Chief, Performance Decrement Function; Survivability/Vulnerability Branch; Radiobiology Division, United States Air Force School of Aerospace Medicine; Brooks Air Force Base, Texas. I received many Outstanding performance ratings, and was promoted rapidly to an administrative position. My office was far removed from the laboratory, from the pain and suffering of non-human animals; I did not have to hear their screams, to see them struggle against the bonds of restraint, to watch them languish in the spotlessly clean stainless steel cages, separated from their fellows as well as from their natural environment. I could order them trained, order their deaths, expect the data print-out sheets which would serve as their death certificates, write a paper and wait for the recognition which would move me another step from the laboratory and another rung up the professional ladder. In the process, I could routinely sign the assurance of compliance with the Animal Welfare Act; just one more signature in a series of bureaucratic procedures. I had no reason to admit to causing the animals pain, so I did not. Who was to know? Who was to care?

As a scientist long committed to the understanding, prediction, and control of biological, physiological, and behavioral events, I have no objection to the animal as a legitimate focus of science. As a parent, a son, a sibling, and the proud recipient of unconditional positive regard from a few special people, I am vitally interested in matters of health and in the most ethically efficient use of available resources. As a member of a species which has evolved sufficiently to allow the relatively broad perspective of a "web of life" and at least a rudimentary concept of altruism, I have laboriously struggled against my individual egoism in an attempt to meet greater responsibilities to my fellow humans, to other animals, and to the planet which gives succorance to us all.

Although I regard the animal as a legitimate focus of research, I am irrevocably opposed to vivisection, the practice of inducing disease or trauma in a healthy animal for the hypothetical sake of another animal or another species of animal, for there is but one rationalization for such research: "The ends justify the means."

The more knowledgeable and astute among you will realize that I've made my point and should retire, but I do not enjoy the luxury of addressing you as tabulae rasa, nor do you have the luxury of being so. I do not insult you; on the contrary, I pay you the respect of challenging you to break the bonds of conditioning to enjoy the freedom of innovative thought.

If I were you, I might be thinking, "This guy is twice as pompous as William F. Buckley and only half as bright!" And you would be half-right at most, for I am not pompous, I am bitterly embarrassed. Embarrassed because I spent 44 years of my life believing that non-human animals were mine to exploit, bitter because no one helped me find my way out of the anthropocentrism which ostensibly justified my behavior, bitter because my own background imbued me with a "conditioned ethical blindness" which is one more disease which cannot be cured through the use of nonhuman animals in the laboratory.

As a very young boy, I was given a bounty for every "varmit" I shot or trapped, for they competed with our livelihood as small farmers during the Depression. I was rewarded for killing "game" for the table, for catching fish for our consumption. I was taught to kill the biggest buck and to catch the most fish. I was taught to raise animals for market, and I was taught to butcher them and enjoy their flesh. Finally, I was taught to use them in experiments, and I spent 16 years watching them suffer and die for my scientific career. I'm embarrassed because I didn't take their pain seriously; I'm embarrassed because I forgot about their needs; I'm especially embarrassed because I wasn't bright enough to see the obvious, that all creatures have the right to life without unnecessary exploitation, that all animals, including human animals, are ultimately dependent upon all other life forms for their very existence.

During the many years I worked for the U.S. government, I was required to attend annual courses designed to promote racial equality. At first, they called these mandatory two- or three-day sessions "Race Relations," but, after a time, even the title seemed racist, so they changed it to "Human Relations." I enjoyed these interludes from the ennui of the laboratory, for the discussions and debate centered around values and conditioned perceptions, and, frankly, I loved to watch the racists squirm in their seats. One day, a full-bird Colonel, obviously irritated that his valuable time should be so wasted, blurted out, "I don't care what color a man is, as long as he's an American!"

Ignoring the sexist implication (to which I was not yet attuned), I responded, "Colonel, isn't it time we began thinking in terms of a world community? Isn't nationalism as dangerous as racism?"

He was furious, but said nothing—to me. He did report me to the Commander, however, who reported me to my Division Chief, who reported me to my Branch Chief, and I eventually heard about it. They thought I was a communist! No! Even worse! They thought I was a "nigger-lovin'" communist!

Is it wrong to think in terms of a world community? Might that community include other species, other life forms, as well as other humans?

I'm no longer seen as a "nigger-lovin'" communist. Now I'm seen as an "animal-lovin'" misanthropist. Is there really any difference?

I'm not an "animal lover." I know very few animals; some of those I love, most I do not. Some are human; some are not. Isn't it "respect" we're really talking about? After all, I don't have to love you to respect your space; do I have to love other animals to respect theirs?

Do you realize we don't even have enough respect for laboratory animals in our country to count them? We're pretty sure we use between 17 and 70 million laboratory animals each year in the United States. Come to think of it, that rather epitomizes the estimates of error surrounding the use of nonhuman animals as surrogates for humans. How many sets of pliers are issued to Department of Defense mechanics in the United States each year? Who knows? Who cares? I care. You should care.

And pliers don't even feel pain. We do. Other animals do.

Are pliers necessary? No, there are many other tools, such as specialized wrenches, which will do the job better—and countless more which have yet to be designed and built.

Are non-human animals necessary as laboratory tools? No, there are many other tools which will do the job better—and countless more which have yet to be conceptualized and used.

But, I've invoked the concept of "necessity." My opponents insist that the use of non-human animals in medical and biomedical research is absolutely necessary. For humans.

There was a time, not very long ago, when the slave owners in America insisted that the institution of slavery was an absolute necessity for the economic viability of our country. It was "necessary" to drive the Indian from his land, inter all Japanese-Americans during World War II, disallow the vote for women, and use children in the sweat shops of the greedy. It was "necessary" for us to fight in Viet Nam and to attack Libya. Even though the World Health Organization tells us there are 220 drugs required for a nation to sustain its citizens' health and we now have ready access to over 20,000 drugs, we are told of the absolute necessity of using non-human animals to develop even more drugs.

My opponents sputter, "You can't do research on blindness in a test tube! We need intact, functioning systems to be able to understand the interactions of those systems."

What presumptuous statements! Such pronouncements assume we understand the underlying mechanisms of physiological systems. We do not. Such statements imply that we can control all important variables which comprise an intact, functioning system. We can not. In the past when I heard these statements, I used to think my opponents were reading too much Lewis Carroll. Remember? "If you say a thing three times, it must be true." I've now decided they haven't read enough Lewis Carroll, for his jocular absurdities have become their realities.

"Nonsense!" we are told, "All major medical advances have come through the use of animals in research!" Keeping firmly in mind that all humans are animals, this statement is irrefutable, for the human is always the definitive subject for any experiment relating to human health. Once again, I must remind you that we are operating on an "ends justify the means" argument here. If consistent, the researcher would grant that 100 human subjects would be more efficient than thousands of non-human subjects. If the ends do, indeed, justify the means, then let's get on with it; let's use humans in our experiments to avoid falling into the inextricable morass of extrapolation.

But, perhaps all these arguments are moot. Let's step back and re-examine our ultimate goal: To improve human health. As Steven Tiger (1986) points out, "Making medical progress is not the same thing as improving human health." On the contrary, medical treatment is a relatively unimportant factor in the equation of human health, trailing far behind heredity, lifestyle, and environmental influences. Sociobiological analyses by McKinlay and McKinlay (1977) and McKeown (1976) support this position. McKinlay and McKinlay conclude: "Indeed...3.5% probably represents a reasonable upper-limit estimate of the total contribution of medical measures to the decline in mortality in the United States since 1900."

A significant number of respected critics have recently published data which deny that we are winning the war against the most serious and widespread forms of cancer (Bailar and Smith 1986). With the exceptions of childhood leukemia and Hodgkin's disease, two of the rarer cancers, survival rates have really not changed over the last 20 years, despite the suffering and death of hundreds of millions of laboratory animals. A recent monograph by Reines (1986) concludes: "Despite the claims of prestigious scientists, there is no evidence that research on animal models of cancer has ever led to a significant advance in the treatment or prevention of human cancer."

We are all aware that some 30% of all cancers can be prevented through the cessation of smoking. Another 30% can probably be prevented by dietary change. Does it not seem logical to revise our priorities, to spend our dollars in preventative education rather than pouring them down the drain with the blood of millions of non-human animals?

Once again quoting from Steven Tiger, "Our research-based sickness-care system is bankrupting the nation, as proven by the fact that pre-set limits to reimbursement are being imposed over the objections of the ultrapowerful medical lobby; there is simply no choice. The medical care industry now costs this nation over $400 billion each year, which is far more than the military budget, and it is still growing out of control, like a cancer. What we need in its place is a prevention-based healthcare system; the nation's health would be improved and we could afford to provide far better care than we now can provide to those who would still need it."

I maintain the only "humane" research possible is clinical research, accomplished for the sake of the individual animal being studied. If an individual animal, human or non-human, is suffering, we have an obligation to attempt to alleviate that suffering. If known techniques prove fruitless, then let us take the most logical experimental path in treatment, and, through proper documentation and publication of results, share our findings with the rest of the world. "This is always done," you say. Is it?

Every one of us has occasion to visit a physician from time to time. The physician diagnoses, prescribes, and follows the course of your malady. These data disappear into your file, usually forever, dying with the patient or being discarded when one changes physicians. As malpractice suits proliferate, the data are pushed deeper and deeper into the file. Here are data based upon human illness and the response to treatment for that illness. Why is such information discarded in favor of data based on artificial responses of other animals? The answer is simple: These data are not profitable. On the contrary, they are potentially damaging, for they may well reflect misdiagnosis and malpractice.

Fewer and fewer human autopsies are performed each year, and yet autopsies have uncovered a wealth of information in the past. Again, the subject is human, the disease usually identifiable, the progress of the disease unmistakable. I expect none of you would deny that routine autopsy would yield valuable data, and yet autopsies are done for only the most specific of reasons, researchers opting instead to work with laboratory animals. Well, laboratory animals cannot sue, nor can their relatives.

While analyses of patient files and routine autopsies are certainly valuable sources of information, the biomedical community steadfastly maintains that no alternatives to the use of non-human animals are available. We all know that simply is not true. Most of you probably remember that only a few short years ago, a rabbit had to die to confirm human pregnancy. A simple litmus test is now readily available at all drug counters and without prescription. I wonder how the medical lobby allowed that source of funds to slip through their fingers.

I have been told repeatedly by representatives of the medical industry that alternatives cannot be gained by funding programs to discover them. "They occur serendipitously, in the course of research with non-human animals," I am told. Only a few short years ago, the Johns Hopkins Center for Alternatives to Animal Testing (CAAT) was established with funds gained by pressure from the animal rights movement. At last count, nine different alternative procedures to the 40-year-old Draize Test are ready for validation. The LD-50 Test has been recognized as invalid and anachronistic and is rapidly being replaced by Limit tests and Up-Down procedures. Scientists at CAAT are studying nonanimal alternatives to eye, skin, kidney, and liver research. There is no question that increased funding would speed the progress of these experiments, but the National Institutes of Health (NIH) have never funded a project specifically seeking alternatives to the expensive and invalid use of non-human animals in the laboratory.

Science, the official journal of the American Association for the Advancement of Science, published a Report entitled, "Physiological Correlates of Prolonged Sleep Deprivation in Rats," in the July 8, 1983 issue. Six rats were kept sleepless until death. The conclusion: " . . . these results support the view that sleep does serve a vital physiological function." Well, here's one experiment for which we don't need to find an alternative. Thousands of equally ridiculous experiments are done each year at a staggering cost to the citizens of this country, costs which cannot be measured in dollars alone, but in the futility and immorality of such blatant assaults upon sentient creatures.

Another recent experiment conducted by Fred Van Dyke and published by the Wisconsin Department of Natural Resources, described the capture of 135 mallard ducks as a subject population. Seventy-four of the ducks had one wing broken; the remaining 61 had one wing tied down with leather straps. All were released into the wild or into a pen. The study showed that all but one of the crippled birds released into the wild—and most of those kept in pens—died from starvation, exposure, attacks by predators, or a combination of factors. The conclusion: Crippled ducks die in the wilderness.

On May 28, 1984, members of the Animal Liberation Front (ALF) gained access to the Head Injury Clinic at the University of Pennsylvania and "liberated" over 60 hours of videotapes depicting innumerable violations of the Animal Welfare Act, the experimental protocol, and any standard of humane treatment. These tapes were copied and an edited version was made available to the general public and to elected representatives. Several Congresspersons were outraged and drafted an amendment to deny funding for this laboratory. In a brilliant preemptory move to avoid congressional intervention, the Department of Health and Human Services, in coordination with the NIH, voluntarily closed the laboratory.

Since that time, several other major laboratories have been found wanting and fines have been levied against them, funding withheld, and, in some cases, the laboratories have been closed, at least temporarily. If the medical establishment had been believed, none of these actions would have been taken, for the public has been consistently assured that all is well within the windowless rooms of the experimental laboratories. Is it merely coincidence that each laboratory visited by the ALF was found to be in severe violation of scores of regulations? I doubt it; independent studies by the Office of Management and Budget and the Office of Technology Assessment verify the inadequacy of the existing system to prevent such abuse.

There are no laws to protect laboratory animals from being subjected to the most horrible experiences. There are guidelines, to be sure, and there are specifications for cage size, availability of food and water and other "housekeeping" standards, but any laboratory non-human animal remains completely at the mercy of the experimental protocol. For example, if the researcher makes a case for withholding anesthetics or analgesics, stating that these palliatives will interfere with the results of the experiment, that request will more than likely be approved. There is no appeal process for the laboratory animal.

My point here is simple: The fox guards the henhouse. This is the same fox, by the way, who cited the Head Injury Clinic at the University of Pennsylvania as one of the finest laboratories in the world, only months before the laboratory was closed. Can we afford to leave it to the Animal Liberation

Front to open public "windows" to these laboratories? I think not; it's time to inform the public, to let them see where their tax dollars go, to take a long and searching look at an archaic and immoral practice.

The state of Florida has established an excellent model with their Sunshine Law. Any citizen of Florida can attend official meetings of state representatives. They may be enjoined from speaking or otherwise disrupting these meetings, but they are authorized to hear all evidence and read all documentation pertaining to the expenditure of public resources.

It was through public attendance at an Animal Care Committee meeting at the University of Florida at Gainesville that we recently learned of a proposal to validate the effectiveness of the Heimlich maneuver for near-drowning victims by anesthetizing, intubating, and nearly drowning 42 random-source dogs. This experiment was approved almost summarily by the Animal Care Committee to be sent forward to the American Heart Association for funding. Another previously-approved proposal was discussed on the same day, this one to suspend cats from pelvic harnesses for periods up to 90 days to study bone remodeling. The public was justifiably outraged; both experiments were subsequently dropped by the University, spokespersons citing reasons other than public pressure. To my mind, all states and the federal government should adopt Sunshine Laws immediately. How can we hope to function effectively as a democratic society if we withhold information from the voters?

"Are there no controls over the researchers?" you ask, and my opponent answers, "Of course! The peer review system is excellent and has long stood the tests of its merit." Peer review, the practice of professionals determining the quality of their fellows' research, has long been a sacred cow to the research industry. But the peer review system is now coming under heavy fire from within. "Even Dr. Stephen Lock, editor of *The British Medical Journal*, who is one of the staunchest defenders of the principle of peer review, is calling for reform of what he regards as serious defects in a system that has occasionally failed to detect fraud, plagiarism, and simple error" (Altman 1986). The word, "occasionally," is somewhat arbitrary here, for we have no idea how well the system of peer review actually works to prevent duplication or unnecessary experimentation. Given a reliance on peer review, we now have more than one fox guarding all the henhouses.

"So, the system has some flaws," you admit. "How can we best correct its deficiencies?"

Yes, the peer review system is flawed, but that's not the point. Patching up this system to allow continued exploitation of non-human (and human) animals is akin to repairing a faulty gas jet in the showers of Dachau, for it's the system itself which is immoral.

How many of you have read *The Case for Animal Rights* by Dr. Tom Regan? Before you make up your minds about this issue and put the facts and concepts you'll glean from this conference into little pigeonholes, take the time to read this book. It's a serious philosophical text, well accepted in the academic world, and, along with Peter Singer's *Animal Liberation*, a major foundation for our ethical position. Dr. Regan establishes a strong case for the inherent rights of all animals to exist beyond the unnecessary exploitation of other species. Professor Singer's utilitarian analysis, while espousing completely different arguments from those of Regan, reaches essentially the same conclusion, i.e., we humans have the obligation to adopt a peaceful and humane ethic, to minimize universal suffering of both human and non-human animals. We stand on firm ethical ground and reject the "ends justify the means" argument.

Just over six years ago, I might well have been at a conference such as this upholding my right and obligation to experiment with non-human animals for the sake of humans. Of course, six years ago, conferences like this were nonexistent, but that's another issue. The fact is, six years ago I was detailed by my boss to defuse the concerns of a young Ph.D. statistician who had recently joined our laboratory, and who was appalled that the monkeys were shocked repeatedly during training. My task, you see, was to determine the post-irradiation behavioral effects of pulsed neutron/gamma energy in an effort to predict the effects of nuclear radiation upon military personnel in an operational environment. I had been engaged in this pursuit for some 15 years and was responsible for the suffering and death of over 1,000 rhesus monkeys, as well as a few baboons.

I conferred with our new statistician, telling him of the importance of this information to the military mission of the United States, assuring him of the absence of alternatives, assuaging his concerns with the shopworn litany I expect to hear later today [at this conference]. I convinced him, but, in the process, my words began to sound hollow; I began to unconvince myself. That statistician and I remain friends today, and I credit his sensitivity with forming the first wedge driven between a dormant ethical concern and a highly conditioned "official" scientific position.

I mentioned the sensitivity of this person. Such sensitivity is relatively unusual in an adult male in our society. The animal rights movement is approximately 85% female, due, I believe, to the fact that our society has allowed females to be empathic and has not conditioned them against feeling or showing sympathy. I remember walking along the San Antonio River about eight years ago and hearing a father trying to comfort his crying son. "Real men don't cry!" he was saying to the boy.

I have already recounted my early conditioning with respect to hunting, trapping, butchering, eating, and wearing non-human animals, but I'd like to say a word about later conditioning. I was taught a fairly rigorous Skinnerian approach to experimental psychology during my graduate days at The Ohio State University. To empathize with the "subject" was a no-no, for that was anthropomorphic and not in keeping with the objectivity required of a real scientist. "Real men don't cry"; real scientists don't either. The animal was a "subject" or a "preparation." Electric shock was a "noxious stimulus" or a "negative reinforcer." Screams were "verbalizations." Starvation was "deprivation."

I can't begin to tell you how glad I am to be back in the real world of emotion, of animals, of recognizing pain and pleasure, and of respecting the validity of my experience without having to label it with a euphemism.

I was proud of the first draft of my thesis, for I felt it communicated well. My advisor wanted more jargon, and, of course, he got it. The second draft was somewhat more difficult to understand, but still not arcane enough. The third, and final, draft was impossibly difficult but readily accepted. I sometimes have trouble remembering the title—which was, *Motive Relevant Perceptual Discriminations of Authoritarians and Non-Authoritarians*, or something like that.

To emphasize this condition of irreality in science, allow me to present a short paragraph from the *New Scientist*, April 10, 1986:

> The current issue of *Life Sciences* contains a paper under the title: "Alternating lateralisation of plasma catecholamines and nasal patency in humans." Our science editor says this means "Why you can breathe up one nostril at a time."

"It's raining pesticides in Hokkaido," is the title of an article in the April 10, 1986 edition of *Nature*. "Japan may be the first country to be suffering from pesticide rain. There is mounting evidence that pollutants found in Lake Manshu, in the northern island of Hokkaido, may have come all the way from the Chinese mainland before being washed out of the sky. Lake Manshu, often claimed to be the cleanest lake in the world, is 7,000 years old. Benzene hexachloride (BCH) levels are rising precipitously, and will probably continue to rise, for the lake has no outlets and the pesticide is virtually non-biodegradable."

The *Washington Post*, April 16, 1986: " A University of California professor has accused the Interior Department of suppressing data suggesting that toxic contamination at a California wildlife refuge is moving up the food chain and could pose a threat to endangered species."

The *Washington Post*, March 4, 1986: "The Agriculture Department listened to testimony from organic farmer, Dick Thompson. Thompson, one of the country's best-known organic farmers, described the non-chemical techniques that have made his Iowa farm a profitable environmental showcase that has taken him out of debt. The secret, he explained, lies in establishing a rotation cycle that helps regenerate soil while blanking out weeds and in using mechanical cultivation at the right time instead of chemicals to combat the weeds that survive. 'This thing is happening everyplace,' Thompson said. 'Both sides—we on the organic side and those on the chemical side—are bending. The ideas of how we have perceived each other are changing.'"

The *Washington Post*, April 15, 1986: "*Hopkins Quits Using Rabbit Tests:* In an effort to address the concerns of animal rights groups, Johns Hopkins School of Public Health in Baltimore announced yesterday that it has developed a program to replace using rabbits in eye tests designed to determine irritation from cosmetics and other substances."

Like Dick Thompson, we see our opposition as bending also. Unlike Thompson, however, we are not satisfied with bending, for each minute sees the irrevocable destruction of between 60 and 100 acres of rain forest and every second of that minute is an eternity of pain to some other animal.

The animal rights movement has progressed far beyond the humane treatment of domestic animals. Listen to the first basic principle of Deep Ecology as defined by Devall and Sessions (1985): "The well-being and flourishing of human and non-human Life on Earth have value in themselves. These values are independent of the usefulness of the non-human world for human purposes."

The July, 1985 issue of *Advances in Nursing Science* includes an article by Crowley and Conners (1985), "Critique of 'The Use of Animals in Nursing Research.'" The authors state: "Those who would harm animals must justify doing so. To produce such a justification, it is not enough to argue that people profit, satisfy their curiosity, or add to scientific knowledge. These facts are not morally relevant."

Science, April 11, 1986, LETTERS: "Directions of Research," by Richard Trumbull: "It might be that the continued inbreeding of researchers under the protective laboratory conditions now afforded by government support has resulted in another laboratory animal that has lost its resistance and resilience. Here again we might face the problem of finding some researchers 'in the wild' for some imaginative crossbreeding to return our stock to one that can deal with problem-solving in the real world."

The animal rights movement is among the most radical in the world today, for we ask for more than simple prudence or improved housekeeping, or even more stringent controls over the approval and implementation of research projects. We ask for a shift in attitude, the adoption of a more humane ethic, a major revision of conditioned thoughts and behaviors. We ask for a peaceful ethic toward all life forms. We ask for ecological, environmental, and personal respect, and we'll accept no less from ourselves and, eventually, from you.

Endnotes

[1] Paper presented at the national conference, "Animals and Humans: Ethical Perspectives," Moorhead State University, Moorhead, MN, April 21-23, 1986.

[2] Director, Washington, DC Office, The National AntiVivisection Society, 112 North Carolina Ave, SE, Washington, DC 20003.

References

Altman, LK. 1986. Peer review is challenged. *New York Times*. February 25.
Bailar, JC, III and Smith, EM. 1986. Progress against cancer? *New Eng. J. Med.* 314: 1226-32.
Crowley, MA and Connors, DD. 1985. Critique of "The use of animals in nursing research." *Adv. Nurs. Sci.* 7:30.
Devall, B and Sessions, G. 1985. *Deep Ecology*. Salt Lake City, UT: Gibbs M. Smith, Inc. p.70.
McKeown, T. 1976. *The Role of Medicine: Dream, Mirage or Nemesis*. London: Nuffield Provincial Hospitals Trust. pp. 143-44.

McKinlay, JB and McKinlay, SM. 1977. The questionable contribution of medical measures to the decline of mortality in the United States in the twentieth century. *The Milbank Memorial Fund Quarterly/Health and Society.* . Summer. p.425.

Reines, B. 1986. *Cancer Research on Animals: Impact and Alternatives.* Chicago: The National Anti-Vivisection Society. p.103.

Tiger, S. 1986. Effective presentation in public debate. *The AV Magazine.* March. p.7.

THE CASE FOR THE USE OF ANIMALS IN MEDICINE[1]

Gary F. Merrill[2]

Introduction

As early as 1500-500 B.C., there are indications of Vedic (Hindu) records of animals being observed by man for medical and scientific purposes. In 300 B.C., in Alexandria, Egypt, Erasistratis placed live birds in closed containers and withheld food to observe the consequences of losing body humors. This is believed to be the first recorded attempt to use live animals for research. The philosopher Aristotle (384-322 B.C.), also a biologist and the son of a physician, founded the sciences of physiology, zoology, and comparative anatomy as a result of his observations of animals. Galen (130-200 A.D.), a physician, anatomist, physiologist, and philosopher, was the founder of experimental physiology. His animal studies were designed to be applicable to humans; their accuracy and completeness greatly improved the understanding of the human body. After the fall of the Roman Empire and during the Dark Ages, the decline that occurred in the arts and sciences was evident in both human and veterinary medicine. The unity that had existed between the two branches of medicine did not reappear until the Renaissance. Thus, the seventeenth and eighteenth centuries saw significant medical advances, many the result of animal studies (Davison 1977).

The use of animals in medical research and teaching, and the public concern this has generated is not a new issue (Visscher 1969). Ever since scientists began using animals to investigate the function of the body in health and disease, there have been those who opposed their work (Fishman and Richards 1982). Whether this controversy is cyclic is not known, but most concerned biomedical investigators agree that the opposition is here to stay. The author shares this opinion, and thus maintains that it is in the best interest of all parties to be properly educated on the issues. Only through the process of education can we accurately evaluate the pros and cons.

The issues to be addressed in this paper entitled "The Case for the Use of Animals in Medicine" include: 1) a selected sample of animals used in biomedical research and teaching and, 2) a selected sample of diseases which have plagued (do plague) man and other animals, and in which animal research has played a major role in minimizing the suffering and early loss of life. The subject is broad and other, more complete, reviews are available (Sechzer 1983).

227

Animals Used in Medical Research

Beginning in June, 1984, the American Physiological Society, one of the many professional biomedical societies dedicated to educating the public and supporting the responsible use of animals in medical research, published a series of articles entitled, "Health Benefits of Animal Research" (Gay 1984). The series, which identifies several animal species currently being used as research subjects in medicine, and the benefits to human health derived from such research, concluded in April, 1985 (King and Yarbrough 1985). These articles are being reproduced as separate chapters in a book entitled, *Health Benefits of Animal Research*, by the National Association of Biomedical Research, Washington, DC. Much of the information presented in the following section of this paper was taken by permission from this material.

Dog. — The dog has been a companion to man since the dawn of history. As a result of this relationship, both the dog and man have experienced advantages and disadvantages. It is quite logical why this animal was chosen by early biomedical investigators as the species of choice for their scientific experiments. In laboratory surroundings, the dog is smaller and easier to maintain than large agricultural animals. Rats and mice at the time were not thought of in terms of human health and welfare, but rather were disdained as transmitters of disease. Stray dogs were generally abundant then, as now, in urban areas.

The dog's gregarious behavior and highly-developed social nature enhance his manageability in the laboratory. This good nature was recognized and appreciated by earlier medical and behavioral scientists, including Pavlov (1927). Consequently, behavioral genetics has been the subject of considerable research with the dog (Scott and Fuller 1965). Amongst the many clinical diseases for which the dog has been an important research subject are: 1) hereditary spinal muscular atrophy (similar to human motoneuron disease [Cork et al. 1979]); 2) arterial diseases and diseases of the heart, including endocarditis (Keys et al. 1972); 3) hepatitis (Morris and Gocke 1971) and ethanol-induced liver cirrhosis (Chey et al. 1971); 4) neoplastic disease (Dewhirst et al. 1983), including the use of chemotherapy to treat infectious cancer; 5) cholera (Sack and Carpenter 1969); and 6) nutritional disease, including a niacin deficiency syndrome similar to human beriberi (Yang and Mickelson 1974). Perhaps the most well known of all such diseases for which the dog has been used experimentally is diabetes mellitus (see "Diabetes" below).

Rat. — The rat is used experimentally in the fields of immunology, reproduction, cancer, behavior, and aging, amongst others. It is also of universal economic importance since it destroys one-fifth of the world's crops annually and transmits a variety of diseases to humans (Canby and Stanfield 1977). The consequences to man of the rat's destructive activities and its major contributions to human health through biomedical research make the rat one of the most important species with which man must continue to interact.

The strain of rats that has been used most for experimental biomedical research is the Norway rat (*Rattus norvegicus*), and its early history has been described by Robinson (1965). In the middle of the nineteenth century, the rat was used for studies in nutrition, physiology, and anatomy. The research rats used in the United States are thought to have originated in Europe. The first inbred lines of rats were developed independently by several groups of investigators at the beginning of the twentieth century. The first studies were in basic genetics and cancer research.

The rat is a major animal for the biomedical investigation of aging. Inbred and randomly-selected strains show major spontaneously-developing, age-related diseases of the heart and kidneys, as well as the nervous, endocrine, skeletal, and immune systems (Baker et al. 1979). Rats are susceptible to the development of: 1) spontaneous, 2) virally-induced, and 3) chemically-induced tumors in all organ systems (Baker et al. 1979). The rat provides models for both diabetes mellitus and diabetes insipidus. The rat is the best animal model currently available for the study of spontaneous, insulin-dependent (juvenile onset) diabetes in humans (Baker et al. 1979).

Rabbit.—Because the general physiology of cells, sera, tissues, and organs of rabbits exhibit a similarity to that of humans, rabbits have been and will continue to be excellent models for many human diseases. The rabbit is large enough to provide adequate quantities of tissue for experimental work without pooling of samples, but is small enough to be more economical than pigs, monkeys, or dogs. Blood samples can be taken without serious harm to the animal from birth to adulthood in sufficient quantity for performance of routine biochemical tests. Rabbits are an excellent mammalian model system for the investigation of placental transfer of drugs, steroids, and other metabolites due to the fact that the rabbit and human have the same type of placenta (hemochorial), which allows the closest contact between maternal and fetal circulations (Fox et al. 1982; Kozma et al. 1974).

Degenerative diseases occur with increasing age in man and all other species. To study the aging process, the National Institute of Aging suggested the rabbit as a species of potential use (CAMRA 1981). There are a number of changes that have been observed in rabbits with increasing age, including: 1) changes in blood composition, 2) cardiovascular changes, 3) endocrine levels, 4) nerve responses to stimuli, and 5) visual parameters.

The author has recently noted the similarity between the response of coronary arteries to the vasoconstrictor effects of histamine in the rabbit (Tozzi and Merrill 1985) and those reported in humans (Ginsberg et al. 1981; Toda 1983). Interestingly, in the rabbit, histamine administration can cause ischemic (lack of coronary blood flow) heart disease (Tozzi and Merrill 1985). This disease causes anginal pain in humans and is associated with high mortality in industrialized nations. Thus, the rabbit might prove to be an acceptable animal model in which to study coronary heart disease.

Subhuman Primates.—Primates share more biological and behavioral characteristics with humans than do any other animal species. These interesting similarities begin with the basic chemical component of life, deoxyribonucleic acid (DNA); ninety- eight percent of the molecular composition of human DNA can be found in subhuman primates.

While anthropologists study subhuman primates for clues to human origins and nature, behavioral scientists examine social organization and interactions in these animals seeking to understand principles governing sexual behavior, aggression, and other social behaviors. In neurobiology, immunology, pathology, reproductive biology, teratology, neonatology, and cardiology, subhuman primates are indispensible biomedical research models for both understanding basic physiological mechanisms and the development of new approaches to the diagnosis and treatment of disease.

Until the 1960s, cardiovascular research was conducted primarily in rabbits, dogs, chickens, rats, and swine. Subhuman primates became major subjects of heart research when it was discovered that atherosclerotic vascular plaques were similar developmentally, microscopically, and biochemically in humans and monkeys (Taylor et al. 1962, 1963a, 1963b). Subhuman primates now are used in cardiovascular research with the following objectives: 1) to identify the mechanisms of atherosclerotic destruction at the organ, cellular, and molecular levels (Ross 1983); 2) to understand the role of diet, blood pressure changes, and other factors in the advancement and stabilization of the disease (Mott et al. 1983); and 3) to define the relative influence of risk factors (tobacco, alcohol, hypertension, obesity, etc.) on disease progression, regression, and prevention (Wissler 1980).

The subhuman primate is also being used in non-cardiovascular disease research. Hepatitis B is a viral disease that annually afflicts 200,000 people in the United States alone. There are approximately 200 million carriers of hepatitis B worldwide. At least three million infants born in Southeast Asia alone will become carriers and develop severe liver damage, including liver cancer (King and Yarbrough 1985). In 1981, the United States Food and Drug Administration approved the world's first hepatitis B vaccine (Maugh 1981). This major advance depended on research with chimpanzees and other subhuman primates, since hepatitis B cannot be transmitted to other laboratory animals.

Other areas of biomedical research in which subhuman primates are being studied include AIDS and vision research. (The chimpanzee is thus far the only laboratory animal that has proven susceptible to the AIDS infectious agents (McClure et al. 1984; *Science News* 1984). One of the most recent contributions of primate research to human clinical treatment has been the studies of Wiesel and Hubel (Wiesel 1982). These studies, for which the authors received the Nobel Prize, involved the development and functioning of the visual system in monkeys. Their results underlie the emphasis on early diagnosis and treatment of human infants with visual disorders.

Cats. —Cats made the cover of *Time* magazine, December 7, 1981 (Reed 1981). They made it for their capacity as people pleasers, their role as the country's most adaptable companion animal, and their potential as a new source of revenue for cartoonists and merchants who have recognized the cat's national popularity. It is estimated that there are 34 million cats inhabiting 24% of the households in America, and another 15 million who are homeless. In the past 30 years, the health of the house-dwelling cat has improved to such a degree that a life expectancy increase of six to eight years allows many felines to reach the age range of 16-20 years. Improvement in the feline's health has been brought about, in great part, by better nutrition, new and improved diagnostic procedures and surgical techniques, and treatment and drug therapies for such conditions as feline urolithiasis, skin disorders, fractures, conjunctivitis, anemia, tumors, and pneumonia, and vaccines for the prevention of many life-threatening diseases such as feline enteritis, rhinotracheitis, pneumonitis, and most recently for feline leukemia (Lewis et al. 1981). This improved quality of life for felines could not have been brought about without the aid of biomedical research—research which often relied on the cat as the subject.

Not dealt with in the *Time* article was the cat's contribution to medical research, perhaps because it is a very unpopular subject with a small segment of the public (Cornelius 1969).

In 1894, Murey, a French physiologist, used the motion-picture camera to illustrate how the cat's body changed position when it was dropped from a height, landing eventually on its feet (McDonald 1960). This observation led physiologists in the early 1900s to examine the physiology of the falling reflex and to come to the conclusion that the brain, and either the eyes or the balancing mechanism (vestibular organs) in the ears, were essential to the cat's landing on its feet. It was not determined, until 1960, that of the two systems, the eyes were more important (Beadle 1977; McDonald 1960).

It is not uncommon for cats that are well cared for to reach the age range of 16-20 years. Thus, this species is suitable for investigating the aging process. It is of interest that many of the procedures used to study aging do not require surgical intervention, nor do they adversely affect the health and wellbeing of the animal. Studies of the aging process in cats has yielded information of relevance to both man and pets, an important factor in view of the increased interest in the role of companion animals, especially for the older population. Although animals (cats, dogs, birds, fish, reptiles) are being prescribed for their therapeutic value with the mentally retarded, physically handicapped, and prisoners, one of their most valuable contributions is with the aged population; cats are the animal most often used in these settings.

Behavioral scientists use the cat for studies of behavioral development, comparative evaluation, and behavioral assessment. Examples of human conditions studied in the cat include neuroses, amphetamine psychosis, learning and memory, socialization, and habit formation. Of the many diseases cats and humans share, probably one of the best known is leukemia. The causative agent of leukemia in the cat is a ribonucleic acid (RNA) virus belonging to the family Retroviridae, a retrovirus. Of great current medical interest are the reports of retroviruses associated with AIDS in humans (Sarngadharan et al. 1984) and SAIDS (Simian Acquired Immune Deficiency) in animals (Marx et al. 1984).

Mice.—Throughout history, mankind has suffered the ravages of disease. Plagues have decimated large segments of the population. Hundreds of millions of people died from the "black plagues" caused by *Yersinia pestis*, a bacterium transmitted from rodents to humans via fleas, or by infectious droplets from pneumonic victims. In the early 1900s during the influenza virus pandemic, bacterial meningitis and pneumonia associated with *Haemophilus influenza* caused countless deaths. As late as 1950, whooping cough caused by *Bordetella pertussis* killed over 1,000 American children. The threat of bacterial disease is not a thing of the past as evidenced by the recent experience in the United States with Legionnaires' Disease caused by a newly recognized bacterium *Legionella pneumophila*.

Parasitic diseases are a major cause of human death and suffering. Schistosomiasis, a blood fluke affecting millions of people, is fortunately of limited public health importance in the United States.

The mouse has played a fundamental role as an experimental animal in many of the infectious diseases just discussed. Mice were, and, in many cases still are an essential part of diagnostic medical microbiology. They are important in viral isolation studies of diseases caused by arthropod-borne vectors and in rabies virus diagnosis. Mice are still a significant investigative model for understanding basic host mechanisms of disease, and for studying susceptibility and resistance to infectious agents (Jonas 1984). A basic understanding of host-parasite relationships is critical if we are to achieve better diagnostic, preventive medical approaches, and therapeutic applications to these problems.

In addition, there is a long list of diseases of humans that are usually thought to be non-infectious but cause immense suffering and in many cases contribute to mortality. Arthritis, diabetes mellitus, central nervous system disorders, immune deficiency diseases, metabolic disorders, hematologic dyscrasias, and cancer are diseases that are all too familiar. The use of mice has provided us with a valuable method to elucidate fundamental biological phenomena associated with these diseases. Through creative basic and applied research studies, we have developed effective therapeutic measures for infectious and noninfectious diseases.

Abnormalities of development are frequently associated with genetic dysfunction. Down's syndrome is a tragic example of a chromosome disorder resulting in a child with a "mongoloid" appearance, mental retardation, cardiac abnormalities, and an increased likelihood of developing leukemia. Through plant and animal investigations, we have gained an understanding of genetic disease that has enabled us to diagnose and, in some cases, treat genetic disorders. Genetic counseling of prospective parents is now commonplace. The use of mice in studying basic genetics at the classical and molecular levels has resulted in development of organ transplantation as well as effective methods to modify the course of immunologic disease. With advances in molecular biology it is now possible to think in terms of modifying genetic errors of metabolism by gene manipulation.

The history of microbiology is intimately interwoven with experimental studies using mice. Pasteur used mice in studying both rabies and anthrax. Mice were at the forefront of experiments dealing with the nature of infectious disease and immunization. The mouse has been and will continue to be an excellent research animal to study the pathogenesis of bacterial diseases. For example, a model for human antimicrobialassociated diarrhea (AAD) was developed using *Clostridium difficile* as a causative organism. This work used the mouse as the experimental animal (Onderkonk et al. 1980).

Developmental analysis of an animal model of a human disease makes it possible to determine what happens metabolically before the actual disease symptoms appear. Not only does this understanding of a disease's progress help design therapy but also allows early detection and treatment. Because so many genes have been mapped in the mouse (Pearson et al. 1982), it is usually possible to identify the presence of a disease-causing gene early by "tagging" it with a closely linked marker gene. One example of this is the diabetes gene (Coleman 1978), which is closely linked to a coat color gene in the mouse called misty (*m*) (Wooley 1943, 1945). By making appropriate matings, it was possible to link the diabetes gene (*db*) with the *m* gene. Thus it is possible to follow the development of diabetes from shortly after birth by simply analyzing those mice with the misty coat color. Analysis of the *db* gene in mice has led to a better understanding of the complex nature of hormonal abnormalities on cellular metabolism.

Mice are also being used to study the structure of immunoglobulins, diseases of the blood-forming elements such as aplastic anemia or leukemia, and in the development of blood assay systems, e.g., there are at least 17 different mutations in the mouse that cause anemias (Russell 1979).

Medical Advances Made Through Animal Research

Diabetes. — The term diabetes mellitus comprises a heterogeneous group of disorders characterized by high blood glucose levels. Approximately 5.8 million people in the United States have been diagnosed by a physician as being diabetic, and an additional four to five million people have diabetes but have not yet been diagnosed (approximately 500,000 new cases are diagnosed annually). There are four main types of diabetes. Insulin-dependent diabetes mellitus (IDDM) constitutes about 510% of all cases of diabetes, and most of the remainder is noninsulin-dependent diabetes mellitus (NIDDM). About 2% of diabetes is the type that is secondary to or associated with other conditions. Gestational diabetes, a transient condition occuring in about 2-5% of pregnancies, is one such malady.

The annual incidence of IDDM is about 12-14 new cases per 100,000 children ages 0-16 years, and by age 20, about 0.3% of persons have developed IDDM. About one-half of the persons with IDDM are currently older than age 20 years. The annual incidence of NIDDM is about 320 per 100,000 persons ages 20 years or older. About 2.35% of the population has been diagnosed as having NIDDM, and at ages 65 years and older, virtually 8.6% have NIDDM.

Diagnosis of diabetes is made either by the presence of the classical symptoms of diabetes accompanied by elevated blood glucose levels, by elevated fasting glucose levels, or by an abnormal oral glucose tolerance test.

In 1982, 34,583 deaths were attributed to diabetes as the underlying cause of death, resulting in diabetes being ranked as the seventh leading underlying cause of death. In approximately 95,000 additional deaths, diabetes was listed as a contributing cause of death. The majority of death certificates (about 75%) listing diabetes also list a cardiovascular condition. Ischemic heart diseases are involved in about 60% of diabetes deaths, and cerebrovascular diseases in about 25%.

Diabetes is not a benign disease. It causes almost 50% of amputations of the foot and leg among adults. It causes about 20% of all cases of kidney failure and 15% of all blindness. Diabetes in the mother causes about 3% of all stillbirths and neonatal and perinatal deaths. The United States ranks among the highest five nations in the world in mortality due to diabetes.

It is of interest that the first major preclinical breakthrough in diabetes-related research was achieved in dogs (Best 1974). It had been shown by a variety of investigators that complete removal of the pancreas always resulted in fatal diabetes. Dogs were the experimental animal. Therefore, it was inevitable that Fred Banting and Charles Best would choose the dog as the experimental animal for their studies which lead to the ultimate discovery in insulin (Best 1974). Both Banting, an expert surgeon, and Best, a medical school student studying experimental physiology, had personal interest in diabetes. Dr. Banting had watched a child in Alliston, Ontario, lose weight and eventually develop diabetic coma and die. An aunt of Charles Best developed diabetes in 1912. The only known therapy at the time was the under-nutrition regimen (Allen et al. 1982). Best's aunt followed the diet faithfully, but died of diabetic coma in 1918 (Best 1984).

When these investigators were provided with research facilities in the Department of Physiology, University of Toronto, in 1921 to study the disease, they were told that only ten dogs would be provided for their proposed studies. The cost per animal in those days was approximately one dollar. When the ten dogs had been used, and suitable results were still lacking, Banting and Best purchased additional dogs with their own funds. In those days, they had no stipends or grants to support their research. In final studies (August and September, 1921), they performed scores of experiments providing convincing evidence that the secretory extracts of beef pancreas (they and others had hypothesized that presence of an anti-diabetic factor) could reverse the symptoms seen in diabetic (pancreatectomized) dogs.

Because of personal interest in the outcome of their experiments, they looked after the dogs themselves; fed them and cleaned their cages. They trained the dogs to put out their paws for withdrawal of blood samples and they made the dogs as comfortable as possible under the circumstances. In the words of Charles Best, "I do not think that in these or subsequent experiments we ever inflicted pain and discomfort on dogs which we would not have been willing to accept ourselves" (Best 1974). The animals that were pancreatectomized before insulin was available and which lived for ten days to two weeks, were necessary as controls for the animals that received insulin.

Banting and Best presented the results of their research before the Physiological Society in Toronto in November, 1921, and before the American Physiological Society in December, 1921. On January 10, 1922, they gave each other injections of their purified insulin (Best 1974). There were no untoward effects except a mild soreness of their arms. On the following day, the material was taken to Toronto General Hospital where a 14-yearold boy, Leonard Thompson, became the first diabetic to receive insulin. The first patient to receive insulin in the United States was James Havens of Rochester, New York. He became a celebrated artist and lived to help raise a fine family.

The travels of Banting and Best in some 70 different countries brought them in contact with thousands of diabetics. Statisticians have told them that figures based on worldwide consumption of insulin indicate that some 130 million diabetics have had their lives prolonged by insulin (Best 1974). Today, of course, the number is even greater.

Fred Banting and Charles Best had two great hopes in 1921 and 1922. The first hope was that the dramatic recovery of diabetic animals could be consistently produced by insulin in human patients. The second was that insulin would become an important agent in metabolic research. Both hopes have been realized. As evidenced by the fact that volumes have been written on these two subjects, it is safe to say that perhaps no other physiological agent has been so instrumental in revealing so much about the intermediary metabolism of carbohydrates, fats, and proteins, as has insulin.

Childhood Communicable Diseases and Vaccines. —A report of "teething paralysis" by Colmer (1843) is thought to be the first description of an outbreak of paralytic poliomyelitis in the United States. For many years, poliomyelitis was primarily manifested as an endemic disease, occurring sporadically among infants and young children, but in the early years of this century, epidemics began to appear, and so-called "infantile paralysis" was soon reported in older children and adults (Paul 1971).

During the 1940s, there was a marked increase in the incidence of poliomyelitis in the United States. For example, between 1920 and 1940, there was an annual incidence of about five cases per 100,000 population. By 1945, the incidence had risen to approximately 15-20 cases per 100,000, and by the early 1950s, it had escalated to an estimated 40-45 cases per 100,000 population (Salk 1980). Similar increases were observed in other developed countries at the time, while developing countries noted an increased incidence and shift from an endemic to epidemic pattern (Paul 1958).

The conquest of polio began in 1909, when Landsteiner and Popper (1909) demonstrated that poliomyelitis could be transmitted from humans to apes and monkeys. Subhuman primates became a key animal model for studies on the pathogenesis of polio and for vaccine development when it was recognized that among animal species, only monkeys and chimpanzees could contract human poliomyelitis.

After a decade of intense, focused research employing mice, cats, chick embryos, and subhuman primates (Jonas Salk, personal communication), a killed poliovirus vaccine (KPV, sometimes referred to as inactivated poliovirus vaccine) was introduced into general use in the United States in 1955. This followed a large-scale, controlled field trial in 1954 (Francis et al. 1955), in which the author, and numerous òther American children became the first recipients of the protective effects of the polio vaccine. Between 1955 and 1962,there was a sharp decline in the incidence of poliomyelitis in the United States, decreasing to fewer than five cases per 100,000 population; a decrement of approximately 95%.

The impact of the polio vaccine on human health, developed totally through the use of animal experimentation, has been nothing less than dramatic. In the peak years of the polio epidemic in the United States, there were an estimated 50,000—60,000 cases reported per year. Many victims died or were left permanently crippled. In the immediate years following a federally-subsidized campaign to vaccinate all American children, there were fewer than ten cases per year (Salk 1980).

Similar results, based on research in animals, have had impacts on other communicable childhood diseases. In 1934, before final development and routine administration of the pertussis ("whooping cough") vaccine, more than 265,000 cases were reported in the United States. More than 7,500 victims died. In 1982, there were fewer than 2,000 reported cases with about four deaths.

In 1980, the World Health Organization announced the worldwide elimination of smallpox. Hence, the development of vaccines against communicable diseases, resulting from animal research, has provided society-wide health benefits. Today, all 50 states require routine vaccination of children against smallpox as a condition for entering school.

Interestingly, the protection against such epidemics cannot be extended to subpopulations who refuse vaccination. Fox et al. (1971) described an outbreak of smallpox in a well-immunized community in eastern Nigeria that was confined entirely to members of a small religious sect who refused vaccination. This group was dispersed throughout the community, but maintained close social ties and relative segregation from the rest of the population.

The 1978 and 1979 outbreaks of poliomyelitis in The Netherlands, Canada, and the United States are strikingly similar to the above example. Since 1961, outbreaks of poliomyelitis in The Netherlands have occurred only in certain Protestant communities with very low vaccine acceptance rates. Between April 15 and October 12, 1978, a total of 110 cases of poliomyelitis (80 paralytic, 30 non-paralytic) occurred in several different Dutch provinces amongst members of a Protestant religious denomination that refuses vaccination. The majority of the cases (99) occurred among 65,000 unvaccinated individuals under 27 years of age (Bijkerk 1979). This outbreak was geographically widespread, but no cases of poliomyelitis occurred among 435,000 unvaccinated individuals under age 27 who were not members of these religious denominations. Also in 1978, several cases of poliomyelitis were reported among members of a religious group in Canada. This latter group is related to the group involved in The Netherlands outbreak (Furesz 1979).

There had been contact between the Canadian group and visitors from The Netherlands prior to the outbreak.

In 1979, poliomyelitis appeared in the largely unvaccinated Amish population of the United States, which includes approximately 75,000 members (CDC 1979). The first paralytic case occurred in January in a Pennsylvania town to which a Canadian family had recently moved from a town in Ontario near an area affected with poliomyelitis in 1978. There are extensive interactions among members of different Amish communities, including contact between groups in the United States and Canada. By mid-spring, it was apparent that wild polio virus type 1 was circulating widely within these communities. The virus was shown to be similar to that circulating in Canada and The Netherlands. By June, 1978, when vaccination campaigns appeared to have controlled the outbreak, a total of 15 paralytic and two non-paralytic cases had been reported.

Cardiovascular Disease.—The story of the diagnosis, treatment, and prevention of cardiovascular disease is the story of animal research. It begins in antiquity with the discovery of the structure and function of the heart, vasculature, blood, and lungs (Fishman and Richards 1982). Of prominence in this story is the discovery by William Harvey (1578-1657) of the circulation of the blood, a feat based on years of observation and experimentation in animals. By an extraordinary chance, the notes of his Lumleiam lecture of April 17, 1616, found their way into the British Museum, where they were discovered in 1886, some 270 years later. Excerpts from these notes, written originally in Latin and English are worth quoting (Franklin 1933):

> It is plain from the structure of the heart that the blood is passed continuously through the lungs to the aorta as by two clacks of water bellows to raise water. It is shown by application of a ligature that the passage of blood is from the arteries into the veins. Whence it follows that the movement of the blood is constantly in a circle, and is brought about by the beat of the heart. It is a question, therefore, whether this is for the sake of nourishment or of heat, the blood cooled by warming the limbs, being in turn warmed by the heart.

Bear in mind that his predecessors were convinced that the heart was the source of body heat production and expelled a mixture of blood and air, but not in a circular fashion (Sarton 1954). Harvey continued such experiments throughout his life although it is not always easy to determine just what these experiments were and when he did them. An elegant experiment in the perfusion of the pulmonary circulation is described in a letter written late in his life to his friend Paul Marguat Slegel of Hamburg (Harvey 1766).

> ...the pulmonary artery and aorta are ligated, the left ventricle opened, and a cannula placed in the vena cava, and water forced in. Quid fit? [What happens?] The right ventricle is vehemently tumefied. Through the opening in the left ventricle, however, not a drop of water or blood escapes....So now (the solution having been predicted), the syringe is introduced into the pulmonary artery, with a ligature around it lest water regurgitate into the right ventricle.

We force the water in the syringe against the lungs and immediately water
with copious amounts of blood leaps out of the cleft in the left ventricle, so
that as much water as is expressed into the lungs, so much flows out of the
hiatus mentioned. You can experiment as often as you like, and know for
certain that this is so.

These and similar experiments designed to prove that blood flows in a
circular fashion, i.e., from right ventricle to lungs, thence through the left
ventricle to the systemic circulation, and back to the right ventricle, are
detailed in Harvey's classic book, *Exercitatio anatomica de motu cordis et
sanguinis in animalibus*, (Guilielmi Harvei, Angli, Francofurti, sumptibus
Guilielmi Fitzeri, anno M. DC. XXVIII [Harvey 1928]). This great work in
experimental physiology stands by itself. In the medical sciences, there is
nothing else in the same class. Its seventeen brief chapters, in 72 pages, are
an inexhaustible source of enlightenment and incontrovertible conclusions.
Its achievement was more than a discovery: it was a revolution (Fishman
and Richards 1982).

Animals, like humans, suffer from many inherited cardiovascular defects,
including patent ductus arteriosus, pulmonary stenosis, persistent right aortic
arch, intraseptal defect, and both tricuspid and mitral valve insufficiencies.
All of the preceding maladies are seen in the dog. Vascular disease, common
in the aging dog, is treated clinically in many veterinary practices with drugs
and other methods of therapy developed during the last couple of decades
for use in humans.

Replacement of heart valves and segments of larger arteries are major
cardiovascular advances resulting almost entirely from dog surgery 30 years
ago. Hemophilia in the dog is an inherited hemostatic defect considered
nearly identical to that seen in humans; it is more frequently reported in
dogs than in most other domestic animals (Gay 1984).

One in every two deaths of Americans today can be attributed to cardiovascular and related disease. The incidence of such diseases, which steadily
increased in the post-World War II years, has been on the decline for the
past couple of decades. This is due in large measure to the promotion of
accelerated activity in biomedical research by the federal government, the
biomedical scientific community, and private enterprise. With such promotion
has come advanced technology that allows us to intervene in the advancement
of cardiovascular disease with new surgical and pharmacological insight that
was thought impossible 20 years ago. For example, the early loss of life due
to ischemic heart disease (secondary to organic vessel disease, i.e.,
atherosclerosis) has been greatly reduced with the development of coronary
bypass surgery. It should be of little surprise that before such technology
was applied to human medicine, it was first thoroughly investigated in
experimental animals. Large agricultural species such as sheep, hogs, and
cattle, and smaller domestic species such as dogs, were routinely investigated
to determine the feasibility of such surgical/pharmacological intervention in
man. Few of us are not familiar with persons whose life expectancy has

been prolonged as the result of coronary bypass surgery. Valve replacement surgery is another intervention routinely performed in man which was first tested and perfected, as far as was possible, in experimental animals.

In November, 1983, the author attended the annual meeting of the American Heart Association held in Anaheim, California. This meeting occurred less than one year after surgeons at the University of Utah Medical Center replaced the first human heart with an artificial counterpart. Drs. William C. DeVries (heart surgeon) and Robert K. Jarvik (heart designer) were both in attendance, as was the wife of Dr. Barney Clark, the first human recipient of a transplanted artificial heart. Dr. DeVries was the keynote speaker. He outlined in considerable detail the years of animal experimentation at the University of Utah and elsewhere which preceded this first human experiment. Dr. DeVries described the experience as follows (DeVries et al. 1983):

> Advances in bio-engineering and animal experimentation have led to the possibility of total artificial heart (TAH) implantation in humans. The Utah TAH consists of two spherical ventricles fabricated of polyurethane which are pneumatically powered by an external driver. The TAH use tilting-disk valves. A 61-year-old man presented with end-stage heart failure due to idiopathic cardiomyopathy which was refractory to drug therapy. With informed consent, the TAH was successfully implanted and allowed survival for 112 days. After implantation, increases occurred in blood pressure, from 85/70 to 124/60 mm Hg, left ventricular ejection fraction, from 10% to 77%, and cardiac output, from 1.5 to 7 l/min. The patient awakened after three hours and was fully alert. He later could be ambulated and exercised with assistance. The postoperative period was complicated by pulmonary insufficiency, intermittent renal failure, seizures (? overperfusion), bleeding complications of anticoagulation, and pseudomembranous colitis with terminal shock. Except for an early valve failure which was corrected, the TAH was functional throughout the clinical course and was intact at autopsy. This initial experience indicates the feasibility of long-term cardiovascular support with a TAH systems, and suggests promise for further development.

At the conclusion of his keynote address, Dr. DeVries described the experience as "inspiring" for all who participated. Other such artificial heart transplants have occurred subsequently and will undoubtedly continue to occur at an accelerated rate as technology improves.

Summary and Conclusions

Francois Magendie, a French physiologist of the early nineteenth century, and considered by many to be the father of experimental physiology, was succeeded at the College of France by his former assistant, Claude Bernard. Bernard was a brilliant physiologist who was described by Pasteur as physiology itself. Every phase of physiology to which Bernard addressed himself was explored with innumerable experiments. Bernard did not lack sensitivity towards the animals he used, nor to the views of his anti-vivisectionist opponents. Yet he was undaunted in his belief that his work was justified by the potential benefits derived for humanity. In his book, *An Introduction to the Study of Experimental Medicine* (Bernard 1957), he asks the question:

> Have we the right to make experiments on animals and vivisect them?...I think we have this right, wholly and absolutely. It would be strange indeed if we recognized man's right to make use of animals in every walk of life, for domestic service, for food, and then forbade him to make use of them for his own instruction in one of the sciences most useful to humanity. No hesitation is possible; the science of life can be established only through experiment, and we can save living beings from death only after sacrificing others. Experiments must be made either on man or on animals. Now I think that physicians already make too many dangerous experiments on man, before carefully studying them on animals. I do not admit that it is moral to try more or less dangerous or active remedies on patients in hospitals, without first experimenting with them on dogs; for I shall prove, further on, that results obtained on animals may all be conclusive for man when we know how to experiment properly.

The author reaffirms the beliefs held by Claude Bernard. If our motivation for using animals in experimental investigation is to improve the welfare of humanity (and other forms of animal life), then by all means it is not only our right but perhaps our obligation to continue such work.

At a time when the use of animals in experimental research is being seriously debated, it is important to step back and review what has been learned and what may lay ahead in the field of biomedical research and medicine. Scientists as well as the lay public are bound by ethical and moral principles to treat living creatures with respect and to protect their environment. By the manner in which we address relevant biological questions, so shall we be judged by our peers and the public. By the manner in which we manage our environment, so will we be judged by future generations. In the final analysis, it is the function of society to determine whether the gain from animal research with its attendant benefit to man and the environment is justified (Jonas 1984).

The use of animals in experimental research is by its very nature an encroachment on those precepts respecting animal life (Jonas 1984). Nonetheless, man is given by divine decree, "...dominion over the fish of the sea, and over the fowl of the air, and over every living thing that moveth upon the earth" (Genesis 1:28). Therefore, it is incumbent not only on scientists who use animals in medical research, but on every person who uses animals for any purpose to do so with care, compassion, and consideration for the animal's well-being.

Endnotes

[1] Paper presented at the national conference, "Animals and Humans: Ethical Perspectives," Moorhead State University, Moorhead, MN, April 21-23, 1986.

[2] Associate Professor, Graduate Program in Physiology, Bartlett Hall, Cook College, Rutgers University, New Brunswick, NJ 08903.

References

Allen, FA. 1982. In: Bliss, M. *The Discovery of Insulin*. Chicago: Univ. of Chicago Press. pp. 105-28.

Baker, HJ, Lindsey, JR and Weisbroth, SH. eds. 1979. *The Laboratory Rat*. Vols. 1 and 2. New York: Academic Press.

Beadle, M. 1977. *The Cat*. New York: Simon and Schuster. Chapt. 13. pp. 166-82.

Bernard, C. 1957. *An Introduction to the Study of Experimental Medicine*. New York: Dover Publications. p. 102.

Best, CH. 1974. A short essay on the importance of dogs in medical research. *The Physiologist*. 17: 437-39.

Bijkerk, H. 1979. Poliomyelitis epidemic in The Netherlands, 1978. *Div. Biol. Stand*. 43: 195-206.

CAMRA. (Committee on Animal Models for Research in Aging). 1981. *Mammalian Models for Research on Aging*. Washington, DC: Natl. Acad. Sci. p. 587.

Canby, TY and Stanfield, JL. 1977. The rat—lapdog of the devil. *Natl. Geo*. 15: 60-87.

CDC. (Center for Disease Control). 1979. Poliomyelitis. *Morbid. and Mort. Wkly. Rep*. 28: 49-50, 229-30, 309, 34546.

Chey, WY, Kosay, S, Siplet, H and Lorber, SH. 1971. Observations on hepatic histology and function in alcoholic dogs. *Am. J. Dig. Dis*. 16: 835-38.

Coleman, DL. 1978. Obese and diabetes: Two mutant genes causing diabetes-obesity syndromes in mice. *Diabetologist*. 14: 141-48.

Colmer, G. 1843. Paralysis in teething children. *Am. J. Mrd. Sci*. 5: 248-50.

Cork, LC, Munnell, JF, Lorenz, MD, Adams, RJ, Griffin, JW and Price, DL. 1979. Hereditary canine spinal muscular atrophy. *J. Neuropathol. Exp. Neurol*. 38: 209-21.

Cornelius, CE. 1969. Animal models—a neglected medical resource. *N. Eng. J. Med*. 281: 934-44.

Davison, FC. 1977. Historical perspectives of biomedical experimentation. In: *The Future of Animals, Cells, Models, and Systems in Research, Development, Education and Testing*. Washington, DC: Natl. Acad. Sci. pp. 7-15.

Dewhirst, MW, Sim, DA, Wilson, S, DeYoung, D and Parsells, LJ. 1983. Correlation between initial and long-term responses of spontaneous pet animal tumors to heat and radiation or radiation alone. *Cancer Res*. 43: 5735-41.

DeVries, WC, Joyce, LD, Anderson, FL, Anderson, JL, Jarvik, RK and Kolff, WJ. 1983. Initial human application of the Utah total artificial heart. *Circul*. 68: III-89, October.

Fishman, AP and Richards, DW. 1982. *Circulation of the Blood: Men and Ideas*. American Physiological Society:Baltimore, Maryland, Waverly Press.

Gay, WI. 1984. Health benefits of animal research. The dog as a research subject. *Physiol*. 27(3): 133-41.

Fox, JP, Elveback, L, Scott, W, Gatewood, L and Ackerman,E. 1971. Herd immunity: Basic concept and relevance to public health immunization practices. *Am. J. Epidemiol*. 94: 17989.

Fox, RR, Muir, H, Bedigan, HG and Crary, DD. 1982. Genetics of transplacentally induced teratogenic and carcinogenic effects in rabbits treated with *N*-nitroso-*N*-ethylurea. *J. Natl. Can. Inst*. 69: 1411-17.

Francis, T, Jr., Korn, RF, Voight, RB, Boisen, M and Tolchinsky, E. 1955. An evaluation of the 1954 poliomyelitis vaccine trials: Summary report. *Am. J. Pub. Health*. 45: 1-63.

Franklin, KJ. 1933. *A Short History of Physiology*. London: Bale.

Furesz, J. 1979. Poliomyelitis outbreaks in The Netherlands and Canada. *Can. Med. Assoc. J*. 120: 905-6.

Ginsburg, R, Bristow, MR, Kantrowitz, DS and Harrison, DC. 1981. Histamine provocation of clinical coronary artery spasm: Implications concerning pathogenesis of variant angina.*Am Heart J*. 102: 819-22.

Franklin, KJ. 1933. *A Short History of Physiology*. London: Bale.

Furesz, J. 1979. Poliomyelitis outbreaks in The Netherlands and Canada. *Can. Med. Assoc. J.* 120: 905-6.

Ginsburg, R, Bristow, MR, Kantrowitz, DS and Harrison, DC. 1981. Histamine provocation of clinical coronary artery spasm: Implications concerning pathogenesis of variant angina.*Am Heart J.* 102: 819-22.

Harvey, W. 1766. Epistola prima Paulo Marquarto Slegelio,Hamburgensis, April 7, 1651. In his *Opera Omnia*. London: Letters collected by Bowyer. p. 613.

—. 1928. *Exercitatio anatomica de motu cordis et sanguinis in animalibus*. With an English translation by CD Leake. Springfield, IL: Thomas. First edition, 1628.

Jonas, AM. 1984. The mouse in biomedical research. *The Physiologist.* 27(5): 330-46.

Keys, TF, Sapico, FL, Touchon, R, Barenfus, M and Hewitt,WL. 1972. Experimental interococcal endocarditis. Description of a canine model. *Am. J. Med. Sci.* 63:103-9.

King, FA and Yarbrough, CJ. 1985. Health benefits of animal research. Medical and behavioral benefits from primate research. *Physiol.* 28(2): 75-87.

Kozma, C, Macklin, W, Cummins, LM and Mauer, R. 1974. Anatomy, physiology and biochemistry of the rabbit. In: *The Biology of the Laboratory Rabbit*. Weisbroth, SH, Flatt, RE and Kraus, AL. eds. New York: Academic Press. pp. 49-72.

Landsteiner, K and Popper, E. 1909. Ubertragung der poliomyelitis acuta auf affen. A *Immunitaestforsch.* 2: 377-90.

Lewis, MG, Mathes, LE and Olsen, RG. 1981. Protection against feline leukemia by vaccination with a subunit vaccine. *Infect. Immun.* 34: 888-94.

Marx, PA, Maul, DH, Osborne, KG, Moody, P and Gardner, MB. 1984. Simian AIDS: Isolation of a type D retrovirus and transmission of the disease. *Science.* 223: 1083-86.

Maugh, TH, III. 1981. FDA approves hepatitis B vaccine. *Science.* 214: 1113-17.

McClure, HM, Swenson, B, King, F, Chermann, JC, Sinousi, J and Macher, A. 1984. Experimental infection of chimpanzees with lymphadenopathy associated virus. *Morb. Mortal. Wkly. Rep.* 33: 442-43.

McDonald, D. 1960. How does a cat fall on its feet? *New Scientist.* 7: 1649-57.

Morris, TQ and Gocke, DJ. 1971. Modified acute canine viral hepatitis—a model for physiologic study. *Proc. Soc. Exp. Biol.* 139: 32-36.

Mott, GE, McMahan, CA, Kelly, JL, Mersinger-Farley, C and McGill, H, Jr. 1983. Influence of infant and juvenile diets on serum cholesterol, lipoprotein cholesterol, and lipoprotein concentration in juvenile baboons. *Atherosclerosis.* 45: 191-201.

Onderdonk, A, Cisneros, R and Bartlett, J. 1980. *Clostridium difficile* in gnotobiotic mice. *Infect. Immunol.* 28: 277-82.

Paul, JR. 1958. Endemic and epidemic trends of poliomyelitis in Central and South America. *Bull. WHO.* 19: 747-58.

—. 1971. *A History of Poliomyelitis*. New Haven: Yale University Press. p. 486.

Pavlov, IP. 1927. *Conditional Reflexes and Investigations of Physiological Activity of the Cerebral Cortex.* Trans. by GU Anrep. New York: Dove.

Pearson, PL, et al. 1982. Report of the committee on comparative mapping. *Cytogenet. Cell Genet.* 32: 208-20.

Reed, JD. 1981. Crazy over cats. *Time.* Dec.: 72-79.

Robinson, R. 1965. *Genetics of the Norway rat*. New York: Permagon Press.

Ross, R. 1983. Recent progress in understanding atherosclerosis.*J. Am. Geriat. Soc.* 31: 231-35.

Russell, ES. 1979. Hereditary anemias of the mouse: A review for geneticists. *Adv. Genet.* 20: 357-459.

Sack, RB and Carpenter, CG. 1969. Experimental canine cholera. Development of a model. *J. Infect. Dis.* 119: 138-49.

Salk, D. 1980. Eradication of poliomyelitis in the United States. I. Live virus vaccine-associated and wild poliovirus disease. *Rev. Inf. Dis.* 2: 228-42.

Sarngadharan, MG, Popovic, L, Bruch, J, Schupbach, J and Gallo, RC. 1984. Antibodies reactive with human T-cell lymphotropic retroviruses (HTLV-III) in the serum of patients with AIDS. *Science.* 224: 506-8.

Sarton, G. 1954. *Galen of Permagon.* Lawrence,KS: Univ. of Kansas Press.

Science News. 1984. First chimps infected with AIDS virus. 126: 121.

Scott, JP and Fuller, JL. 1965. *Genetics and the Social Behavior of the Dog.* Chicago: Univ. of Chicago Press.

Sechzer, JA. 1983. The role of animals in biomedical research. *Annals of the New York Acad. of Sciences. Vol. 46.*

Taylor, CB, Cox, GE, Morale-Estrella, P and Southworth, J. 1962. Atherosclerosis in rhesus monkeys. II. Arterial lesions associated with hypercholesterolemia induced by dietary fat and cholesterol. *Arch. Pathol.* 74: 16-34.

Taylor, CB, Patton, DE and Cox, GE. 1963a. Atherosclerosis in rhesus monkeys. IV. Fatal myocardial infarction in a monkey fed fat and cholesterol. *Arch. Pathol.* 76: 404-12.

Taylor, CB, Manalo-Estrella, P and Cox, GE. 1963b. Atherosclerosis in rhesus monkeys. V. Marked diet-induced hypercholesterolemia with xanthomatosis and severe atherosclerosis. *Arch. Pathol.* 76: 239-49.

Toda, N. 1983. Isolated human coronary arteries in response to vasoconstrictor substances. *Am. J. Physiol. (Heart Circ. Physiol.).* 14: H937-H941.

Tozzi, CA and Merrill, GF. 1985. Evidence of histamineinduced myocardial ischemia: Reversal by chlorpheniramine and potentiation by atherosclerosis. *Cardiovas. Res.* 19: 74453.

Visscher, MB. 1969. A half century in science and society. *Am. Rev. Physiol.* 31: 108.

Wiesel, TN. 1982. Postnatal development of the visual cortex and the influences of environment. *Nature (London).* 299: 583-91.

Wissler, RW. 1980. Perspectives on cardiovascular research in primates. In: Kalter, SS. ed. *The Use of Nonhuman Primates in Cardiovascular Diseases.* Austin, TX:Univ. of Texas Press. pp. 15-32.

Wooley, GW. 1943. "Misty," a new coat color dilution in the mouse *Mus musculus. Genetics.* 28: 95-96.

—. 1945. Misty dilution in the mouse. *J. Hered.* 36: 269-70.

Yang, MG and Mickelson, O. 1974. Laboratory animals in nutrition research. In: Gay, WI. ed. *Methods of Animal Experimentation.* Vol. 5. New York: Academic Press. pp. 136.

THE CASE FOR INTENSIVE FARMING OF FOOD ANIMALS[1]

Stanley E. Curtis[2]

Introduction

Great strides have been made in recent decades in applying principles of biology and engineering in animal agriculture. Sophisticated animal production systems have been developed in response to numerous pressures. But many technological riddles still remain. Fortunately, our seemingly insatiable curiosity about the nature of things and our inexorable drive to apply what we know guarantee both our continuing search for new knowledge about relations between animals and their environments, and our rapid use of that knowledge to upgrade and fine-tune animal production systems for the benefit of animals, the consuming public, and the agricultural industries alike.

Ecology has always been at the heart of animal farming. After the glaciers receded and crop production was established, animal production ascended. The animals recruited for domestication by early farmers—the same species farmers all over the world raise nowadays—differed from their cousins that have been left in the wild in that those domesticated were adaptable to a wider range of environments. Agricultural animals are relatively unfinicky and tolerant. During the millennia when farm animals were kept in the natural environment, or at best shielded poorly from climatic rigors, the relative ease with which the animals appeared to adapt to their surroundings led husbandmen to give the environment a low spot in the hierarchy of production factors.

The situation has changed greatly during the last 40 years. The advent of widespread intensivism in animal agriculture—together with ever smaller profit margins and our relentless search for ways to increase food-production efficiency—increased the relative significance of animal-environmental relations. Now, in addition to paying close attention to the nutritional and health-care needs of the animals and to increasing the genetic fit of the animals to the environment, we try to meet the animals' needs by modifying the several facets of their surroundings. Of course, all of these efforts have been made possible by the increase in our knowledge base that has resulted from experimental research on animals.

It is not likely that animal farms are today as we remember they were yesterday. It is likely that they never were. Our notions of how things used passage of time. Be that as it may, to appreciate the modern, intensive systems of animal agriculture that have sprung up in many parts of the world, we must consider them in the context of the human cultures in which they arose.

Our world is still a hungry place. At the same time, the number of people worldwide who grow food for themselves continues to dwindle. Most U.S. citizens have never set foot on a farm or harvested one mouthful—let alone a lifetime's worth—of daily bread. Yet our farmlands and climates and our agricultural and food industries are this nation's ultimate resources. By increasing productivity, our farmers and the scientific and business endeavors that support our nation's food production, processing, and distribution have proved to be able and reliable husbands of these precious resources. But make no mistake: The challenge to increase food production in step with increasing demands is a huge one. It requires managing numerous elements of nature which are recalcitrant at best, while coping with others which are manageable or unpredictable, or both. Our agribusinesses have made the task simple; plenty of safe, wholesome, inexpensive food is available in our groceries every day. Again, make no mistake: Making this so has not been a simple task. It is incumbent upon severe critics of American animal agriculture to recognize and appreciate this and to make their retrospective judgments accordingly.

Stress

Before discussing intensive management of food animals in the United States, a few words should be said about stress in general. Any animal, in the wild or on a farm, is usually responding to several stressors at once. Stress is the rule, not the exception. And nature has endowed animals with a marvelous array of reactions to stress. The animal must maintain a steady bodily state despite fluctuating external conditions. By means of dozens of negative-feedback control loops, the animal tries to regulate within narrow limits the environment in which its individual cells reside and operate.

An environmental adaptation is any functional, structural, or behavioral trait that favors an animal's survival or reproduction in a given environment. A stress is any environmental situation that provokes an adaptive response. Stress can occur when an animal's environment changes so as to trigger some homeokinetic response (as when environmental temperature falls below some critical point) or when the animal itself changes in relation to its surroundings (as when shearing reduces a sheep's cold tolerance).

The scientific literature contains reports of hundreds of experiments purported to measure the environmental adaptability of agricultural animals. It is a relatively simple task to subject experimental animals to a controlled stressor and measure a resultant change in some physiological, immunological, anatomical, or psychological characteristic. But an objective index of stress in terms of animal health, productivity, and overall well-being has been elusive.

Still, it is a fundamental tenet of modern animal agriculture that environmental stress generally alters animal performance. The stress provokes the animals to react, and this reaction can influence the partition of resources amoung maintenance, reproductive, and productive processes in at least five ways (Curtis 1983). (1) The reaction might alter internal functions. Many bodily functions participate in productive and reproductive processes as well as in reactions to stress, which are of higher priority. (2) The reaction might divert nutrients. When an animal responds to stress, it in effect diverts nutrients from productive and reproductive processes to uses in higher priority maintenance processes. (3) The reaction might reduce productivity directly. The animal's response sometimes comprises an intentional reduction in productive processes in order to free some nutrients for maintenance use. (4) The reaction might increase variability in productive performance. Individual animals differ from one another in the ways they respond to the same stressor. The complements of mechanisms used often differ in the energy expenditure they require, so the amount of metabolizable-energy expenditure which must be diverted from productive processes to mainte- nance differs. The result is that the amount of variation in individual perfor- mance tends to be directly related with the environmental adversity to which a group of animals is subjected. (5) The reaction might alter disease resistance. Because an animal's reaction to stress can affect disease resistance, that stress reaction can influence the frequency and severity of infectious diseases.

That stress in agricultural animals must be optimized in terms of both animal welfare and economic ramifications should be obvious from these general relationships. This is an important issue, and it will pervade all that follows.

Intensive Farming of Food Animals

Hundreds of millions of Americans must have food but choose not to grow it for themselves. Food production is a business and subject to the same economic forces as any business (Halcrow 1980). The chances of a turnaround in the trend to fewer, larger, more intensive animal farms are akin to those of a return to mom-and-pop grocery stores in the residential areas of every city and an independent fast-food restaurant on the main street of every town.

Intensive dairy, livestock, and poultry farms came on the scene soon after World War II. The movement of agricultural animals from dirt lots and pastures to confinement facilities accelerated markedly during the 1950s in the poultry and dairy industries and the 1960s in livestock production. It continues to this day. The most important reason for this did not revolve around the well-being of the animals. Admittedly, although there have been significant side-benefits of intensivism for the animals, there have been new problems, too (Curtis 1983).

One major force leading to intensivism in animal agriculture had to do with responsible land management. Rearing animals extensively requires tremendous acreages, and in many parts of the United States it not only constitutes unsound stewardship of the soil, but it has proved economically unfeasible as well.

Another critical factor was labor. With the family farm goes the force of cheap workers upon which this kind of farming was based. Also, animal caretaking is a seven-day-a-week job, so to attain a living standard similar to that of society as a whole, outside help was needed. Today's poultry, livestock, and dairy producers increasingly need to hire workers from the general labor pool to do chores formerly assigned to family members. Of course, prevailing wages must be offered if workers are to be attracted. Despite relatively high rates of unemployment in many rural communities in recent years, the farmer often has had to provide unusual incentives to employees, because the work is hard, and in some respects, unappealing. Thus, animal producers have had to expand and specialize their operations to the extents necessary to justify increased outlays for hired help.

A third factor has been animal waste. Farm animals produce tremendous amounts of feces and urine. For example, one hog puts out as much waste as three adult humans. Of course, the magnitude of the waste-management task rises in parallel with the size of operation. Because of the keen interest in environmental protection over the past two decades, regulations have been put in place which in effect preclude animal production on many of the hills and in many of the valleys these animals roamed in days past. For practical purposes, waste containment is achievable only with a confinement-production facility.

Land, labor, and waste—these have been the principal socioeconomic forces behind the widespread adoption of intensive animal-production systems. The changes that have resulted from these forces have had impacts on the animals' welfare. At this point, let us mention those changes that have been beneficial for the creatures. For one thing, seasonal production cycles have been dampened considerably. It is easier to manage newly born or hatched animals—and juveniles and adults, too, for that matter—the year around in houses than in either natural surroundings or rudimentary artificial shelters typical of extensive production. This has been good for the animals. And the resultant changes in dairy, poultry, and livestock marketing increased economic efficiencies in food production, processing, and distribution. The ultimate beneficiaries of these efficiencies in our free-enterprise economy are the consumers of food products of animal origin.

More pluses have to do with biological management, with the animal's life per se. (1) Providing steady supplies of a well-balanced diet and sanitary water is easier in confinement than on range. (2) Predation of young and small animals by wild and feral carnivores is a tremendous problem in many parts of the United States. Intensive animal facilities such as sheep folds have been used to foil this aspect of the web of life since biblical times. (3) The perforated floors commonly used in animal facilities separate the beasts and birds from their own excreta, thus preventing them from practicing some unhygienic, obnoxious habits such as coprophagy and wallowing in their own excrement. Because enteric infections are major causes of disease and even death in all species of farm animals, the perforated floor improved the living conditions of these creatures greatly. (4) Caretakers can observe individual animals more thoroughly when they are close at hand, held singly, or in small groups. Injuries and disease can be detected more readily and remedial measures implemented more easily as a result.

Interestingly, despite technological changes, managers of large, intensive animal farms still consider sound animal care the keystone of profitability in animal production. Can anything else be imagined by anybody? Excellent animal husbandry is the sine qua non of successful animal production.

The advent of larger units also made it possible to upgrade management quality. On many farms, animal production is no longer a sideline activity or one of several enterprises competing for the manager's attention. More and more, managers of animal-production operations are multitalented professionals who devote all of their time to a single species. Demand for well-educated and -trained managers has led to the establishment of special curricula in intensive animal production.

Finally, with increasing size of operation come economies of purchasing and marketing in large lots, with more or less continuous flow (Halcrow 1980). While this generally enhances the profitability of an individual enterprise, again consumers of foods of animal origin are the ultimate beneficiaries in our kind of economy.

In agriculture, it is not sufficient to be interested only in physiological, behavioral, immunological, and anatomical indices of animals' environmental adaptability. The next question is: How much decrement in production is associated with residing in a particular environment? To learn the quantitative effects of a given environment on animal performance, we still must measure the productive traits themselves. An animal exhibiting obvious reaction to stress, as mentioned above, is generally assumed to be having depressed performance. But the performance loss may be reversible only by a modification of the environment that cannot be repaid in terms of increased animal productivity. Further, visible strain in an animal signifies that it is trying to compensate for an environmental impingement. These attempts might succeed, and they might interfere with performance only slightly. Of course, the question remains as to whether the stresses imposed by a certain production system comprise an unacceptable environment in terms of the animals' overall welfare, a point to be expanded upon later.

Abuse, Neglect, and Deprivation

Animal production resembles other professions in that there are (in terms of humane treatment) good animal farmers and poor ones. When critics of animal farms cite examples of cruelty to animals, they are referring to farms run by poor producers. Inhumane treatment leads to unhealthy, unproductive animals, and consequently, financial losses. Poor stockmen are among the first animal farmers to go out of business in times of economic crisis.

It has been suggested that any suffering an animal experiences at the hands of a farmer falls into one of three categories: abuse, neglect, or deprivation (Ewbank 1981). Abuse refers to obvious, active cruelty, such as beating an animal with a stick. Neglect is obvious, passive cruelty; for example, confining an animal and then not providing it one or more vital resources, such as food or water. Everyone would agree that abuse and neglect are cruel, and state and federal legislation outlawing both was passed many years ago. Progressive animal producers neither condone nor encourage such cruelty, and any representation to the contrary comprises a calumny. Further, abuse and neglect constitute or lead to severe stress and thus are clearly counterproductive; their practice by farmers would be just as clearly irrational.

Deprivation is the most subtle form of cruelty, and thus the most difficult to assess. It involves the denial of relatively less vital resources, the actual requirements for which mostly have yet to be established. Whether or not farm animals in certain living situations in intensive production systems are suffering from deprivation is a major issue being discussed by humane activists, farmers, and scientists. If so, economical and practically feasible means of alleviating the deprivation will need to be discovered and developed for adoption by farmers. While it might be tempting to speculate anthropomorphically as to the stress perceived by animals when they are prevented by the nature of the environment in which they reside from performing some specific behavior, both humane and economic aspects of environmental design and management are better served when the scientific approach to needs identification and fulfillment is taken.

Needs: Physiological, Safety, and Behavioral

It is axiomatic that, when an animals' needs are not being met, its welfare is more or less jeopardized. But here again it must be remembered—and this idea also will be expanded upon later—that a particular welfare decrement does not necessarily place the animal in an ethically unacceptable environment; perhaps the animal simply experiences less—but still an ethically acceptable amount—of well-being.

In any case, it has been suggested that agricultural animals have a hierarchy of needs along the lines of Abraham Maslow's scheme for humans, and that animals' basic needs are being met in most intensive production systems (Curtis 1984). First and most basic are farm animals' physiological needs; for feed, physical and biological elements of the environment, and health care. These are already relatively well understood and fulfilled.

Intermediate are the animals' safety needs. Although the needs to be protected from harmful environmental elements are important, these safety needs are tended somewhat less rigorously than are the physiological needs. Weather accidents,predation, and poorly designed, manufactured, and operated equipment and facilities still exact reducible tolls in terms of both animal welfare and financial profits (Curtis 1984).

Last in the hierarchy are the animals' behavioral needs. The question among most scientists is: Is there reasonable evidence supporting the existence of any behavioral need in any agricultural animal? Indeed, no such need has been established, although many scientists believe that they well might exist, however difficult they may be to elucidate (Hughes 1980). Of course, fundamental to assessing welfare in a farm animal are answers to two questions, the second of which is proving to be exceedingly difficult to answer: (1) Does the animal have subjective feelings? (2) What indicators reveal any such feelings? (Duncan and Dawkins 1983). Knowledge of animals' mental activities can be gained only via indirect experimental evidence at this time, hence any conclusions must be considered tentative.

Assessing Welfare in Farm Animals

Attempts to quantitatively evaluate suffering or its antithesis, welfare, in animals residing in various farm environments have proved futile so far (UFAW 1979; Brown 1980; Bessei 1982; Curtis 1982; Baxter et al. 1983; Duncan and Dawkins 1983; Smidt 1983a; Tarrant 1984; Moberg 1985; Wiepkema 1985). There is a consensus that the welfare of farm animals eventually will be best assessed by an integrated system of indicators from four categories: (1) reproductive and productive performance, (2) pathological and immunological traits, (3) physiological and biochemical characteristics, and (4) behavioral patterns (Duncan 1981; Curtis 1982; Smidt 1983b). At present, potential pitfalls notwithstanding (Duncan and Dawkins 1983), health, reproductive, and productive traits continue to be the most readily measurable, most practically useful indicators of fit between agricultural animals and the environments in which they reside (Curtis 1982).

Welfare Plateau, Economics, and Production Environments

On any animal farm, achieving the highest level of animal welfare possible, consistently, is still a vague exercise. It most likely will be so for several years. C.D. Hardwick formulated the idea that an ethically acceptable level of animal welfare exists over a range of conditions provided by a variety of agricultural production systems, not only in one ideal set of circumstances (Duncan 1978). This acceptable range of environments, Hardwick said, comprises a "welfare plateau" (figure 1). The word, "plateau," may be misleading; in the acceptable range, with improvements in environment, the animals' total welfare increases, too. Indeed, increasing welfare is the basis for identifying environmental "improvement." But Hardwick's notion was this: Any point on the welfare plateau is ethically acceptable in terms of animal welfare. In other words, on the welfare plateau, a relatively small environmental change might improve subtly an animal's overall well-being, but *anywhere* on the welfare plateau the animal is as *free of suffering* as possible.

The concept of the welfare plateau is profoundly relevant to discussions of environmental design for animal farms. At the lower limit of the welfare plateau might stand one or more production systems that are marginally acceptable in terms of the animal welfare they engender, while beyond this gray zone stand systems that more or less fail to support the animals' needs sufficiently well to be considered ethically acceptable.

The farmer as businessperson recognizes that, already in the zone of marginally acceptable production systems, the law of diminishing returns (Halcrow 1980) has ensued; returns to investments in environmental improvements are not sufficient to pay for the improvements. In contrast to this, in the range of unacceptable environments, small environmental improvements result in returns that are more than adequate to pay for the improvements. Thus, there is the logical tendency for the production systems adopted most

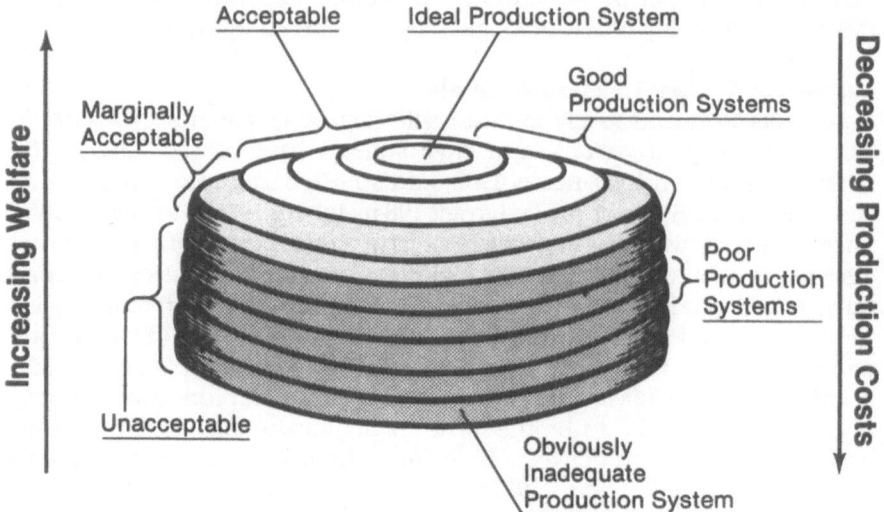

Welfare Plateau

Figure 1. Scheme relating respective animal production systems to the animal welfare they engender. Scheme incorporates C.D. Hardwick's idea of a "welfare plateau" (modified after Duncan 1978).

widely to be located at the upper end of the marginally acceptable zone and the lower end of the acceptable zone. Said another way: The shape of the HardwickDuncan scheme is determined mainly by (1) decreasing frank suffering by the animals as the environment is improved within the unacceptable zone; (2) flexion in the marginally acceptable range, where the law of diminishing returns sets in; and (3) inadequate returns to investment on the welfare plateau, throughout which (a) frank suffering is minimal and (b) those small additional returns that do occur owe to increased animal well-being alone.

The animal producer is thus faced with the necessity of compromising welfare for profit. As a humane person, the producer strives to provide the animals an existence as free of suffering as possible. This limits the possibilities to the upper region of the marginally acceptable range or the welfare plateau itself. As a businessperson, the producer strives to adopt the production system that will be economically optimal for prevailing conditions. In view of the law of diminishing returns, this latter constraint tends to locate adoptable systems in the upper region of the marginally acceptable range or in the lower part of the welfare plateau. To locate nearer the humane ideal would be an unwise business decision. Still, as long as the system adopted lies in or very near the acceptable zone, the producer's ethical obligations have been satisfied.

Summary and Conclusions

Alas, farmers face an animal-welfare dilemma. They must decide on animal-production systems while constrained by humane concerns—both their own as well as those of the general citizenry—on one hand and by the realities of doing business in a free-enterprise milieu on the other. And the dilemma will be resoluble only if and when we know much more than we now know about animal suffering and thus about animal well-being. The question is not whether animals have feelings; there is general agreement up and down the line that they do. The question is: How does the animal feel, living in this production system or that? Ian Duncan and Marian Dawkins (1983) believe that there are "...indicators that with careful experimentation we may be able to accumulate indirect evidence about animals' subjective feelings. This should be our ultimate aim. There are many problems but they are not insurmountable."

How can these problems be surmounted? How will it come to pass that we learn once and for all whether certain production systems cause farm animals to suffer? How will economically feasible, more socially acceptable systems of farm-animal production be discovered and developed? The answer: We can learn these things only from research.

The time is ripe for humane activists to support in all ways possible bona fide scientific investigations of farm animal welfare. This suggestion is not heretical, naive, or ridiculous. My reasoning follows, in the form of a brief recapitulation and juxtaposition of earlier points with a couple of new ones, together with pragmatic analysis and synthesis.

1. Consumer demand for human foods of animal origin is strong, and it will continue to be so for decades. The vast majority of consumers decide whether or not to eat these foods on the bases of nutritional factors, convenience, and flavor, not on the basis of ethical questions. It is folly to hope that animal farms will disappear from the U.S. scene. Those of us who want farm animals to experience as little suffering and as much wellbeing as possible ought to do what we can to ensure that these animals' needs and feelings are understood, and that the needs are fulfilled, the feelings protected.

2. Food-animal production is a business. As such, it is constrained by economic factors.

3. Society—including animal producers—requires that food animals not be caused to suffer in any way. Therefore, food animal production is also constrained by humane factors.

4. Economic and humane factors do not always work in tandem. Compromise between humane and economic constraints is inevitably necessary in terms of animal-production-system design. This compromise occurs at the juncture of the welfare plateau and the range of marginally acceptable production systems.

5. Animal agriculture quickly adopts appropriate technologies, especially when the benefit/cost ratio is favorable.

6. Animal producers are at least as humane as members of society in general. Any representation to the contrary comprises a calumny.

7. If animal producers have adopted inhumane production technologies, it has been because they and those who advise them have been ignorant. Any such ignorance owes to lack of scientific evidence, not lack of concern for the animals' general well-being.

8. Those of us who care about animals and want to try to improve the welfare of food animals ought to do everything we can to learn more about what these animals need and how they feel. At the same time, we can be searching for improvements in terms of production equipment and facilities and husbandry systems designed to fulfill the animals' needs and support favorable feelings. Basic and applied research along these lines deserves the complete support of all who want to engender the highest level of welfare possible in food animals.

Endnotes

[1] Paper presented at the national conference, "Animals and Humans: Ethical Perspectives," Moorhead State University, Moorhead, MN, April 21-23, 1986.

[2] Professor, University of Illinois at Urbana-Champaign, College of Agriculture, Department of Animal Sciences, 126 Animal Sciences Laboratory, 1207 W. Gregory Dr, Urbana, IL 61801.

References

Baxter, SH, Baxter, MR and MacCormack, JA. eds. 1983. *Farm Animal Housing and Welfare.* Boston, MA: Martinus Nijhoff Publishers.

Bessei, W. ed. 1982. *Disturbed Behaviour in Farm Animals.* Stuttgart, FRG: Verlag Eugen Ulmer.

Brown, PL. ed. 1980. *Proc. First European Conf. on the Protection of Farm Animals. Anim. Reg. Stud.* 3(1,2): 1174.

Curtis, SE. 1982. Measurement of stress in animals. In: Woods, W. ed. *Proc. Symp. Management of Food Producing Animals. Vol I.* West Lafayette, IN: Purdue Univ.

—. 1983. *Environmental Management in Animal Agriculture.* Ames, IA: Iowa State University Press.

—. 1985. What constitutes well-being? In: Moberg, GP.ed. *Animal Stress.* Bethesda, MD: American Physiological Society.

Duncan, IJ. 1978. An overall assessment of poultry welfare. In: Sorensen, LY. ed. *Proc. First Danish Seminar on Poultry Welfare in Egglaying Cages.* Copenhagen, Denmark: National Committee on Poultry and Eggs.

—. 1981. Animal rights—animal welfare: A scientist's assessment. *Poul. Sci.* 60(3): 489-99.

Duncan, IJ and Dawkins, MS. 1983. The problem of assessing "well-being" and "suffering" in farm animals. In: Smidt, D. ed. *Indicators Relevant to Farm Animal Welfare.* Boston, MA: Martinus Nijhoff Publishers.

Ewbank, R. 1981. Alternatives: Definitions and doubts. In: *Alternatives to Intensive Husbandry Systems.* Potters Bar, United Kingdom: Universities Federation for Animal Welfare.

Halcrow, HG. 1980. *Economics of Agriculture.* New York: McGraw-Hill Book Co.

Hughes, BO. 1980. The assessment of behavioural needs. In: Moss, R. ed. *The Laying Hen and Its Environment.* Boston, MA: Martinus Nijhoff Publishers.

Moberg, GP. ed. 1985. *Animal Stress.* Bethesda, MD: American Physiological Society.

Smidt, D. ed. 1983a. *Indicators Relevant to Farm Animal Welfare.* Boston, MA: Martinus Nijhoff Publishers.

—. 1983b. Advantages and problems of using integrated systems of indicators as compared to single traits. In: Smidt, D. ed. *Indicators Relevant to Farm Animal Welfare.* Boston, MA: Martinus Nijhoff Publishers.

Tarrant, PV. ed. 1984. *Evaluation Report 1979-83, Farm Animal Welfare Programme.* EUR918EN. Luxembourg, LX: Commission of the European Communities.

UFAW (Universities Federation for Animal Welfare). 1979. *The Welfare of the Food Animals.* Potters Bar, UK: Universities Federation for Animal Welfare.

Wiepkema, PR. 1985. Abnormal behaviours in farm animals: Ethological implications. *Neth. J. Zoo.* 35(1,2): 27999.

THE CASE AGAINST INTENSIVE FARMING OF FOOD ANIMALS[1]

Linda D. Mickley[2] and Michael W. Fox[3]

Overview

Intensive poultry and livestock husbandry practices, which developed in the United States shortly after World War II, are part of the "revolution" in American agriculture. This revolution, however, is not leading to the flourishing of American farming, but to its demise. Our once labor-intensive food-production system has become increasingly capital intensive, and dependent upon machinery, automation, and petrochemical-based fertilizers and pesticides. This now over-capitalized industrialization of agriculture has reaped enormous profits (for a few), and agricultural economists are quick to point out that not only do Americans pay proportionally less for their food than any other country, our farms also help feed a hungry world. Yet, there are many hidden costs, costs not directly reflected by or computed into the price of our food.

The sociological and economic costs of the U.S. agricultural system are evident in such problems as rural unemployment, bankruptcy of family farms, and chronic overproduction of commodities such as milk, that are buoyed up by price support programs. The large-scale operations or "super" farms are benefitted by capital-intensive buildings, automation, and drugs and feed additives, as well as the economies of scale, while small- to medium-sized farmers must borrow against their land, crops, or animal products. Such inequities are further compounded by tax structures that favor large farms.

The ecological cost is measured in terms of irreparable damage to our farmlands due to soil erosion, depletion of trace minerals, soil humus, and deep water aquifer reserves, along with pesticide and chemical fertilizer pollution of ground waters. Such are the consequences of imposing an inappropriate technology and industrial paradigm upon the delicate biological balance of agriculture.

This same paradigm has been applied to farm animals in the industrial-scale production of meat, eggs, and dairy products. Such application is changing animal *husbandry* into animal *technology* wherein the health and well-being of the animals, like the health and well-being of the land, are sacrificed in the name of efficiency and productivity. The enhancement of efficiency and productivity of farm animals is achieved through the feeding of high-energy, low-fiber feedstuffs (such as corn, soy, and food industry by-products), selective breeding for rapid growth and weight gain, and housing in varying degrees of confinement.

The feeding of high-energy, low-fiber diets has been linked to metabolic and production-related diseases such as ketosis and laminitis in dairy cattle (Webster 1986; Harvey 1983; Fox 1983; Van den Bergh 1976), and the rumenitis-liver abscess complex in beef cattle (Fox 1984). Farm animals are also harmed by being fed crops and by-products that are contaminated with residual pesticides and other hazardous chemicals such as drugs (Long 1985; Peterson 1986; Somogyi 1985), and which are nutritionally deficient as well (a problem in part attributable to depletion of trace minerals in the soil). Improper nutrition is one of the factors that contributes to the suffering of intensively-housed farm animals, which, like the improper use of nutrient fertilizers, is linked with poor viability and higher pest susceptibility of crops (Chaboussou 1980).

Humans have long exercised control over animals and plants by means of selective breeding. In the case of farm animals, however, selective breeding for rapid growth (e.g., broiler chickens and hogs) and high productivity (e.g., laying hens and dairy cows) contributes to a host of production-related ("domestigenic") diseases and/or increases in susceptibility to infection (Siegel 1983; Fox 1984). The term "agricologenic" is applicable to those unintentional or undesirable side effects of crop production systems, such as greater susceptibility to disease and pests due to selection for greater productivity in various hybrid strains (Hodges and Scofield 1983). Selective breeding alone of crops for high yield does not guarantee such such yields, as it has been noted that U.S. crops produce, on the average, only 20-25% of their genetic potential, and are prevented from reaching that potential by adverse physical (abiotic) environments, diseases, arthropod pests, nematodes, and weeds (Cook 1986).

Confinement housing, especially in the case of veal calves, poultry, and hogs, is the third tool used by modern U.S. agriculture to increase animals' efficiency and productivity. It is often stated by agribusiness advocates that animals in intensive housing would not produce if their well-being and health were truly compromised. This belief is only a half-truth: In reality, productivity (or performance) is not an absolute guarantor of welfare (Fox 1984). While few farmers are deliberately cruel, the economic treadmill on which modern farmers find themselves forces them to increase stocking density, which in turn forces them to jeopardize or ignore individual animal welfare in order to maximize overall production (Fox 1984).

Such increases are justified by the confinement unit producers, as they purportedly allow for greater productivity per unit of building space. These producers may also argue that less heat must be supplied to the units due to body heat generated by the animals, and therefore stock requires less feed in colder months.

This latter reason for overstocking is not legitimate; it is false economy at best, as the animals will suffer heat stress in the summer, and the negative effects of high relative humidity and noxious fumes from excreta if ventilation is inadequate.

In addition, animals in overcrowded, restrictive environments may develop learned helplessness as a result of their having no control over or escape from their immediate environments. It is theorized that learned helplessness leads to immunosuppression, reduced stress resistance, and increased disease susceptibility (Dantzer and Mormède 1983; Fox 1984). It must be reiterated that in spite of these known problems that affect animals' productivity, it has not been cost-effective to rectify them. It is more expedient instead to treat the symptoms with drugs. An analogous situation exists in the use of pesticides on crops.

These nutritional and genetic factors, in combination with environmental influences, account for the etiological bases of most of the complex multi-factor diseases and attendant suffering of farm animals. It is simplistic thinking to blame viruses, bacteria, and other pathogens solely for the infectious diseases that afflict farm animals. The presence and abundance of pathogens (as also occur in monocultures of corn, oranges, and other crops) are symptoms of improper husbandry and a consequence, in part, of selective breeding and feeding for high productivity. Hence, the over-reliance today on pesticides in crop production and on vaccines, antibiotics, and other drugs in livestock and poultry production, is an over-reliance that profits neither the farmer nor the consumer—and can harm both (see figure 1).

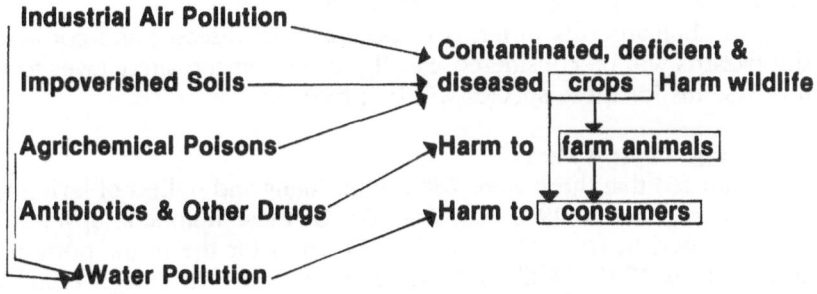

Figure 1. Multi-factor scenario of farm animal health problems.

The ultimate high technology corrective—genetic engineering—is the next capital-intensive and monopolistic innovation of agribusiness that has been shown to favor large and super farms (OTA 1985). Monsanto's genetically-engineered bovine growth hormone and herbicide- and pesticide-resistant seeds herald the next agricultural revolution. The future also holds the possibility of transgenic animals—pigs endowed with the genes of cattle, for example. All these newly engineered life-forms are patentable, and as Doyle (1985) has shown, may well lead to global corporate monopoly of germ plasm and of agricultural practices. A paradigm shift in agricultural and farm animal husbandry practices is imperative, lest genetic engineering result in adverse impacts on the environment and farm animal welfare and health, resulting in this new technology becoming a Pandora's box rather than a cornucopia.

It is perhaps too simplistic and judgmental to conclude that the many domestigenic diseases that afflict farm animals arise because the animals are treated like unfeeling production machines. What has taken place, in essence, is a substitution of empathetic and compassionate husbandry by animal management techniques designed to maximize the overall productive efficiency of the entire system, rather than optimizing the production, health, and well-being of animals on an individual basis. In other words, the ethical principles of humane husbandry have become subordinate to two higher values: first, the economic imperative of maximizing profit margins, (which is essential considering the high capital investment of confinement systems for veal, poultry, and hogs); and second, the subordination of ethics to the ideology of industrialized efficiency. Animals have come to be regarded as simply the living parts of the "factory" that is modern farming.

The well-being of American agriculture is indeed a complex subject, fraught with interrelationships, predictions, recriminations, and at times, high emotionalism. In this paper, we will be concentrating on several aspects of the entire picture that we feel are fundamental to the issue of animal welfare in modern agriculture. First, we will take a brief look at two farm animal species maintained in very restrictive systems, that is, battery-caged laying hens and tethered and/or crated brood sows.

Next, the human costs in terms of occupational diseases and consumer health hazards will be considered. Finally, some humane alternatives to the factory systems for these species will be presented.

Battery-Cage Laying Hens

It is estimated that there were 280 million hens and pullets of laying age in the United States in 1982 (USDA 1983). Of these numbers, upwards of 95% are housed in restrictive wire battery cages for the major portion of their lives (Fox 1984). While such intensive systems do indeed eliminate certain stressors (e.g., predators, some climatic extremes, and parasites), one must undoubtedly question the humaneness of systems in which cannibalism, feather and vent picking, and stereotypic pacing are common occurrences (Fox 1984; Sambraus 1985). The production-related behavioral problems seen in laying hens (table 1) are perhaps better understood when one considers that chickens evolved from ancestors who built nests, were secretive in their laying habits, and lived in small, male-dominated flocks (Kilgour 1985).

The two parameters of the battery system, cage size and design and stocking density, are to be held accountable for the production-related behavioral and

Table 1. Behavior or husbandry problems in poultry.

Behavior or husbandry problem	Possible cause(s)	References[1]
disease	social stress	1
hysteria	monotonous environment	2, 3
head ticks, head-flicking, and hyperactivity	confinement	4
cannibalism	nutritional deficiency (arginine), overcrowding	5, 6, 7, 8
feather pecking (picking)	dietary imbalance (methionine deficiency) and overcrowding, socially facilitated "vice," lack of fiber, eating diet too fast, boredom	5, 6, 7, 8
pick-out	associated with cloacal prolapse in battery layers, possibly "vice" or related to overcrowding	8, personal observation (M.W. Fox)
pre-laying pacing	lack of adequate stimuli or site for nesting	3, 9, 10
self-multilation	visual isolation from other birds	8, 11
redirected aggression and stereotyped behavior	crowding stress, feeding frustration	12, 13
aggression and social stress	overcrowding, unstable grouping	14
egg-eating	boredom	8
tonic immobility	fear response to novel situations	3

[1] Data obtained from: 1) Gross 1976; 2) Ferguson 1968; 3) Craig and Adams 1984; 4) Levy 1944; 5) Allen and Perry 1975; 6) Duncan and Hughes 1974; 7) Ewbank 1969; 8) Sambraus 1985; 9) Bareham 1975; 10) Mills and Wood-Gush 1985; 11) Wood-Gush 1973; 12) Duncan and Wood-Gush 1972; 13) McBride 1966; 14) McBride 1968.

physical maladies of the laying hen. The size of the cage (25 × 38 cm or 38 × 51 cm), coupled with stocking density (three to five birds per cage), denies the bird of the chance to perform many of its "natural" movements, such as wing flapping and stretching, dust bathing, ground scratching; movements that undoubtedly have an important place in the behavioral repertoire of the laying hen (Fox 1984). Frustration of pre-laying activities, such as distancing from the flock and nest building, also occurs in the barren battery cages, and manifests itself in the form of stereotypic pacing. The incidence of stereotypic pacing should be considered indicative of frustration from the welfare viewpoint, and is undesirable for the following economic reasons as well: 1) considerable energy is expended, and this may in turn reflect an increase in feed consumption; 2) there is an increased risk that birds will suffer feather loss, skin abrasions, or even death from trapping while pacing; and 3) some birds will actually lay while pacing, resulting in broken eggs (Mills et al. 1985).

Important research has been conducted in an effort to more fully understand the needs of the laying hen and how fulfillment of such needs can be met in modern systems. Gross and Siegel (1982), for example, discovered that chicks that were habituated (socialized) to humans had more than a 60% reduction in the occurrence of death and pericarditis compared to ignored birds. Perhaps even more exciting, these researchers found that socialized birds had improved feed efficiency, increased antibody response to both *Mycoplasma gallisepticum* and *Escherichia coli*, and increased resistance to the effects of environmental stresses. In reality, Gross and Siegel may have quantitatively proved what all good animal caretakers know—that tender, loving care pays off. Tauson (1984) enriched the caged hen's environment with the addition of a perch, and found that it is possible to get the same number of eggs in cages with perches as in the standard cage at the same stocking density used in Sweden at the time of his study (480 cm^2 cage floor area). The Gleadthorpe or "get-away" battery cage does provide means for the hens to meet some of their behavioral needs, with the provision of a nest box, and separate feeding area (Fox 1984). Reductions in aggression, feather picking, foot problems, nervousness, and stereotypic behaviors were noted in this system, as was an increase in production over the standard battery cage (Fox 1984). The use of the shallow cage has been extensively reviewed by Hughes (1983) in terms of productivity, mortality, and feed efficiency. Hughes pointed out that while some benefits did indeed occur with the use of these cages, such improvements should not distract from the crucial issue of total space requirements of the hens, "by merely providing more space for one particular, albeit, important, activity."

Total space required by hens is probably the single most important parameter of welfare in the issue of the laying hen. In the United States and much of Europe, living space for these animals is in the range of 230-300 cm^2 (Fox 1984). Yet, a West German study (Bogner et al. 1979) determined that a hen requires 538 cm^2 to perform a restricted wing stretch, 528 cm^2 to body stretch, 506 cm^2 to preen, 497 cm^2 just to ruffle its feathers, and 424 cm^2 when resting. In September, 1985, the United Kingdom government-appointed Farm Animal Welfare Council (FAWC) released its assessment of the laying hen systems currently in use in the European Economic Community (EEC). The FAWC states that it does not approve of either the battery-cage system nor the wire- or slatted-floor loose houses where no litter is provided. Bearing the above space requirements and the FAWC decision in mind, it is difficult to fathom the reasoning behind the decision by the E.E.C. Council of Ministers to set the minimum cage size at 450 cm^2 per bird. The reasons are undoubtedly political, and undoubtedly the hens will continue to suffer in E.E.C. countries until that time when sufficient pressure can be brought to bear on the Council to set a date for the phasing out of battery cages in those countries (*Agscene* 1985, 1986). All countries would do well to follow the enlightened example set by Switzerland, which has outlawed the battery cage. All Swiss egg producers must provide their hens with comfortable compartmentalized housing by 1990 (*New Scientist* 1986). Swiss researchers (Huber et al. 1985) have already begun work on determining the hens' preferences in nesting materials.

Tethered and/or Confined Brood Sows

Intensive livestock husbandry systems attempt to maximize production as a response to upwardly spiraling costs for stock, feed, equipment, medication, and labor. One way in which hog producers are attempting to maximize production is by restraining brood (gestating) sows in narrow stalls, either for just farrowing, or for both gestation and farrowing. Although this paper will concentrate on the concerns raised about the tethering of sows, the confinement of sows to narrow crates that offer no room in which to turn around present many problems similar to those encountered in tether systems. Injuries, obesity, stereotypic behavior (Fox 1984), and lowered prefarrowing sow weights and reduced farrowing rates have been reported (*Pig America* 1984). It is estimated that there were 10 million sows farrowing in 1982 (USDA 1983), and if even a conservative estimate of 1% of the total (100,000) are tethered sows, and 510% (500,000 to one million) are confined to gestation crates, surely such numbers dictate that welfare of the brood sow be addressed.

Tether stalls are usually narrow, devoid of bedding, and concrete slat-floored, with the animals tied either to the floor or to the front of their crates by a short tether attached to a neck- or shoulder-collar or girth harness. This method of restraint appeals to the modern producer for the following reasons: 1) low investment, 2) minimum use of floor space and equipment, and 3) ease of maintenance and monitoring of animals (Becker et al. 1985). Although the foregoing reasons may make sense from a purely technocratic viewpoint, the tethering of sows must be analyzed primarily from an animal welfare viewpoint. The pig is one of the most intelligent domestic animals, with a highly developed repertoire of social and exploratory behaviors, yet it has been subjected to some of the most intensive systems of animal agriculture (Kilgour 1985). The tethering of brood sows should be considered to be particularly deprivational to the animals, as close confinement in a gestation stall, with subsequent placement into a farrowing crate, will often condemn the sow to severe restriction for her entire reproductive life.

In recent years, many scientific studies have been conducted in an attempt to determine if sow welfare is indeed compromised in tether systems (see, e.g., Becker et al. 1985; Barnett et al. 1985; Cronin 1985; Gustafsson 1983; Ekesbo 1981; and Vestergaard 1981). An evaluation of welfare of tethered sows necessarily includes investigation of such factors as the physiological indicators of stress, production parameters such as breeding rate and piglet mortality, and incidence of stereotypic behavior.

Physiological Indicators of Stress.—As noted by Dantzer and Mormède (1981), exposure to physical or psychological stressors elicits a wide range of physiological changes in the organism that can be more or less easily detected. One such change important in the evaluation of stress in pigs is the change in levels of serum corticosteroids; animals exhibiting higher than normal plasm corticosteroid levels are claimed to be in a state of stress (Dantzer and Mormède 1981; Moss 1981). Two recent studies strongly suggest that gilts kept in tether systems develop chronic stress response. Barnett et al. (1985) demonstrated that pregnant gilts kept in tethers showed a 76-82% increase in corticosteroid levels "at rest" over other systems (2.2 ng/ml for tethered gilts vs. 1.4 ng/ml for those housed in stalls and indoor or paddock groups). Becker et al. (1985) found that at the end of four weeks' tethering, gilts exhibited higher concentrations of serum cortisol in the morning. The authors suggest that because morning is a time of greater activity for swine, the physical restrictions imposed upon them by tethering may account for this higher concentration. It is also suggested that such restriction may induce chronic stress response in these animals (Becker et al. 1985).

Production-Related Problems.—Reports of such problems as sow illness at farrowing (Ekesbo 1981), increased piglet mortality (Gustafsson 1983), and lower mating rates and irregular estrus (Becker et al. 1985) indicate that the keeping of brood sows in tethers or stalls is, in fact, counter-productive to the maximization of production.

Vestergaard (1983/84) reports that tethering during pregnancy and/or farrowing-lactation resulted in an increased duration of farrowing itself, and that tethered sows showed much restlessness in the last 24 hours prior to giving birth; he interpreted this restlessness as thwarted nesting behavior. Modern producers often respond to longer farrowing times by administering prostaglandin, which induces labor. As this hormone drastically increases the nest-building motivation, sows restricted in a narrow farrowing crate may be extremely stressed (Fox 1984), and perhaps even more so if tethered as well. A synopsis of behavioral, health, and husbandry problems documented in confined sows is presented in table 2.

Stereotypies.—Although stereotypies as related to welfare have been reported in tethered sows in previous studies (see table 2), Cronin (1985) has presented a comprehensive look at these behaviors. One study was designed to describe the development of stereotypies in sows after tethering (see Cronin 1985, chapter 3). The authors concluded that environmental-directed stereotypies (directed towards chains, drinkers, bars, etc.) develop as

Table 2. Behavior, health, and husbandry problems in tethered or confined sows.[1]

Behavior, health, or husbandry problem	Possible cause(s)	References[2]
traumatic or physical injuries	poorly designed crates and/or flooring, residual urine and dung	1, 2
infertility or low fecundity	social isolation/ confinement	3, 4, 5, 11
oral stereotypies: mouthing, champing, polydipsia, vacuum chewing, bar-biting	boredom, lack of bedding, feed-directed activity, low-bulk feed	1, 6, 7, 8, 9, 10, 11, 12, 13
arteriosclerosis	social isolation	8
subclinical disease	close confinement	1
"mourning behavior"	boredom, lack of bedding	9, 11, 14
increased farrowing time, complications at farrowing	lack of exercise due to confinement	1, 2, 12, 15
increased piglet mortality	fetal development and farrowing illness due to confinement	1, 2
lameness	lack of wear on toes due to flooring	12

[1] Confined in this context refers to farrowing and/or gestation crates.
[2] Data obtained from: 1) Ekesbo 1981; 2) Gustafsson 1983; 3) Kiley-Worthington 1977; 4) Becker et al. 1985; 5) Fox 1984; 6) Fraser 1974; 7) Ewbank 1969; 8) Fraser 1975; 9) Vestergaard 1981; 10) Sambraus 1985; 11) Cronin 1985; 12) Barnett et al. 1985; 13) Rushen 1985; 14) Hall 1984; 15) Vestergaard 1983/84.

a result of frustration/conflict at being restrained, and the sows' consequent loss of control over their environment.

A study was also formulated to test the hypothesis that endorphins (endogenous opiates) play a role in the development and performance of stereotypies by tethered sows (see Cronin 1985, chapter 4). Tethered sows were treated with the specific opiate antagonist, naloxone, while performing stereotypies. While saline injections did not effect the behaviors of the sows, naloxone caused severe disruption of the stereotypies, but not of the normal behaviors of the sows. Cronin summarized these results as follows:

> The results strongly suggest that endorphins may be the factor underlying the development and performance of stereotypies. Endorphins are released in response to stress, and in time, sows may learn to self-stimulate the release

through the performance of stereotypies. Stereotypies probably function to reduce the perception of the negative aspects of the real environment, over which tethered sows have no control, and "rebuild" a new and possibly much reduced environment that they control through the performance of stereotypies. The results suggest that sows perceive tethering in a very negative way. (p. 140)

It was also suggested by Cronin that stereotypies have direct influences on sow productivity. Many tethered sows become highly active through stereotypy performance (even though they are unable to locomote), and these animals have a higher metabolic rate and poorer feed conversion efficiency/growth than less active sows. Such activity may contribute to the "thin sow" syndrome. Cronin points out that not all sows can adequately cope with the stress of being tethered; such animals are more likely to be culled at an early age as poor producers. To sum these observations, Cronin states:

Stereotypies are indicators of a poor environment and thus lower welfare status. It should be the aim of all pig producers therefore, to achieve better welfare for their animals. Better welfare will undoubtedly result in higher profits, but also in reduced public displeasure at the current intensive husbandry systems which disregard the welfare of the sows. (p. 135)

In a positive move for the welfare of dry sows, the Standing Committee of the European Convention for the Protection of Animals Kept for Farming Purposes is considering draft recommendations concerning pigs. The Farm Livestock Specialist Group of the Scientific Advisory Panel, World Society for the Protection of Animals, has submitted the following:

That the confinement of dry sows to individual stalls, with or without tethering, is a serious welfare problem, inevitably leading to severe restrictions on the animal's freedom of movement thus denying normal exercise that can give rise to patterns of abnormal behaviour and commonly causes injuries and leg weakness (WSPA 1985).

It is to be hoped that this statement will be taken under serious consideration and a precedent thereby set for the humane treatment of sows.

Occupational and Consumer Health Hazards

Although the American public generally envisions the farmer as working out-of-doors, breathing fresh air, and being subjected to few of the stresses endured by his urban and suburban neighbors, nothing could be further from the truth for the majority of the farmers engaged in intensive animal agriculture. Today's animal farmers are exposed to toxic fumes from the herbicides and pesticides used on the crops, noxious dust from on-farm grain storage centers, poisonous gases from the animal confinement units, and physical danger from the powerful, high-speed equipment. It has been reported that 200,000 disabling injuries and 2,000 annual accidental deaths befall farmers; they share the three highest rates for "industrial accidents" with miners and construction workers (*Houston Post* 1984). It has been estimated that up to one million American farmers may now work in livestock confinement buildings alone, and as many as 70% of Iowan farmers may have respiratory difficulty at any given time (Donham et al. 1984).

In a study of swine confinement units, Clark et al. (1983) found levels of carbon monoxide, carbon dioxide, ammonia, and hydrogen sulfide to be in excess of threshold limit values for occupational exposure. The same study reported that airborne concentrations of total and gram negative bacteria in poultry and swine units were as high or higher than those found in wastewater treatment plants, solid waste/sludge composting plants and cotton card rooms where microbiologically contaminated organic dusts were also present. Judging from the ramifications of being housed in such conditions, it is little wonder then that modern farmers feel it necessary to maintain their intensively housed stock on subtherapeutic levels of antibiotics, and on growth promotants such as hormones.

Such drugs constitute the second health hazard, that of consumer risk from eating animal products from factory-farmed livestock. There are several recent developments implicating that intensive animal agriculture may jeopardize consumer health. Consider that of the nearly 30 million pounds of antibiotics produced annually on the United States, one-half of them are fed to farm animals, and that the Food and Drug Administration estimates that 80% of swine, 60% of cattle, and 30% of chickens are fed antibiotic-laced feeds (Allman 1984). The sobering aspect of these facts is that animals reared for consumption are often fed the same antibiotics (i.e., tetracycline and penicillin) in subtherapeutic doses that are used to treat bacterial infections in humans. It is now known that microorganisms can become resistant to these drugs (Holmberg et al. 1984; Dixon 1986), and that such resistance can be transferred between microbes (O'Brien et al. 1984). The implications of these findings is that subtherapeutic dosing of farm animals may effectively select *for* organisms resistant to the antibiotics. The use of antibiotics to combat stressful and crowded environmental conditions may well be considered an irresponsible and/or dangerous practice; one that renders tetracycline and penicillin useless against human illness. Unfortunately, in November, 1985, the U.S. Department of Health and Human Services chose to ignore these implications and refused to ban the use of these two drugs in animal feeds (Rhien and Siwolop 1986).

It is interesting to note that while livestock usage of drugs and feed additives amounted to a $2 billion a year expenditure for producers, veterinary care for these animals was disproportionately low, as shown in table 3 (Rheines and Siwolop 1986; Charles and Charles 1983). The figures in the table are based on a survey conducted in 1982, and it is evident that the majority of households contacted never obtained veterinary care for their farm animals.

Table 3. Percent of households not getting veterinary care.

Pet and/or animal	% of households not obtaining veterinary care
household pets	
dog	26.4
cat	52.8
fish	97.8
caged bird	92.5
rabbit	89.2
hamster	93.4
guinea pig	79.8
gerbil	97.4
other small rodent	84.6
reptile	95.8
other fowl	97.1
all others	97.5
horse	57.4
agricultural animals	
dairy cattle	52.7
beef cattle	59.9
swine	75.3
sheep	72.1
poultry	97.5
goat	76.9

Note: Percents based off of number of households which owned each type during year.

From: Charles, Charles and Associates, Inc.: *The Veterinary Services Market Study* prepared for the American Veterinary Medical Association, July 1983.

Alternatives and Economics

A comprehensive treatment of alternatives to factory farming and the differences in production parameters, health maintenance expenditures, and cost-benefit ratios is beyond the scope of this paper. Suffice it to say at this time, alternatives are, fortunately, being developed and tested. The alternatives presented below do attempt to meet at least some of the following basic rights or needs that must be met by animal husbandry systems (Carpenter 1980): 1) freedom to perform natural physical movement; 2) association with other animals, where appropriate, of their own kind; 3) facilities for comfort activities, e.g., rest, sleep, and body care; 4) provision of food and water to maintain full health; 5) ability to perform daily routines of natural activities; 6) opportunity for the activities of exploration and play, especially for young animals; and 7) satisfaction of minimal spatial and territorial requirements, including a visual field of "personal" space. Deviations from these principles should be avoided as far as possible, but where such deviations are absolutely unavoidable, efforts should be made to compensate the animal environmentally (Carpenter 1980).

The fact that high-technology swine units are expensive to build and maintain (and often do not meet production expectations) has been acknowledged by the pork industry (Vansickle 1984). Such considerations, coupled with welfare concerns, have prompted alternatives research and implementation. A "Family System" for hogs that endeavors to show how pig housing conditions can be designed according to basic ethological requirements is being developed (Stolba 1982; Wood-Gush 1985/86). Other alternatives to the narrow crates include A-frame huts on pasture (Hohmann 1985), and roomy (5 × 7 feet), tilted indoor crates which allow both sow movement and piglet protection (McClinton 1985). A turn-around gestation crate has been designed (McFarlane and Curtis 1983) which may be used by producers interested in offering their sows more diversity in and control over their environment.

An aviary system for laying hens has been developed in Switzerland (Fölsch et al. 1983) and is reported to be comparable in productivity and economics to the deep litter and cage systems. Mason (1985) notes that a 1978 study by the Swiss Centre for Poultry collected data on the laying performance of 65,000 hens in 38 Swiss flocks (two-fifths on litter, remainder in cages) and there were no differences in laying performance between the two systems.

The standard economic arguments for factory farming are rapidly becoming passè in light of the farm economic crisis, consumer health risks due to residual chemicals and drugs, pollution, and depletion of soil and water resources. To quote Mason (1985):

> ...the financial benefits of factory farming are exaggerated, and furthermore, that they produce unhealthy animals and poor-quality products: to offset these effects factory farmers must employ an arsenal of antibiotics, hormones, drugs, chemical additives, colouring agents and other substances that may threaten human health. When one considers the potential magnitude of these health problems and the social cost of dealing with them, the food produced by factory methods may well be too expensive—regardless of its price at the market.

Endnotes

[1] Paper presented at the national conference, "Animals and Humans: Ethical Perspectives," Moorhead State University, Moorhead, MN, April 21-23, 1986.

[2] Research Associate, The Institute for the Study of Animal Problems, 2100 L St, NW, Washington, DC 20037.

[3] Scientific Director, The Humane Society of the United States, and Director, The Institute for the Study of Animal Problems, 2100 L St, NW, Washington, DC 20037.

[4] The fact that we present in-depth consideration of these two species is reflective solely of a limitation of space, and in no way implies that laying hens and brood sows are the *only* food animals kept in deprivational systems detrimental to their physical and psychological well-being. Indeed, the welfare of crated veal calves is of major concern, requiring urgent attention from animal scientists, animal welfarists, and the public alike. It is ethically unconscionable to continue to maintain veal calves in crates for the reasons of tradition and "psychology" (i.e., consumers expect veal to be white); the refinement of alternative systems make it unnecessary as well.

Recent scientific studies document that the crate system is deleterious in terms of health, behavior, and production (see Friend et al. 1985; Fox 1984; Sambraus 1980, 1985; Saville and Webster 1981; Dantzer et al. 1983; and Webster 1986). Mason (1985) also reports that 40% of the members of the National Association of Veal Producers in Great Britain have adopted loose housing in groups for economic reasons; housing costs are halved, calves are healthier, and veterinary bills are reduced by 65%.

References

Agscene. 1985. The big sell-out. 81: 6-7.

—. 1986. The wrong directive. 83: 4.

Allen, J and Perry, GC. 1975. Feather-pecking and cannibalism in a caged layer flock. *Brit. Poult. Sci.* 16: 441-52.

Allman, WF. 1984. Drugs in feed: Fatter cattle, fitter bacteria. *Science 84.* 5(10): 16.

Bareham, JR. 1975. Research in farm animal behaviour. *Brit. Vet. J.* 131: 272-83.

Barnett, JL, Winfield, CG, Cronin, GM, Hemsworth, PH and Dewar, AM. 1985. The effect of individual and group housing on behavioural and physiological responses related to the welfare of pregnant pigs. *Appl. Anim. Behav. Sci.* 14: 149-61.

Becker, BA, Ford, JJ, Christenson, RK, Manak, RC, Hahn, GL and DeShazer, JA. 1985. Cortisol response of gilts in tether stalls. *J. Anim. Sci.* 60(1): 264-70.

Bogner, H, Peschke, W, Sed, V and Popp, K. 1979. Berliner und Münchener Tierärztliche Wochenschrift. 92: 340.

Carpenter, E. 1980. *Animals and Ethics.* London: Watkins.

Chaboussou, F. 1980. *Les Plants Malades des Pesticides: Bases Nouvelle d'une Prevention Contre Maladies et Parasites.* Paris: Deband.

Charles, Charles and Associates, Inc. 1983. *The Veterinary Services Market Study.* Prepared for the American Veterinary Medical Association, July 1983.

Clark, S, Rylander, R and Larsson, L. 1983. Airborne bacteria, endotoxin and fungi dust in poultry and swine confinement builidings. *Am. Ind. Hyg. Assoc.* 44(7): 537-41.

Cook, RJ. 1986. Interrelationships of plant health and the sustainability of agriculture, with special reference to plant diseases. *Am. J. Alt. Agric.* 1(1): 19-25.

Craig, JV and Adams, AW. 1984. Behaviour and well-being of hens (*Gallus Domesticus*) in alternative housing environments. *WSPA J.* 40(3): 221-40.

Cronin, GM. 1985. *The Development and Significance of Abnormal Stereotyped Behaviours in Tethered Sows.* Doctoral Thesis: Agricultural University of Wageningen. 1985.

Dantzer, R. and Mormède, P. 1981. Can physiological criteria be used to assess welfare in pigs? In: Sybesma, W. ed. *The Welfare of Pigs.* London: Martinus Nijhoff.

—. 1983. *J. Anim. Sci.* 57: 6-18.

Dantzer, R, Mormède, P, Bluthe, RM and Soissons, J. 1983. The effects of different housing conditions on behavioural and adrenocortical reactions in veal calves. *Reprod. Nutr. Develop.* 23(3): 501-8.

Dixon, B. 1986. Overdosing on wonder drugs. *Science 86.* May. 7(4): 40-44.

Donham, KJ, Zavala, DC and Merchant, JA. 1984. Respiratory symptoms and lung function among workers in swine confinement buildings: A cross-sectional epidemiological study. *Arch. Env. Health.* 39(2): 96-100.

Doyle, J. 1985. *The Altered Harvest.* New York: Viking Press.

Duncan, IJ and Hughes, BO. 1974. Some emotional factors influencing feather-pecking in growing birds. *Brit. Vet. J.* 129: 503-4.

Duncan, IJ and Wood-Gush, DG. 1972. Thwarting of feeding behaviour in the domestic fowl. *Anim. Behav.* 20: 444-51.

Ekesbo, I. 1981. Some aspects of sow health and housing. In: Sybesma, W. ed. *The Welfare of Pigs.* London: Martinus Nijhoff.

Ewbank, R. 1969a. Social behaviour and intensive animal production. *Vet Rec.* 85: 183-86.

—. 1969b. Behavioural implications of intensive animal husbandry. *Outlook Agric.* 6: 41-46.

Ferguson, W. 1968. Abnormal behaviour in poultry. In: Fox, MW. ed. *Abnormal Behaviour in Animals.* Philadephia: Saunders.

Fölsch, DW, Dolf, C, Ehrbar, H, Bleuler, T and Teijgeler, H. 1983. Ethologic and economic examination of aviary housing for commercial laying flocks. *Int. J. Stud. Anim. Probs.* 4(4):330-5.

Fox, MW. 1983. Animal welfare and the dairy industry. *J. Dairy Sci.* 66: 2221-5.

—. 1984. *Farm Animals: Husbandry, Behavior, and Veterinary Practice.* Baltimore, MD: University Park Press.

—. 1985. Philosophy and ethics in ethology. In: Fraser, AF. ed. *Ethology of Farm Animals.* World Animal Science Series. A. No. 5. Amsterdam: Elsevier.

Fraser, AF. 1975. *Farm Animal Behaviour.* London: Bailliere Tindall.

Fraser, D. 1974. Behavior at three weeks. *Pig Farming Suppl.* Oct.: 61, 63, 71.

Friend, TH, Dellmeier, GR and Gbur, EE. 1985. Comparison of four methods of calf confinement. I. Physiology. *J. Anim. Sci.* 60(5): 1095-1101.

Gross, WB. 1976. Plasma steroid tendency, sound environment and *Eimeria necatrix* infection. *Brit. Poul. Sci.* 55: 1508-12.

Gross, WB and Siegel, PB. 1982. Socialization as a factor in resistance to infection, feed efficiency, and response to antigen in chickens. *Am. J. Vet. Res.* 43(11): 2010-12.

Gustafsson, B. 1983. Effects of sow housing systems in practical pig production. *Trans. ASAE.* 1181-93.

Hall, WF. 1984. Stress and the farrowing sow. *Hog Farm Manage.* July: 13-14.

Harvey, G. 1983. Poor cow. *New Scientist.* Sept. 29: 940-43.

Hodges, RD and Scofield, AM. 1983. Effect of agricultural practices on the health of plants and animals produced: A review. In: Lockeretz, W. ed. *Environmentally Sound Agriculture.* New York: Praeger.

Hohmann, K. 1985. Extensive A-frames. *Hog Farm Manage.* July: 26-27.

Holmberg, S, Osterholm, M, Senger, K and Cohen, M. 1984. Drug-resistant *Salmonella* from animals fed antimicrobials. *New. Eng. J. Med.* 311(1): 617-22.

Houston Post. 1984. Health watch: Ailing farmers. Dec. 28.

Huber, HU, Fölsch, DW and Stähl, U. 1985. Influence of various nesting materials on nest site selection of the domestic hen. *Brit. Poult. Sci.* 26: 367-73.

Hughes, BO. 1983. Conventional and shallow cages: A summary of research from welfare and production aspects. *WSPA J.* 39(3): 218-28.

Kiley-Worthington, M. 1977. *Behavioural Problems in Farm Animals.* London: Oriel.

Kilgour, R. 1985. Management of behaviour. In: Fraser, AF. ed. *Ethology of Farm Animals.* World Animal Science Series. A. No. 5. Amsterdam: Elsevier.

Levy, DM. 1944. On the problem of movement restraint, ticks, stereotyped movements, hyperactivity. *Am. J. Orthopsychiatr.* 14: 644-71.

Long, C. 1985. Dirty dozen. *Agscene.* Nov./Dec. 81: 13.

Mason, J. 1985. Is factory farming really cheaper? *New Scientist.* 28 March (1449): 12-15.

McBride, G. 1966. The conflict of crowding. *Discovery.* 27: 16-19.

—. 1968. Social organization and stress in animal management. *Proc. Ecol. Soc. Austral.* 3: 133-38.

McClinton, D. 1985. "Happier" animals, higher profits. *The Furrow.* July/Aug. 8-11.

McFarlane, JM and Curtis, SE. 1983. Behavior of mated gilts in a turn-around gestation crate. Cooperative Extension Report, 1983-16. Univ. of Illinois at Urbana-Champaign.

Mills, AD and Wood-Gush, DG. 1985. Pre-laying behaviour in battery cages. *Br. Poult. Sci.* 26: 247-52.

Mills, AD, Wood-Gush, DG and Hughes, BO. 1985. Genetic analysis of strain differences in pre-laying behaviour in battery cages. *Br. Poulty. Sci.* 26: 187-97.

Moss, BW. 1981. The development of a blood profile for stress assessment. In: Sybesma, W. ed. *The Welfare of Pigs*. London: Martinus Nijhoff.

New Scientist. 1986. Swiss chickens choose home comforts. 109(1493): 33.

O'Brien, T, Hopkins, J, Gilleece, E, Medeiros, A, Kent, R, Blackwood, B, Holmes, M, Reardon, J, Vergeront, J, Schell, W, Christenson, E, Bissett, M and Morse, E. 1982. Molecular epidemiology of antibiotic resistance in *Salmonella* from animals and human beings in the United States. *New Eng. J. Med.* 301(1): 1-6.

OTA (Office of Technology Assessment). 1985 *Technology, Public Policy, and the Changing Structure of American Agriculture: A Special Report for the 1985 Farm Bill*. Washington, DC: U.S. Congress.

Peterson, C. 1986. Food supply called unprotected: FDA accused of allowing improper use drug use in farm animals. *Washington Post*. Jan. 13: A5.

Pig America. 1984. Universities dispute value of gestation stalls. March: 16-17.

Rhein, RJ and Siwolop, S. 1986. Drugs in animal feed: Now the FDA is drawing heavy fire. *Business Wk*. Feb. 3: 70F.

Rushen, JP. 1985. Stereotypies, aggression and the feeding schedules of tethered sows. *Appl. Anim. Beh. Sci.* 14: 137-47.

Sambraus, HH. 1980. Humane considerations in calf rearing. *Anim. Reg. Stud.* 3: 19-22.

—. 1985. Mouth-based anomalous syndromes. In: Fraser, AF. ed. *Ethology of Farm Animals*. World Animal Science Series. A. No. 5. Amsterdam: Elsevier.

Saville, C and Webster, JF. 1981. Basic necessities for ensuring the welfare of veal calves in various housing systems. *Appl. Anim. Ethol.* 1: 382-83.

Siegel, HS. 1985. Effects of intensive production methods on livestock health. *Agro-Ecosystems*. 8: 215-30.

Somogyi, A. 1985. Drug residues. *World Health*. July: 26.

Stolba, A. 1982. A family system of pig housing. *Proc. Symp. Alternatives to Intensive Husbandry Systems*. Potters Bar, England: Universities Federation for Animal Welfare.

Tauson, R. 1984. Effects of a perch in conventional cages for laying hens. *Acta Agric. Scanda*. 34: 193-209.

USDA (United States Department of Agriculture). 1983. *Agricultural Statistics*. Washington, DC: U.S. Govt. Print. Off.

Van denBergh, SG. 1976. Abnormal lipid metabolism and production diseases. In: *Proc. of the Third Internatl. Conf. on Production Diseases in Farm Animals*. Wageningen, The Netherlands. Sept. 13-16, 1976.

Vansickle, J. 1984. High-tech units swallow profits. *Natl. Hog Farmer*. March 15: 80,82.

Vestergaard, K. 1981. Influences of fixation on the behaviour of sows. In: Sybesma, W. ed. *The Welfare of Pigs*. London: Martinus Nijhoff.

—. 1983/84. Are tethered sows stressed: A behavioural comparison of tethered and loose sows. *Appl. Anim. Ethol.* (Abstr.) 11: 81-82.

Webster, J. 1986. Health and welfare of animals in modern husbandry systems—Dairy cattle. *In Practice*. May.: 85-89.

Wood-Gush, DG. 1973. Animal welfare in modern agriculture. *Brit. Vet. J.* 129: 67-74.

—. 1985/86. The attainment of humane housing for farm livestock. In: Fox, MW and Mickley, LD. eds. *Advances in Animal Welfare Science 1985/86*. Washington, DC and Dordrecht, The Netherlands: The Humane Society of the United States and Martinus Nijhoff, respectively.

Wood-Gush, DG, Duncan, IJ and Fraser, AF. 1975. Social stress and welfare problems in agricultural animals. In: Hafez, ES. ed. *The Behaviour of Domestic Animals*. London: Bailliere Tindall.

WSPA (World Society for the Protection of Animals). 1985. *Tethering of Dry Sows*. by the Scientific Advisory Panel. May.

THE CASE FOR HUNTING[1]

William L. Robinson[2]

There are serious ecological problems in the world today. Lead from the exhaust of California automobiles is found in the ice of Greenland, and sulfur from Ohio industries acidifies lakes in Michigan and New York and fish die. Toxic wastes once buried out of sight near the Great Lakes have unexpectedly returned from their graves, causing gulls and cormorants to be born with deformed bills. Tropical rainforests are being clearcut at a rate of 250,000 square kilometers per year, destroying entire species of animals before they are even described by scientists.

People who understand and care about the biosphere—its beauty and health (which I believe are closely related)—have joined together to oppose the crush of humanity which threatens to overwhelm this planet, manipulating its ecological functions with technologies whose effects are at best unknown, and at worst, harmful to the long-term interests of man and nearly all living things.

The greatest threats to wildlife on earth are widespread loss of habitat, chemical poisons, and uncontrolled or improperly regulated exploitation. Regulated hunting, as practiced in the Unites States, Canada, and most of Europe, is but a minor worry to nearly all ecologists.

Nevertheless, there are numerous local citizens' groups, and large organizations such as Friends of Animals, Fund for Animals, and The Humane Society of the United States, for whom opposition to hunting constitutes a major public relations, lobbying, and legislative activity. On the pro-hunting side are thousands of hunting-oriented local sportsmen's groups, and on a larger scale, the National Rifle Association, the Wildlife Management Institute, and the National Wildlife Federation. The National Audubon Society maintains a neutral stance on hunting. I believe that a common goal of all these groups is to maintain populations of wild animals, with individuals among those populations provided adequate opportunity for survival and reproduction. The hunting debate has had an unfortunate polarizing effect among people sharing this goal.

My purpose at this symposium is to present the case for hunting. I am a wildlife ecologist by training and profession, and I am also a hunter. As a hunter, I am sensitive to criticisms of this pursuit, as any hunter should be. Some people question how, with knowledge of the nature and functioning of ecological systems, I can go out with a gun and kill grouse, ducks, and deer. I respond that, indeed, my understanding of ecology and the nature of man enhances my enjoyment of hunting.

Human Attitudes Toward Wildlife

Significant progress in understanding attitudes of people toward animals has been made in the past decade, largely through the work of Steve Kellert (1978, 1980) of Yale University. After polling a cross-section of Americans, Kellert identified ten types of attitudes. These are summarized in table 1.

Table 1. Attitudes toward animals (from Kellert 1976).

Attitude	Characteristics
Naturalistic	Desires personal contact with natural habitats, concern for wildlife
Ecologistic	Intellectual understanding of interactions of wildlife and environment
Humanistic	Strong personal affection for animals, especially pets
Moralistic	Ethical concern; vigorous opposition to inflicting suffering or death in animals
Scientistic	Interest in animals as objects of study
Aesthetic	Interest in physical and symbolic attractiveness of animals
Utilitarian	Animals are valued for tangible usefulness to man
Dominionistic	Interest in mastery and control of animals
Negativistic	Desire to avoid animals; fear and alienation

Anti-hunters are most likely to exhibit humanistic and/or moralistic attitudes. The humanistic attitude attributes human qualities to animals, and possessors of that attitude are frequently interested in animals as pets, often treating them as they would other humans. While other animals share similarities in our basic senses, they are not human. In many ways, they are superior to us. A beaver, for example, can gnaw down an aspen tree 10 inches in diameter and live by eating its small branches. A lion can pursue and kill a zebra without the use of a rifle or a Land Rover, and my dog possesses such a sense of smell that she lives in a world so rich with odors it is unimaginable to me. But the beaver, the lion, and the dog are unable to comprehend the complex ideas that we are dealing with in this conference; and many mammals never comprehend the dangers of highways and frequently run into the front of an oncoming car rather than away from it. Unlike humans, the reproductive rates of birds and mammals, particularly game species, are high and keyed to a normally high rate of mortality. All this is not to imply that we should not respect other animals, only to show that they are not humans.

People with a moralistic attitude feel that causing pain or death to any animal capable of knowing pain is immoral and should not be done by humans (Singer 1975). A difficulty that people espousing this view have is in determining what organisms "know pain," that is, drawing a line between which animals may be killed morally, and those which cannot. I believe that wherever one draws this line is artificial and uncharacteristic of the physiological and behavioral inheritance of our own species. I should point out that I have no objection whatever to people choosing their own behavior toward animals and their own diets (which invariably involve killing something). What I do object to is their attempts to make their preferences and beliefs mandatory for all others.

Hunting as a Human Tradition

It is my contention that a propensity to hunt is a part of our humanity or "humaneness," using my *Webster's Dictionary* (Mish 1983) definition of humane as "characterized by or tending to broad humanistic culture," and that while hunting may not be an entirely necessary component of modern human culture, it has value in keeping a perspective on man's role as an active participant in the community of life.

Hunting, I claim, is part of our human inheritance. Our dentition and our digestive system, both unaffected by learning, tell us that we are adapted as omnivores, like bears, pigs, and raccoons. We have incisors, canines, and molars. Our digestive tract is relatively short without a compound stomach or a caecum typical of herbivores. We do not do well on a vegetarian diet unless we apply the most sophisticated nutritional information and select items grown in various parts of the continent and shipped to us. Our natural diet includes meat; the middens of our ancestors through the ages attest to this. And just as we have inherited a digestive system adapted partially for meat, we also have inherited behavioral traits that enhance our ability to capture and kill other animals.

On a summer day two years ago, I sat at a Toronto city park on the shore of Lake Ontario and watched a three-year-old boy try to catch gulls. As his parents spread the lunch, he chased the birds, took a lunch break, then went after them again, without success. I doubt very much that his gull chasing was a learned behavior. He simply wanted to catch a gull and probably wouldn't have known what to do with one had he done so. His predatory drives were expressing themselves and he was having a good time.

Most ethologists recognize play as an inherent behavioral trait of many animals, especially those with more complex brains. Play behavior among young animals is frequently interpreted as a form of practice for catching prey and escaping predators. Kittens and puppies instinctively play, crouching behind the living room couch, pouncing upon a passing sibling, rolling about, then exchanging roles. Skills of hiding, escape, speed, and angles of approach are honed.

In these games there are three basic elements: prey, predator, and cover. A kitten can be the predator during one encounter and the prey for another. The couch or a chair serves either as cover to hide the predator or to protect the prey from attack. I first encountered these views expressed by the British ecologist, Charles Elton (1939) in an article entitled "On the Nature of Cover." Elton also explained human athletic contests as expressions of our inherent interest in predator, prey, and cover relationships. In American-style football, for example, the object of the game is for one team to carry or throw the ball into the end zone despite the great physical efforts of the opposing team to prevent it. The end zone represents cover, the ball carrier represents a prey animal, and the defensive team a pack of predators. Players, even amateur players, strive mightily on both sides and frequently get hurt in the melee. We could civilize football a great deal through a gentlemen's agreement before the game. One team might decide to let the other team carry the ball untouched into the end zone, if the other team in turn will permit them to do the same. This would be humane football: no one would get hurt and the score would end in a tie. But who would play or watch such a game?

Baseball, hockey, basketball, soccer, and even tennis and golf can be explained by reference to the three basic elements of predator, prey, and cover. Why do millions of people play, and even watch and become emotionally involved in these games? This real and vicarious participation is a result of our inherent interest in predator-prey relationships.

Some people will argue that they themselves are not in the least interested either in hunting, fishing, or athletic events. Genetic variation as well as environmental influences operating among individuals explains this attitude, and we should not expect all humans to possess an interest in hunting any more than we expect all humans to look alike.

Whereas athletic competitions are extremely ritualized versions of predator-prey-cover relationships, with elaborate rules for scoring and exchange of predator-prey roles, hunting and fishing are merely more primitive expressions of human predatory nature. Instead of pretending we are predators, as hunters or fishermen we actually are predators. I agree that the tendency to hunt can be "civilized" out of us; and it has been done, to some extent. We have rules of the game; there are closed seasons, protected species, bag limits, and restrictions on the type of equipment permitted to protect animal populations from over-exploitation. Opponents of hunting argue that we should put hunting as a sport behind us. But is it necessary or desirable to prohibit hunting for "humane" reasons? Do humans have a right to kill an animal for sport and food? Is there something that gives the fox a right to kill and eat a rabbit but denies that right to humans? Do people who have little interest in hunting have the right to deprive those who do from pursuing their sport?

Attitudes of Hunters

As all who oppose hunting do not have the same reasons for doing so, hunters do not all hunt for the same reasons. Kellert (1978) found that there were three prevalent attitude types among hunters: utilitarian (making up 44% of hunters), dominionistic (38% of hunters), and naturalistic (18% of hunters). The dominionistic or "sport hunter" views hunting primarily as a contest between the hunter and the game. This hunter sees the hunt as a challenge of skill in stalking or marksmanship, and is frequently interested in taking trophy animals. The utilitarian hunter or "meat hunter" hunts primarily for meat, and may be little concerned with ritual or method of attaining that meat. The naturalistic or "nature hunter" hunts primarily to participate in the community of life, viewing himself or herself as a predator in this community. Such hunters frequently hunt alone and often with primitive equipment such as bows and arrows or muskets. I classify myself among the nature hunters, with a secondary utilitarian interest.

Hunting and Genetics of Wildlife Populations

There are suggestions that hunters remove the strongest, and genetically superior individuals, while other predators ("natural" predators) remove the less genetically desirable individuals. While several studies do indicate that non-human predation does take the weaker individuals, in many cases these are simply young, sick, or old animals, of which none may be genetically inferior. I know of no research that supports the conclusion that hunting selects out the most fit individuals. In fact, theoretically speaking, a conclusion that hunters remove the most genetically fit animals by taking the largest individuals from the population and therefore weaken it, denies the very nature of selection. If these animals are being selectively removed they are then by definition less fit for survival against hunting. It remains to be shown that these animals are genetically superior to the less-sought-after smaller animals.

There is some evidence that hunting has selected against cottontail rabbits that run some distance above ground rather than dash into the nearest burrow. In such cases, survival against dogs and hunters of the hole-seeking rabbits is enhanced. There are also indications that ring-necked pheasants with a genetic propensity to run rather than fly have a higher survival rate, and are more likely to pass on their ground-hugging qualities to their offspring, thus more successfully avoiding hunters. Research needs to be done to determine whether holeseeking rabbits and running pheasants are less genetically adapted for surviving the impacts of other stresses such as malnutrition, diseases, and non-human predators.

Raveling (1978) found that increasing numbers of Canada geese are remaining in northern parts of their range rather than migrating farther south. Those which remain in the north, assisted in many cases by being fed in parks, have a higher survival rate than their counterparts that go south, primarily because they escape hunters. In this case, it appears that hunting is selecting for the less migratory geese. Whether this hunter-induced trait is bad for the geese remains unknown.

Geist (1986) reviewed the work of a German chemist named Vogt who showed quite convincingly that nutrition rather than genetics is primarily responsible for large body size and large antler growth. Vogt, by providing European and American deer from normal stock with high quality diets, was able to produce stags with weights and antler growth equivalent to the largest trophy animals taken in Europe. Thus nutrition, rather than genetic capabilities, was shown to be the major factor in producing trophy animals. Nevertheless, there is room for further study of the effects of hunting on genetics and fitness of animal populations. Should trophy hunting be shown to have a detrimental effect on the genetics of big game populations, measures may be taken to protect larger members of the population without eliminating hunting entirely. Such regulations are now used in management of sport fisheries.

Hunting as a Sacred Ceremony

I feel more comfortable defending the nature hunter and the utilitarian hunter rather than the trophy hunter, although I am not suggesting outlawing the dominionistic hunter. Several years ago, Dennis Olson (1980), who was then teaching at the Environmental Learning Center in northern Minnesota, published a searching article superbly expressing his views as a hunter. He describes his thoughts and experiences as he hunts a deer with bow and arrow on an early October morning:

> I recall discussions with friends, sensitive people. Preservation logic is one thing that binds us. Wilderness teaches us. My friends have expectations of a naturalist:
> "We look to a teacher of the woods for inspiration, for a spirit of integration with plants, animals, and Earth. His sensitivity must be manifest in a deep humanity toward animals. Knowledge of nature's nuances must lead him to respect and preserve life."
> But I am a hunter and they are not.
> "A hunting naturalist is a contradictory creature," they argue. "With his own hand, he takes the very life he respects. Can respect be shown in making a corpse?"
> "The gossamer of life, so tenuous and fragile," the poet says, "can scarcely withstand this violent betrayal. How can anyone kill except in need?"
> Hunter. Like it or not, I belong to a group.
> The arguments turn in my mind as I drive to the woods I know from childhood. I park and get out, slowly adjusting to the darkness...
> A sharp-shinned hawk flashes between the trunks. A twist, and he continues his erratic maneuvering—toward me! Perhaps he caught a blink of my eye. Expecting a close fly-by, I watch in admiration. He lands on my bow! My excitement sends slight tremors through the bow limbs, but he perches and scrutinizes the leaf litter. Mouse movement spins his head and he is off again in wild flight. A predator. A killing efficiency honed by the millennia.
> An hour passes and I still wait.
> Humanity is a curious invention. The constraint of positive emotions, love and care, is placed upon human potential for carnage. The rest of nature is simply indifferent. Plants and other animals don't need moral control because they don't have our omnipotence. We feel we should be humane to other animals.
> Fairness is humane. The bow and arrow I hold are more fair than cannons. A wolf would use every means it has to make the kill and, if it could, would think me too generous. It doesn't understand my power. Maybe I don't either.

A small buck approaches. I hold my breath.

He moves closer, weaving and stopping. Closer. My heart pounds. Closer. Shaking. Closer. Draw. Closer—broadside. Heart shot!

The string whips. The arrow is a resolute line, a decision made. To kill.

I feel I should have been here before. I speared the mastodon with chipped stone and seasoned maple, thrusting desperately at its heart. The last magnetic forces twitched violently from its muscles. I smelled warm blood, oozing between pebbles toward thirsty rootlets. A choked gasp, and life passed to scavenger, maggot, and me. Life unto life, only through death.

It isn't painless. The deer runs the hunched, tail-down death panic. In three seconds his brain is bloodless. He weaves, collapses, kicks, and is still. A brown eye peers, unseeing.

The twist of opposing emotions clenches my throat. This feels right to me.

Can I ever resolve the cardinal question? Is my euphoria from the "fun" of killing or the joy of participating in a natural system? I do feel sorrow, but never enough to quit hunting. Is it more humane to munch steak or soybeans and feel nothing at all? Where is the real insensitivity? Who feels more removed from the natural world—the participant or the watcher? I can't answer, because I am both....

One friend will eat meat, but only wild meat. She seeks honesty, wishing to escape the anonymity of meat counter, cellophane, and slaughterhouse. Our common world is respect for life-and-death poetry. There is little honesty in a fast-food hamburger. The styrofoam wrapper insulates the reality that it died for me. I know where my venison comes from. I watch and feel it die...

There are two kinds of kills: one dulls, one sensitizes. The former is tragedy. The latter is a sacred celebration, as old as time. Every living being kills life. Some firsthand.

Olson quoted his friends as seeking a spirit of integration with plants, animals, and Earth. Integration implies an intermingling—something that is impossible unless we consider ourselves a part of the ecosystem, and not apart from it.

Ecology and Economics of Hunting

Are there ecological arguments for prohibiting hunting? Where populations are endangered, hunting should unquestionably be stopped. But in North America, such situations in this century are rare. Hunters have for years shown their interest in maintaining game populations through license fees and taxes on sporting goods earmarked for game research and management, and by supporting legislation and regulations that are designed to protect against overharvest. The response has been quite gratifying. The wood duck, brought nearly to extinction in the early 1900s by overshooting and nesting habitat loss, now numbers in the millions, and sustains annual hunter harvests of hundreds of thousands. Wild turkeys, once reduced to small remnants of their original range, have in the past two decades been reestablished in nearly all of the suitable remaining turkey range. At the end of the nineteenth century, only about 40,000 elk remained in existence in North America. Now there are over 400,000. In the United States, the white-tailed deer, the most popular big game animal in North America, increased, with properly managed hunting and habitat, from about 500,000 in 1900 to 12 million in 1980. Financial contributions of hunters, however, do not necessarily entitle them to ownership of the wildlife, nor even to any more say in its management than other citizens. In North America, resident wildlife belongs to the people.

I am not a great believer in economic arguments to justify the existence of anything. Hunting is unquestionably of great economic benefit to many people, but so is smuggling heroin. If something is morally or ecologically wrong, even if it generates a lot of money, it should not be done. I will therefore not present the arguments about how hunters pay for wildlife habitat management to favor the existence of hunting.

There are accusations that habitat management for game species may be carried out at the expense of other species. This is true in some cases. Research is being conducted on many national forests and refuges to determine whether non-game species are being neglected. New management plans must take these species into account, and much more work needs to be done in this area.

Can Wildlife Be Managed Without Hunting?

Despite the claims of some pro-hunters, it is possible to manage wildlife without shooting. We do so now for many species. But shooting animals does give wildlife managers another means for controlling numbers of some species that are no longer under control of natural predators and in places where natural predators cannot practically be reintroduced. An example of what must be considered neither sound management nor humane treatment of animals occurred within the past few years on Angel Island, California. On this small island park near San Francisco, blacktailed deer multiplied to a point where there was clearly not enough natural food to sustain them. Artificial feeding would only worsen the problems as fed deer would continue to reproduce. The California Department of Fish and Game proposed shooting deer to bring the population back within its carrying capacity. This was opposed by the San Francisco Society for Prevention of Cruelty to Animals (SPCA). A suggestion by wildlife ecologist Dale McCollough of the University of California Berkeley to release coyotes on the island to prey on the deer was also opposed by the SPCA. A plan was then developed under court order for the California Department of Fish and Game to live-trap and release 214 deer from Angel Island into a recently burned area on the California mainland which supposedly could sustain a growing number of deer. Radiotelemetry of released deer showed that within a year, 85% were dead, more than half within three months. The cost of this operation was $20,000 (O'Bryan and McCullough 1985). While death by a bullet may not always be the best answer to overpopulations of deer and elk, suitable humane and practical alternatives have not been forthcoming.

As one trained in science, I have found it somewhat frustrating not to be able to express my humanity in terms of means, standard deviations, and statistical probabilities. Instead I must resort to describing my feelings as a hunter, and as a human being who differs genetically and by experience from all others, and also as one who shares many qualities with other humans. The pleasure I get from hunting is something that many do not understand. Likewise, others experience some pleasures that I do not understand. I have little interest in horseback riding, Disneyland, automobile racing, or grand opera, yet I am not advocating the abandonment of those pursuits. Anti-hunters have not come up with sufficient ecological, moral, or humane arguments to deprive me and millions of other citizens the opportunity to hunt—to participate actively and directly in the community of life of which we are all members.

I believe strongly that the welfare and survival of wild animals on the earth will depend not on protecting them from regulated hunting, but upon our success in defending them against broadscale destruction of their life-support systems. I am talking about the devastation of rainforests in Central and South America, wetlands in North America, savannas in Africa, and lakes in Scandinavia. Anti-hunters, non-hunters, and hunters would do well to join together in fighting the forces of human population growth and destructive exploitation of resources that affects not only ourselves, but also our fellow creatures.

Endnotes

[1] Paper presented at the national conference, "Animals and Humans: Ethical Perspectives," Moorhead State University, Moorhead, MN, April 21-23, 1986.

[2] Professor, Department of Biology, Northern Michigan University, Marquette, MI, 49855.

References

Elton, C. 1939. On the nature of cover. *J. Wildl. Manage.* 3: 332-38.

Geist, V. 1986. Super antlers and pre-World War II European research. *Wildl. Soc. Bull.* 14: 91-94.

Kellert, SR. 1978. Attitudes and characteristics of hunters and anti-hunters. *Trans. N. Am. Wildl. and Nat. Res. Conf.* 43: 412-23.

—. 1980. America's attitudes and knowledge of animals. *Trans. N. Am. Wildl. and Nat. Res. Conf.* 45: 111-24.

Mish, FC. 1983. *Webster's Ninth New Collegiate Dictionary.* Merriam-Webster, Inc. Springfield, MA.

O'Bryan, MK and McCullough, DR. 1985. Survival of blacktailed deer following relocation in California. *J. Wildl. Manage.* 49: 115-19.

Olson, D. 1980. When a naturalist goes to kill, he raises some hard questions. *Milwaukee J.* 6 January. Discover Section: 1, 9.

Raveling, DG. 1978. Dynamics and distribution of Canada geese in winter. *Trans. N. Am. Wildl. and Nat. Res. Conf.* 43: 206-25.

Singer, P. 1975. *Animal Liberation.* New York: Avon Books.

THE CASE FOR HUNTING ON NATIONAL WILDLIFE REFUGES[1]

Harvey K. Nelson[2]

Introduction

Public land management agencies are faced with greater challenges today than ever before in responding to the recreational needs of society. As Will Rogers so aptly stated, "Land, they make so little of it nowadays" (Steinhart 1986). The U.S. Fish and Wildlife Service (FWS) also must face these challenges in management of national wildlife refuges (NWRs). There is a growing demand by the American people to utilize and enjoy NWRs in a variety of ways. Managers are faced with the dilemma of determining how much and what kind of management and utilization of natural resources is appropriate without compromising the mandates and integrity of the 434 NWRs they administer (see figure 1). They must accept the fact that, in our complex society with sharply conflicting interests and philosophies, it is impossible to conduct management programs that will be acceptable to all the people all of the time. Hunting on NWRs is a prime example of this conflict.

In the last decade, recreational hunting has been challenged, largely on philosophical grounds. Some citizen's groups maintain that hunting is inhumane, irresponsible, and violates the rights of individual animals. Pro-hunting groups counter that hunting is a sound tool in wildlife management, that hunting license money has paid for much of the wildlife habitat that remains today, has provided the incentive to retain significant additional wildlife habitats, and that wild animals increase in value when they are used by man.

Since 1960, a number of new refuges have been opened to hunting, and the trend has alarmed some groups. The Humane Society of the United States has charged that FWS is not complying strictly to Congressionally established mandates governing refuges. FWS takes the position that hunting on NWRs is an acceptable, traditional form of wildlife-oriented recreation that can also be used to manage wildlife populations.

This paper describes the historical, legal, social, and biological bases for the FWS position.

FIGURE 1

NATIONAL WILDLIFE

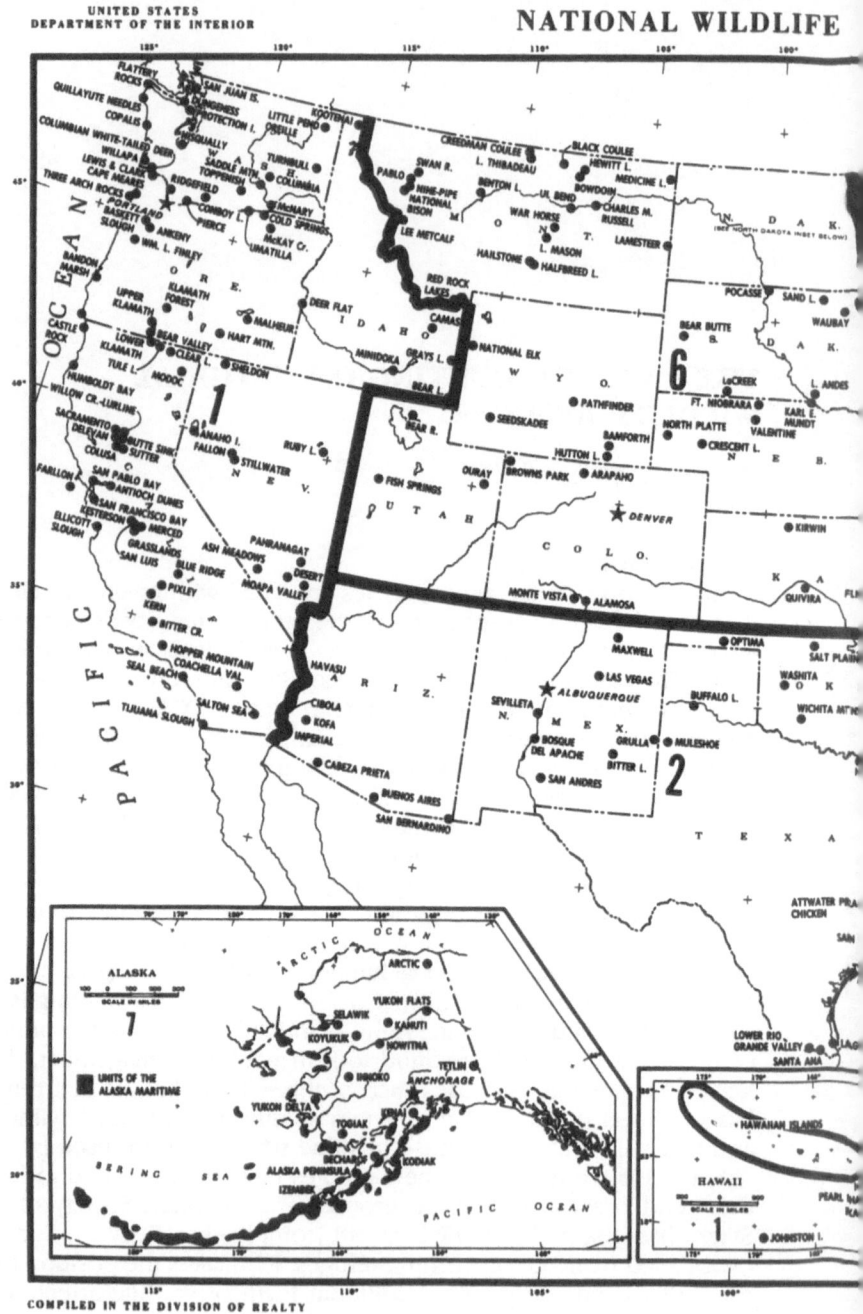

COMPILED IN THE DIVISION OF REALTY

WASHINGTON, D.C.   SEPTEMBER 30, 1985

REFUGE SYSTEM

UNITED STATES
FISH AND WILDLIFE SERVICE

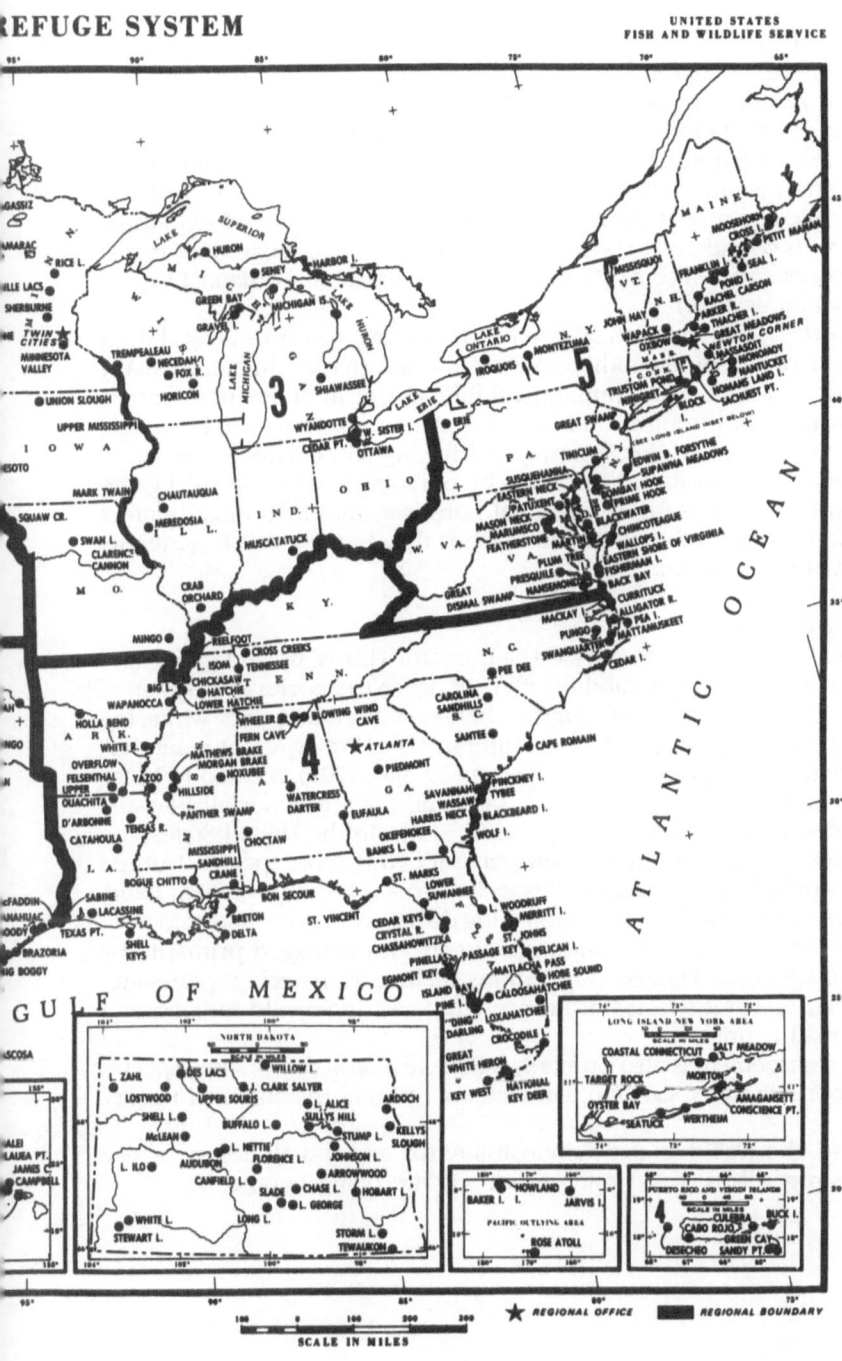

SCALE IN MILES

★ REGIONAL OFFICE   ▬ REGIONAL BOUNDARY

Definitions

Types of Hunting

Hunting can be viewed in historic, legal, or management perspectives.

Subsistence hunting is the taking of animals for food, shelter, or other personal uses essential to survival. For example, Peterle (1977) reported that the Eskimo hunters of Baffin Island acquire about 83% of their annual food by hunting marine mammals. Native people of the Yukon-Kuskokwim Delta in Alaska have relied on migratory birds for about 4% of their annual food supply, and this source is considered to be important seasonally (James Bartonek, Pacific Flyway Representative, Office of Migratory Bird Management, Portland, Oregon, personal communication).

Market hunting, the commercial harvest of wildlife, was at its peak during the latter part of the nineteenth century. Although market hunters usually made a portion of their living hunting wildlife, this term differs from subsistence in that most of the birds and animals taken are sold to someone else for profit. Some illegal market hunting and fishing still occurs today.

Sport hunting is generally construed to be pursuing and killing legal game animals with gun or bow for recreational purposes. Although most hunters consume the meat from the birds and animals they harvest, this is generally not the primary motivation.

Hunted Populations

Hunted populations are categorized as migratory birds, upland game, and big game. Management responsibility differs between categories as follows:

Migratory birds, species that migrate between breeding and wintering habitats are Federally protected by the Migratory Bird Treaty Act, as amended, by treaties with Great Britain (and Canada) (39 Stat. 1702), Mexico (50 Stat. 1311), Japan (concluded March 4, 1972), and Union of Soviet Socialist Republics (concluded November 19, 1976 [16 USC 715j]). In the United States, the FWS has primary responsibility for management of migratory species. Hunted populations include most waterfowl species, coots, sandhill cranes, band-tailed pigeons, whitewinged doves, mourning doves, woodcock, and snipe.

Upland game are resident, non-migratory species managed primarily by State wildlife agencies. Hunted populations include ring-necked pheasants, bobwhite quail, several species of grouse, gray partridge, wild turkeys, and numerous small mammals.

Big game management is also primarily a State responsibility. Hunted species include moose, elk, deer, caribou, wild sheep, pronghorns, mountain goats, and bears.

In any case, the FWS has the responsibility for any and all access or use of the NWRs regardless of the category of wildlife being hunted.

Biological Criteria

Soundness.—A hunting program is considered biologically sound if a predetermined number of game animals can be removed from the population without adversely influencing the desired base population. Most wildlife species have the reproductive capacity to quickly fill a void in the population caused by the removal of individuals from natural mortality or harvest.

Population Control.—Where wild predators are absent, a common occurrence for big game species, normal productivity can lead to crowding, disease, malnutrition, and habitat destruction. There are numerous examples of these problems occurring with white-tailed deer. Similar problems are occurring today with certain populations of Canada geese that are increasing in numbers. Hunting is used as a management tool to reduce and hold the populations at levels that are within the carrying capacity of their environments. In the case of migratory birds, that carrying capacity includes an element of the tolerance of farmers to suffer crop depredation by those birds.

Impact of Migratory Bird Hunting Stamp Act on Transition from Sanctuary to Management Area Concept

Various publications (Reed and Drabelle 1984; Drabelle 1985) present accounts of the early establishment of the National Wildlife Refuge Systems (NWRS) and the evolution of hunting on those areas. The first NWRs were established as sanctuaries by President Theodore Roosevelt to protect colonial nesting birds heavily exploited by market hunters. In 1909, Congress authorized the first use of Federal funds to purchase land for wildlife. An area comprising 12,800 acres in Montana, now known as the National Bison Range, was acquired to protect habitat for the American bison, an animal that John Audubon predicted would become extinct without immediate protection. Thus, the first NWRs were established as inviolate sanctuaries against the onslaught of market hunting and habitat destruction. Reed and Drabelle (1984), reported:

> Hunting was late to entrench itself in the National Wildlife Refuge System. Until 1949, it was permitted on only a few refuges, where it had been a traditional activity prior to the area's inclusion in the system and its continuation was a political necessity—no hunting, no refuge. This was the case with the first refuge opened to hunting, the Upper Mississippi River Wildlife and Fish Refuge, which Congress established in 1924 on condition that hunting be permitted there.

In 1934, passage of the Migratory Bird Hunting Stamp Act provided a new source of revenue for acquiring waterfowl refuges. Every waterfowl hunter the age of 16 or over was required to buy a duck stamp. When the FWS increased the price of the duck stamp from $1.00 to $2.00 in 1949, it agreed to open up to 25% of the acreage on certain refuges to hunting. Wildlife management areas open to hunting increased to 40% in 1958, when the duck stamp price was increased to $3.00. Since 1960, the increase in migratory bird hunting on NWRs correlates closely with the increase in duck stamp revenues as shown in figure 2 (Drabelle 1985; Superintendent of Documents 1981).

The National Wildlife Refuge System Administration Act of 1966, as amended, further expanded the Secretary of the Interior's authority for initiating hunting programs on NWRs. This Act also authorized the hunting of upland game and big game on refuge lands (Committee on Merchant Marine and Fisheries 1975).

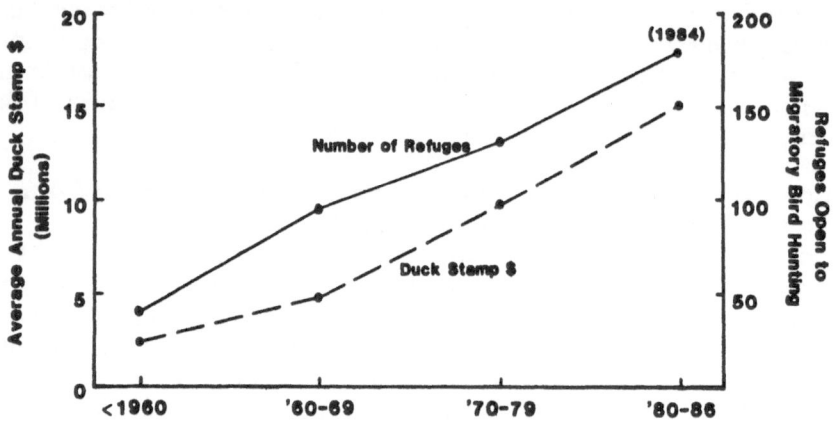

Figure 2. The number of refuges open to hunting for migratory birds has increased at the same rate as revenues from duck stamps.

Thus, hunting on NWRs is not a new phenomenon; it has been gradually increasing since 1935. In some instances, the provision of limited public hunting opportunity has been a condition of State approval for proposed Federal land acquisition.

Refuge Acquisition and Funding

The various refuges were created primarily by three means: (1) withdrawal from public lands via Executive Order, Proclamation, or Public Land Order; (2) purchase; or (3) legislative action. The majority of refuges created by the first mode, land withdrawals, could properly be opened to hunting under regulations established by the Secretary of the Interior through a series of executive orders and Presidential proclamations issued during the period 1934-40 (Presidential Proclamation 2287, June 6, 1938, Executive Order 2416, July 25, 1940).

The second mode of acquiring refuges was by purchase. Acquisition funds for land in NWRS come from two sources—the Migratory Bird Conservation Fund (MBCF) and the Land and Water Conservation Fund (LWCF). The MBCF depends primarily on duck stamp revenue and Congressional advances from the wetland loan fund. It is used to purchase production, migration, and wintering areas for migratory birds. A total of 186 refuges have been purchased with MBCF money and of these, 71, or 38%, are open to hunting. LWCF money comes from a variety of sources, such as off-shore oil and gas revenues, tax receipts from motor boat fuel, user fees, and sale of surplus government property, and has generally been used to purchase habitat for species other than migratory birds.

A third mode of early refuge acquisition was by direct legislative creation, and Congressional support of hunting is clear. Upper Mississippi River NWR and Bear River NWR are examples of early refuges where regulated hunting was allowed.

The provision for hunting on refuges—whether created by withdrawal, purchase, or legislation—thus reflects a long standing policy, enunciated by both Congress and the Executive Branch throughout the last 60 years. The refuges system have been created with an understanding that regulated public hunting is not, per se, barred and is often an important factor supporting the establishment of new refuges.

Policies and Procedures Governing Refuge Hunting Programs

The Secretary of the Interior is authorized by the National Wildlife Refuge System Administration Act of 1966, as amended, and the Refuge Recreation Act of 1962 to permit public recreational use, including hunting, on any refuge within the NWRS, providing that such recreational use will not interfere with the primary purposes for which the area was established and that funds are available for the development, operation, and maintenance of these permitted forms of recreation.

The dual needs for new recreational authorization and for control over possible recreational impairments of refuge purposes formed the background of the Refuge Recreation Act of 1962. Several concepts occurred throughout the history of this legislation: (1) hunting is an accepted use of refuges; (2) observation; and (3) concern that such uses might direct land purchased with duck stamp funds to benefit non-hunter recreationist who made little or no contribution to this fund.

Procedural Requirements for Opening New Refuges to Hunting

The *Refuge Manual* (U.S. Fish and Wildlife Service 1982) provides the specific management planning guidance for refuge managers who may be considering a hunting program. A specific hunting plan must be prepared.

During the planning process, a highly structured decision-making procedure is used to determine compatibility of hunting with refuge objectives. Conflicts between biological resources, management activities, and other public uses are resolved. Throughout the process, the public is encouraged to participate.

There are specific steps required before an NWR can be opened to hunting.

Compatibility.—First and foremost, refuge hunting programs must be compatible with the major purposes for which the refuge was established. The compatibility of hunting to refuge purposes has to be reviewed periodically. This is required by the NWRS Administration Act.

Funding. —Funds must be available to develop, operate, and maintain a hunting program, as required by the Refuge Recreation Act of 1962. Hunting programs have, traditionally, been one of the least costly management tools available to refuge managers.

Environmental Impacts. —Any plan that will affect resources must be given environmental review. All hunting plans developed by the FWS have an Environmental Assessment, or, if necessary, an Environmental Impact Statement. These documents are prepared as required by the National Environmental Policy Act of 1969, as amended.

Endangered Species. —All hunting programs are reviewed to determine if they have impacts on threatened or endangered, sensitive species, or habitats. The refuge manager initiates appropriate consultation procedures in accordance with Section 7 of the Endangered Species Act.

Consultation and Coordination. —The affected state conservation agency is involved in the development or revision of any refuge hunting program and State approval in writing is required. Whenever possible, hunting regulations on refuges are the same as the State regulations. Refuge regulations may, on occasion, be more restrictive but may not be more liberal than State regulations. This approach fosters a spirit of State-Federal cooperation and minimizes the number of regulations required for refuge programs.

Refuges proposing new hunting programs or significant changes in existing programs must provide for public involvement in the decision-making process. This may range from informal consultation with interested individuals or groups to formal public meetings or workshops.

Requirements of the Administrative Procedures Act

When the planning effort for a proposed hunting program on a given refuge is completed and the necessary documentation has been compiled, reviewed, and approved, the proposed opening is published in the *Federal Register.* The Administrative Procedures Act requires a proposed rulemaking to be published, with a public comment period. Following the comment period, responses to all comments received are prepared and a final rulemaking incorporating the comments is published. If the decision is to proceed with the proposed opening, 30 days following the publication of the final rulemaking the refuge is listed in *Title 50 Code of Federal Regulations* (50 CFR) as open to hunting.

There are problems in dealing with refuge specific regulations on an annual basis. In 1960, the Fish and Wildlife Service implemented a provision for annual review of hunting plans and subsequent publication of annual hunting regulations for each refuge in the *Federal Register.* Because of the large number of refuges involved, this procedure became a costly administrative burden and the timely issuance of these regulations was jeopardized. The FWS recently modified this procedure and replaced the annual hunting regulations with refuge specific regulations codified in 50 CFR 32. These regulations remain in effect on a given refuge until they are amended. Amendments must go through the steps discussed, concluding with publication in the *Federal Register.*

The Humane Society of the United States filed a lawsuit in November, 1984, charging that the FWS had abandoned the annual review of hunting programs on refuges. While there has been no decision received,[3] the FWS takes the position that it now requires closer review of hunting programs than previously. Existing compatibility and Section 7 determinations must now be reviewed annually and confirmation of existing determination or a new determination placed on file in the refuge office. Based on the author's recent experience, the codified refuge specific regulations which have been published for refuges in the FWS North Central Region are much improved in comparison to those published annually five years ago.

Other Guidelines for Conducting Refuge Hunts

Hunting must be conducted to minimize conflicts with other refuge programs. Managers must consider the social impacts of hunting on other refuge visitors engaged in non-hunting activities. Often zoning the refuge ensures a quality experience for everyone.

Current demands for hunting in the vicinity of a refuge must be considered to determine if negative impacts on overall hunting opportunities in the surrounding area will occur as a result of the refuge hunting program. The potential for excessive demand must also be carefully evaluated to determine if hunter numbers need to be limited to maintain a quality hunting program and provide for participant safety.

Refuge hunting programs are planned, supervised, conducted, and evaluated to promote positive hunting values and hunter ethics. Refuge hunts are designed to provide participants with reasonable harvest opportunities, uncrowded conditions, and limited interference from, or dependence on, mechanized aspects of the sport. A responsible hunting program on refuges functions as an important management tool, while it also improves State and Federal cooperation and public enjoyment of Federal lands.

Present Day Hunting on NWRs

The NWRS presently includes 434 units encompassing approximately 88 million acres, of which 77 million are in Alaska (see figure 1). In addition, FWS administers more than 1.6 million acres of waterfowl production areas (WPAs) located in 152 counties in Iowa, Michigan, Minnesota, Montana, Nebraska, North Dakota, South Dakota, and Wisconsin.

Although the number of refuges open to hunting has increased, harvest of migratory birds and upland game has declined since the 1976 estimates were published in the *Final EIS on Operation of the National Wildlife Refuge System*. At the same time, refuge waterfowl production has increased by 120% (table 1).

In summary, during 1985, hunting was allowed on 236 refuges, or about 54% of the total NWRs. This includes all refuges in Alaska designated as being open to hunting by the Alaska National Interest Lands Conservation Act of 1980 (P.L. 96-487). In spite of increased numbers of refuges open to hunting, only about 1.6% of the annual U.S. waterfowl harvest was taken on the 179 refuges open to waterfowl hunting during 1984.

Table 1. Comparison of hunting harvest and waterfowl production on national wildlife refuges between 1976 and 1983-84.

Harvest	1976 EIS	1983-84 Estimate
migratory birds	296,000	275,000
upland game	228,000	100,000
big game	10,200	16,000
waterfowl production	1.5 million	3.3 million

Conclusion

Congress has for 60 years expressed a strong and unequivocal national interest in maintaining hunting opportunity on national wildlife refuges. There is a strong base of legislative and executive action, policy, program direction, and public interest to support regulated hunting on refuges.

The term "refuge" is often misunderstood to be synonymous with "sanctuary;" that is, a place where animals are protected from danger and distress. Thus, allowing hunting on refuges causes some confusion and concern among the public as to the intended purposes and management rationale for the NWRs. This is understandable, and clearly the question of what constitutes a "refuge" gets at the core of the issue.

Many wildlife species residing on refuges may occur in considerable numbers throughout thousands of acres of habitat. With the exception of some endangered species (such as the whooping crane), refuge managers, of necessity, must deal with animal populations, not individuals.

Congressional and legal mandates authorizing the establishment of individual refuges often prescribed that compatible forms of public use be permitted. Under these conditions, a more concise description of a "wildlife refuge" might be: "An area set aside for the protection and management of wildlife populations and their habitats and for the benefit of people." The FWS subscribes to the idea that it is possible to remove a given number of individuals of a population from a refuge by hunting, while still protecting, and, in some cases, improving the overall welfare of that population, as well as other species. Further, FWS believes that under given circumstances it is prudent management, and in the public interest, to not open certain refuges to hunting if a given species or group of species need complete protection.

The FWS will continue to defend its position on the appropriate role of hunting on NWRs where the required compatibility determinations have been met. At the same time, the agency recognizes the important role of citizen's groups that hold differing viewpoints. Constructive criticism of government programs by citizens ensures that agencies comply with their legal and moral responsibilities to manage resources for the best interests of the public at large.

The FWS will strive to improve the quality of refuge hunting programs in the future. Former FWS Director Greenwalt (1978) summarized this policy quite well:

The quality of the hunting on refuges will increase as the Fish and Wildlife Service efforts to this end are increased. More hunts will be controlled; that is, limits will be imposed upon the total numbers of hunters present at any one time on certain refuges ... and, consequently, the possible conflicts between hunters and those refuge visitors who prefer not to hunt will be reduced.

Peterle (1977) predicted that the proportion of the total human population participating in hunting will continue to decrease in the future. However, the demand to hunt on refuges may continue to increase because of fewer opportunities to hunt on private lands. Concurrently, the demand for non-hunting uses of NWRs is expected to continue to increase. In the future, managers must place greater emphasis on the social aspects of refuge hunting in terms of its relationship to, and perception by, other users of refuge resources.

Acknowledgement
Special thanks to Robert Drieslein, Jan Eldridge, Calvin Gale, James Gillett, and Robert Oetting, FWS staff members who provided assistance and editorial review of this manuscript.

Endnotes
[1] Paper presented at the national conference, "Animals and Humans: Ethical Perspectives," Moorhead State University, Moorhead, MN, April 21-23, 1986.

[2] Regional Director, U.S. Fish and Wildlife Service, Federal Building, Fort Snelling, Twin Cities, MN 55111.

[3] On July 25, 1986, the United States District Court for the District of Columbia ruled that The Humane Society of the United States lacked standing to bring the case in behalf of its members. The Court ordered that summary judgment shall be entered in favor of the defendants on all of the Humane Society's claims, but excepted an individual plaintiff's claim based on defendant's alleged failure to prepare environmental analyses for hunt openings occurring since 1980. The Court allowed further proceedings on that single issue.

References
Carey, J. 1986. Changing face of America, 50 years of land use. In: Strom, J. ed. *National Wildlife.* Washington, DC: National Wildlife Federation. pp. 8-27.

Committee on Merchant Marine and Fisheries. 1975. *A Compilation of Federal Laws Relating to Conservation and Development of our Nation's Fish and Wildlife Resources, Environmental Quality, and Oceanography.* Washington, DC: U.S. Govt. Print. Off.

Drabelle, D. 1985. The National Wildlife Refuge System. In: Di Silvestro, RL. ed. *Audubon Wildlife Report — 1985.* Washington, DC: National Audubon Society.

General Services Administration. 1984. *Code of Federal Regulations,* Title 50. Washington, DC: U.S. Govt. Print. Off.

Greenwalt, L. 1978. The National Wildlife Refuge System. In: Brokaw, H. ed. *Wildlife in America.* Washington, DC: U.S. Govt. Print. Off.

Peterle, T. 1977. Hunters, hunting, anti-hunting. *Wildlife Soc. Bull.* 5(4): 151-61.

Reed, N and Drabelle, D. 1984. *The United States Fish and Wildlife Service.* Boulder, CO: Westview Press, Inc.

Superintendent of Documents. 1981. Duck stamp data. Washington, DC: U.S. Govt. Print. Off.

United States Code Annotated. 1982. Title 16. Conservation 461 to 760. St. Paul, MN: West
    Publishing Co.
U.S. Department of the Interior. 1976. *Final Environmental Statement: Operation of the National
    Wildlife Refuge System.* Washington, DC: U.S. Govt. Print. Off.
—. 1979. *Final Recommendations on the Management of the National Wildlife Refuge System.*
    Washington, DC: U.S. Govt. Print. Off.
U.S. Fish and Wildlife Service. 1982. *Refuge Manual.* Washington, DC: U.S. Govt. Print. Off.

PROVIDING HUMANE STEWARDSHIP FOR WILDLIFE: THE CASE AGAINST SPORT HUNTING[1]

John W. Grandy[2]

It is a pleasure for me to be here today to address this important symposium on an increasingly controversial topic: sport hunting of wildlife. But we are talking about more than sport hunting. We are implicitly discussing the need for humans to provide humane stewardship for wildlife both for the sake of wildlife and for the sake of ourselves. One way to begin this is to challenge the killing of wild animals for fun, that is, sport hunting, in America at large and on National Wildlife Refuges.

Hunting in America Today

Initially, it is important to define the terms that we are discussing, by viewing hunting in the context in which it exists today. In that regard, we must recognize that we are not talking about hunting as it existed for the pilgrims at Plymouth Rock. We are not discussing hunting as it existed in the time of Abraham Lincoln. We are not even discussing hunting as it existed at the turn of the century, or during the Depression. Nor are we discussing a remote sort of "wilderness hunting" that "pits man against beast."

Rather, we must recognize hunting as it exists for most of those hunting in America today. It is a recreational pastime. Sometimes, as in some forms of "waterfowl hunting," it is little more than shooting at animated targets. It is pursuit of big game "trophies." At its best, it is a form of pleasure from which some people derive some meat that is either eaten or discarded. However, hunting today is not necessary in any sense. As a television reporter said to me not long ago, "we're talking about Joe Sixpack." And while that image is not precise, it is very descriptive. Hunting or killing animals is a form of mostly macho, comraderie-based human recreation. In general it is not romantic, it is not meat hunting, it is not necessary. It is a form of fun. The question is whether society will continue to permit and glorify modern day hunting, the killing of mammals and birds for human pleasure, as an acceptable form of fun.

Objections to Sport Hunting

The objections to sport or recreational hunting are not new. Most basically, hunting inflicts needless undeniable cruelty—pain, suffering, trauma, wounding, and death—on living, sentient creatures. And it does so for nothing more than sport.

Methods of killing animals are archaic, without even minimal attempts to ensure either a lack of pain and suffering, or a quick, humane kill. Hunters use bow and arrow even though it is well known that the bow and arrow causes orders of magnitude more wounding and suffering than might be caused by another kind of weapon. "Primitive" muskets are used, although they are subject to the same kinds of objections as bow and arrow. Not only do officials continue to allow the use of muskets and bow and arrow, but state Fish and Game agencies across the country have established special seasons so that people interested in using bow and arrow and/or musket will not be interfered with by other hunters (sort of a special season which legitimizes the increased cruelty from these weapons). Finally, hunters use lead shot in shotguns. Lead shot not only kills the animals which get shot, but also causes the death by prolonged, excruciating poisoning of more than three million ducks, geese, swans, doves, and other seed-eating birds each year. This needless destruction continues because the hunting community refuses to use steel shot, in spite of the mass poisoning and despite the fact that steel shot is an effective substitute for lead.

Next we must look at what messages we are leaving for our children, young people, and society. The ethic—if it may be called that—of killing for fun teaches callousness, disrespect for life, and the notion "might makes right." It certainly causes a numbing of sensitivity and empathetic responses to animals, including humans. Are these the kinds of values that a civilized society wants to pass on as its legacy? I hope not.

Sport hunting also has destructive impacts on the animals themselves. Obviously, millions of animals are killed outright. These are animals for whom death is at best a "clean kill," but who often suffer a slow, lingering death that includes wounding, suffering, pain, and fear. Representative numbers are fifteen million dead doves each year, twenty million dead ducks and geese, millions of deer and scores of millions of such animals as squirrels, rabbits, bobwhite quail, prairie grouse, turkey, woodchucks, foxes, bobcats, crows, magpies, and myriad others. Hunting also has other destructive impacts. In big game animals, for example, hunters attempt to kill the largest (and presumably genetically superior) animals from the population. Reportedly, this has had serious impacts in terms of reducing the survivability of herds of bighorn sheep. Similar impacts are likely in deer, elk, and moose. Also, hunting removes from natural systems the animals that other animals— predators and scavengers—utilize for food. Nature wastes nothing. Doves which are not killed for human fun, are food for hawks, owls, falcons, and others. The removal of wildlife by sport hunting deprives ecosystems and their animal inhabitants of food.

And, lastly, there is the ghastly "incidental" toll from sport hunting. Each year hunters across this land in pursuit of what they describe as "good clean fun" shoot cows, sheep, goats, horses, pets, highway signs, homes, and endangered and protected species. And even if this is all accidental or incidental, it is still death and destruction that occurs as a by-product of a highly objectionable form of recreation: killing other animals for fun. And, finally, how should society respond to the fact that thousands of *people* are killed or wounded in sport hunting accidents every year?

Someone is sure to say that many of the problems I have pointed out can be alleviated by "cleaning hunting up." By teaching hunters to identify their targets, by instilling "respect" and "good hunting manners," by eliminating lead shot, or by enacting and enforcing strong regulations. But still, the salient fact is that we are talking about "cleaning up" the killing of animals for fun. That, I submit, is a contradiction in terms: you can not clean up a sport based on inflicting needless pain, suffering, and death to innocent animals.

Objections to Hunting on Refuges

For the other part of my presentation, the organizers of this symposium requested that I discuss the case against hunting on National Wildlife Refuges. I am, I admit, positively underwhelmed. Having presented in excruciating detail the case against sport hunting in general, it is an insult to our collective sanity to discuss the "merits" of sport hunting on Refuges.

National Wildlife Refuges were set up to be the one "inviolate sanctuary" for wildlife in this nation. That is why they were called refuges. They were to provide safe haven and sanctuary, and that is the public view of their purpose.

Sport hunting programs assault the very concept of "refuge." It is no less than the rape of the refuge system. There are numerous other impacts as well. There are the objections to sport hunting in general which I mentioned earlier. Hunting programs result in the death of half a million refuge animals each year. Hunting scares wildlife, making it hard for people to see animals on refuges. Hunters kill endangered species on refuges. Destroy habitat. Deposit toxic lead shot. Cause suffering, wounding. And... waste public tax dollars on programs that are destructive of the very purpose for which refuges were established. Finally, as if all that were not enough, refuge officials are now conducting predator control programs: that is, killing predators like foxes and raccoons so that hunters can have more ducks to shoot on refuges.

The whole system is an affront to the public as well as wildlife. Hunting simply has no place on National Wildlife Refuges.

Arguments Used in Favor of Hunting

So with all of that, what, you may ask, are the reasons offered by the hunting community as justification for continuing their sport? First, and most interestingly, is the assertion that hunting is a near-religious—even mystical—experience, that it brings man closer to nature and to his roots. This argument is exemplified by Dr. William Robinson's paper [see this volume]. In the portion of his paper where he is quoting from a bow hunter's description of the kill, we are implicitly invited to hear and see the very origins of man.

With all due respect to Dr. Robinson, however, such hunting is an anachronism. It is a throwback to a time gone by... long... long ago! Dr. Robinson would ask the public to accept sport hunting and killing for fun because a few people continue to try to practice hunting as a semi-religious experience. Dr. Robinson, however, can not avoid the salient facts of hunting in America today. It is a recreational pastime that has nothing to do with feeding families and getting people closer to nature. It is a form of recreation that thrives on the killing of innocent beings.

It is also important to note that the move to end sport hunting in America is a part of social evolution that Robinson's arguments deny. Hunting may once have been a mystical experience as Robinson would portray it today. However, as society has grown and matured, the place of hunting and the reasons for hunting have changed. Society now finds itself reevaluating the ethics of sport hunting and the need for it in a civilized twentieth century culture. Robinson's arguments suggest that we should ignore this social evolution and try to return to the days of our forefathers. This is totally unrealistic as well as counterproductive to society's ethical and moral growth.

Aside from Robinson's naturalistic arguments in favor of hunting, Nelson's paper at this symposium [see this volume] alludes to other purportedly modern-day "justifications" for sport hunting. These are arguments concerned with wildlife management, and I would like to examine them for a moment.

Nelson suggests that sport hunting on National Wildlife Refuges is necessary to prevent starvation of wild animals. Let's examine that assertion. First, I certainly agree that an animal killed by sport hunting is *prevented* from starving to death! However, sport hunting does not in any sense eliminate starvation. Hunters attempt to kill the largest animals, not those animals which are small, weak, or about to starve to death. Moreover, wildlife itself would be the loser if hunting could prevent starvation, because many wild animals (omnivores and carnivores) depend for their food on animals that die of starvation or are weakened by a lack of food.

Mr. Nelson also suggests that hunting is generally permitted on Refuges in order to limit growth of wildlife populations. This is simply not true. Indeed, fall hunting for species like whitetailed deer stimulates birth, by tending to make more food available to each deer that remains. No better example of the fact that hunting seasons are not aimed at limiting deer populations can be given than to note that if wildlife managers were truly interested in just limiting populations they would open a season for females only, but they never do. One might also ask: if deer hunting is really just done as a form of benevolent population limitation, would hunters favor other more humane forms of population limitation that would not involve sport hunting? And the answer to that question is almost certainly "No." The facts are, we must keep reminding ourselves, that hunting on National Wildlife Refuges and elsewhere is not being conducted and promoted to limit wildlife populations but rather to provide "fun" for a small percentage of the people.[3] In fact, Refuge managers have no idea what the population levels of most hunted animals are, much less how much, if any, the population of such animals will be reduced by sport hunting. A representative list of animals hunted in recent years on National Wildlife Refuges includes:

mergansers	partridge	badger
blackducks	grouse	bobcat
other ducks	sage grouse	mountain goat
geese	scaled quail	bighorn sheep
rails	Gambel's quail	whitetailed deer
crows	turkey	black bear
ravens	willow ptarmigan	grizzly bear
woodcock	grey squirrel	Kodiak bear
snipe	red fox	rabbit
doves	grey fox	hare
magpies	arctic fox	opossum

Mr. Nelson also suggests that hunting is necessary to keep populations of wild animals healthy, or to free them from disease. However, these assertions are simply wrong. At best for the wildlife populations, the effects of hunting are random. That is to say, animals are killed randomly regardless of health, genetic makeup, age, sex, or experience. Often, I suggest, the effects of sport hunting programs are far worse because hunters shoot the largest, strongest, most genetically superior animals, thereby removing them from the population.

Finally, Mr. Nelson suggests that hunting should be allowed on Refuges because hunters buy duck stamps and duck stamp monies have been used to purchase land for wildlife refuges. First, I will examine the details of Mr. Nelson's assertion. True, some refuge lands have been purchased by duck stamp funds. According to the 1985 Fish and Wildlife Service Realty Survey, a maximum of 2.9 million acres of land (or 3.3% of the Refuge system) have been purchased with funds derived from duck stamp sales. In contrast, the refuge system contains some 89 million acres of land, nearly all of which has been withdrawn from the public domain or purchased with Land and Water Conservation Fund monies. Thus, if the claims of hunters to hunt on Refuge lands could be said to be legitimately based on purchases of land with duck stamp funds, hunting would be scarce indeed.

And it is important to examine the implications of the Land and Water Conservation Fund purchases as well. The Land and Water Conservation Fund consists of monies derived in large part from the revenues received by the United States government as a result of offshore (continental shelf) oil and gas leases. Funds from the Land and Water Conservation Fund have been used to purchase National Wildlife Refuge lands, city parks, state parks, and a host of other public recreation areas. However, just because the Land and Water Conservation Fund was derived from oil and gas leases does not give the oil and gas industry the right to drill in city parks or on National Wildlife Refuges. By the same token, the fact that some refuge lands have been purchased with duck stamp monies does not confer any implicit or explicit right to hunters to hunt on those lands. Indeed, society might more properly consider revenues from duck stamp purchases as a minimal form of damage payment to society at large for the loss of tens of millions of ducks and geese each year. And, I might incidentally note, the time has come for society to evaluate whether such a "bargain" is a good one.

The bottom line about sport hunting is that it is no management tool at all. It is, as I have said repeatedly, a form of human recreation—fun—that brutally sacrifices the life of innocent wild creatures. Moreover, it must be viewed in another way. It is, in fact, a commercial enterprise. Commercial, you say? Someone is bound to mention that this is not commercial hunting; commercial hunting has been outlawed. But it is commercial in two important senses. First, hunting is commercial in that state fish and game agencies receive substantial portions of their salary monies and operating funds by virtue of the sale of hunting licenses. In essence, there is a strong commercial incentive for state fish and game agencies to encourage and promote sport hunting just to stimulate the sale of hunting licenses and resultant income.

And, hunting is a commercial enterprise in another important sense as well. Pick up an issue of *Field and Stream, American Rifleman, Outdoor Life*, or another outdoor/hunting magazine. Those magazines and the advertisements therein promote the sale of guns and ammunition, outdoor clothing, and other associated products. Readers of these publications are being encouraged to become hunters and buy these products. The message is "be tough, be strong, be in, be a hunter . . . buy." Promoting hunting is a way of stimulating increased sales of guns, shoes, wilderness "adventures" and numerous other commercial products.

Thus, in a very real way, two forces in twentieth century America are manipulating people to hunt through slick advertising and promotional programs. The sooner people recognize and reject this manipulation and resultant destruction, the better off both wildlife and the public at large will be.

Conclusion

Sport hunting has no place on the National Wildlife Refuges of this nation. To even consider it is an affront to the concept of a Refuge, the right of wild animals to safe haven, and the wishes of society. The question of sport hunting in society at large is slightly more complex because society, its thoughts and values, are evolving. Thankfully, we are moving more and more to a view that wildlife should be treated with the same dignity, respect, and freedom from avoidable cruelty that we would ask for ourselves. That process can be moved miles ahead if we eliminate sport hunting—killing for fun—now.

Endnotes

[1] Paper presented at the national conference, "Animals and Humans: Ethical Perspectives," Moorhead State University, Moorhead, MN, April 21-23, 1986.

[2] Vice President, Wildlife and Environment, The Humane Society of the United States, 2100 L St, NW, Washington, DC 20037.

[3] Sport hunting by the public at large is conducted by eight to ten percent of the public; on Refuges, sport hunting is conducted by less than one-half of one percent of the public.

All manuscripts submitted and solicited are subject to prior review and approval by members of the Manuscript Review Committee. Authors may submit manuscripts to: Dr. Michael W. Fox, The Humane Society of the United States, 2100 L St, NW, Washington, DC 20037. Manuscripts not adhering to the style outlined below will be returned to the author.

Manuscripts: Footnotes, references, tables, and figure legends are to be typewritten, double-spaced on 8½ × 11 inch bond paper leaving generous margins. Manuscripts must be in English using the preferred spelling in the *Webster's Third International Dictionary.* Submit original and two (2) copies.

Manuscript organization: Title page (p.1) containing title of the manuscript, author(s), and author(s) affiliation, present address where proofs are to be sent; text (beginning p. 2), with introduction, methods, procedures, results, discussion, conclusion, acknowledgements, references, tables, and figure legends. Special instruction to the copy editor or printer should be affixed to the original copy.

Abbreviations and units: Standard dictionary abbreviations are generally accepted. Other abbreviations should be explained when first mentioned in the text. SI units are preferred.

References: The Author-Date system, not a numbering system, is to be used for the citation of references in the text, e.g., Jones (1977) or (Jones and Smith 1977) or (Jones et al. 1977) for more than two authors of one citation. When citing the authors of such papers in the reference list, *all* authors must be listed for citations using et al. in the text. Where more than one paper by the same author(s) has appeared in one year, the reference should be distinguished by "a," "b," "c," etc. (e.g., 1971a). The list of references should be arranged alphabetically by authors' last names, and chronologically per author.

Titles: Journals should be abbreviated in accordance with the *Chemical Abstract Service Source Index.* References to books/monographs should include editor(s), edition/volume number, city, state, publisher, and relevant page numbers. A paper in press may be referenced if it has been accepted for publication. References to personal communications and unpublished works are permitted in the text only. *Proceedings* and talks given at meetings are to be treated as published material.

Sample references (in accordance with *The Chicago Manual of Style*)
Journal article:
Smith, J. 1979. The effect of heat stress on hogs. *J. Anim. Sci.* 5(1): 127-35.
Book:
Smith, J. 1979. *Animals and Environment.* New York: Academic Press.
Chapter or paper in book:
Smith, J. 1979. The effects of farrowing system on piglet health. In: Jones, A. ed. *Animal Husbandry.* New York: Academic Press.
Tables: These should be concise and double-spaced throughout.

Figures: Submit two sets of glossy prints (no negatives) with identifying arrows and letters contrasting sharply with the background. Indicate on the back the author's name, figure number, and "top." *Do not* write this information on the back of the print with felt-rip or ball-point pen; a separate typewritten sheet attached to the back is preferable. Prints will not be returned to the author (s).

Figure legends: Captions should contain sufficient information allowing the figures to be clearly understood without reference to the text.